BIRDS OF INDIA

BIRDS OF INDIA
A Literary Anthology

Edited by
Abdul Jamil Urfi

OXFORD
UNIVERSITY PRESS

YMCA Library Building, Jai Singh Road, New Delhi 110 001

Oxford University Press is a department of the University of Oxford.
It furthers the University's objective of excellence in research, scholarship,
and education by publishing worldwide in

Oxford   New York
Auckland   Cape Town   Dar es Salaam   Hong Kong   Karachi
Kuala Lumpur   Madrid   Melbourne   Mexico City   Nairobi
New Delhi   Shanghai   Taipei   Toronto

With offices in
Argentina   Austria   Brazil   Chile   Czech Republic   France   Greece
Guatemala   Hungary   Italy   Japan   Poland   Portugal   Singapore
South Korea   Switzerland   Thailand   Turkey   Ukraine   Vietnam

Oxford is a registered trade mark of Oxford University Press
in the UK and in certain other countries.

Published in India
by Oxford University Press, New Delhi

© Oxford University Press 2008

The moral rights of the authors have been asserted
Database right Oxford University Press (maker)

First published 2008

All rights reserved. No part of this publication may be reproduced,
or transmitted in any form or by any means, electronic or mechanical,
including photocopying, recording or by any information storage and
retrieval system, without permission in writing from Oxford University Press.
Enquiries concerning reproduction outside the scope of the above should be
sent to the Rights Department, Oxford University Press, at the address above

You must not circulate this book in any other binding or cover
and you must impose this same condition on any acquirer

ISBN-13: 978-0-19-568945-7
ISBN-10: 0-19-568945-3

Typeset in New Caledonia 10.5/12.2
by Eleven Arts, Keshav Puram, Delhi 110 035
Printed in India at De-Unique, New Delhi 110 018
Published by Oxford University Press
YMCA Library Building, Jai Singh Road, New Delhi 110 001

For my mother,
Mehjabeen Jalil

Contents

Preface xiii
Introduction xv

Birds and the Human Mindscape

BIRDS IN THE WRITER'S IMAGINATION

Panchatantra　3
The Heron that Liked Crab-meat
The Shrewd Old Gander

Joseph Rudyard Kipling　7
Rikki-Tikki-Tavi and Darzee

Douglas Dewar　12
The Cock-a-doodle-doo

Thomas Bainbrigge Fletcher
and Charles McFarlane Inglis　16
How the Hoopoe Got its Crest

DELIGHTFUL DISTRACTIONS

Nuruddin Salim Jahangir　21
The Quail and the Hawk
The Story of a Pair of Cranes

Khushwant Singh　24
Sher Singh Shoots a Crane

Mark Twain　28
In Praise of the Indian Crow

Abul Kalam Muhiyuddin Ahmed
(Maulana Azad)　30
Sparrows of Ahmad Nagar Fort Prison

Jim Corbett   53
Nightjar Eggs

M. Krishnan   58
Rescuing a Fledgeling

R. Lokaranjan   60
Delightful Distractions

D.A. Stairmand   62
Castaway with Birds

Simon Barnes   65
Falling in Love Again

Jawaharlal Nehru   71
'Foreword' to The Wildlife of India

Sport, Entertainment, and Falconry

HUNTING AND SPORT

Jim Corbett   77
Hunting at Bindukhera

Sálim Ali   80
Bird Holocausts at Bharatpur

Mary Anne Weaver   82
Hunting with the Sheikhs

ENTERTAINMENT

Zahir-ud-din Mohammad Babur   102
Parrot

Nuruddin Salim Jahangir   103
Carrier Pigeons

Abdul Halim Sharar   104
Bird-fighting and Pigeon-flying in Lucknow

Edward Hamilton Aitken   114
Peter and His Relations

William Dalrymple   118
Partridge Fight in Delhi

FALCONS AND FALCONRY

T.C. Jerdon   124
Falcons of India

R.S. Dharmakumarsinhji   128
Falconry Flights

Naturalists on the Prowl

Edward Hamilton Aitken   143
The Common Birds of Bombay
Bird Nesting

Douglas Dewar   167
The Great Himalayan Barbet
The Spotted Forktail
The Naturalist in a Railway Train

R.S.P. Bates   177
A South Indian Heronry

B.B. Osmaston   185
Edible Birds' Nests

Thomas Bainbrigge Fletcher and Charles McFarlane Inglis   187
The Common Hawk-Cuckoo or Brain-fever Bird

C.H. Donald   189
The Flight of Eagles

Natural History and Science

Zahir-ud-din Mohammad Babur   197
Birds of Hindustan

Nuruddin Salim Jahangir   203
Koel
A Quail with a Spur
Windpipes of Bustard and Crane
Close Observations on Bustards
The Nesting Cranes

Allan Octavian Hume and Charles Henry
Tilson Marshall   207
The Snow-wreath or Siberian Crane
The Pink-headed Duck
The Mute Swan

C.W. Mason and Harold Maxwell-
Lefroy   228
Economic Ornithology

Hugh Whistler   231
Suggestions on How to Run a Bird Survey

John Burdon Sanderson Haldane   236
The Non-violent Scientific Study of Birds

Birdwatching and Beyond

Sálim Ali   245
*Stopping by the Woods on a
    Sunday Morning*
The Nesting of Baya

Shivrajkumar (of Jasdan), Ramesh
M. Naik, and K.S. Lavkumar   253
*A Visit to the Flamingos in the Great Rann
    of Kutch*

Malcolm MacDonald   259
Green Parakeets in a Delhi Garden

Edward Pritchard Gee   276
The Breeding Birds of Bharatpur

R.A. Stewart Melluish   286
Notes from Madras

Nissim Ezekiel   290
Poet, Lover, Birdwatcher

Thomas Gay   291
An Evening at Pashan Lake, Poona

Peter Jackson   293
A Day's Worth of Delhi Birds

Philip Kahl 296
The Courtship of Storks

Madhav Gadgil 304
Ornithology in Bandipur

Hamida Saiduzzafar 315
Some Observations on the Apparent Decrease in Numbers of the Blue Jay

Prakash Gole 317
The Pair Beside the Lake

Asad Rafi Rahmani 322
The Greater Adjutant Stork

Otto Pfister 325
Cranes of Hanley

Mark Cocker 330
Rare Bird of the Mountains

S. Theodore Baskaran 332
The Haflong Phenomenon

Zai Whitaker 335
Misty Binoculars and Other Strategies for Survival among Birdwatchers

Personalities and Controversies

REMEMBERING SÁLIM ALI

Zafar Futehally 341
Remembering Sálim Ali

Bharat Bhushan 346
The EmPee Saar in Andhra Pradesh

SÁLIM ALI ON OTHERS

Sálim Ali 349
Colonel Meinertzhagen
Revealing Excerpts from the Diaries of Colonel Richard Meinertzhagen, D.S.O.

Edward Pritchard Gee   352
The Maharaja of Kutch

CONTROVERSIES

Madhusudan Katti   356
Are Warblers Less Important Than Tigers?

Sálim Ali   361
The Bird Ringing Controversey

R. Manakadan, J.C. Daniel, A.R. Rahmani, M. Inamdar, and Gayatri Ugra   365
Common English Names of the Birds of the Indian Subcontinent

Michael Lipske   371
Forest Owlet Thought to be Extinct is Spotted Anew after 113 Years

Sources   375
Acknowledgements   380
Further Reading   382

Preface

This book had its origins more than a decade ago when the late Ravi Dayal, Zafar Futehally (founder-editor of the *Newsletter for Birdwatchers*), and I toyed with the idea of bringing together the best articles published in the *Newsletter* as a volume. Unfortunately, this project did not materialize, although I continued to mull over the idea of compiling an anthology of writings on Indian birds. Late Shama Choudhary, a dear friend, who besides being an accomplished writer and novelist, was also a keen birdwatcher, helped in developing this idea further and provided a number of useful suggestions.

A number of friends and associates helped me, in one way or the other, while I was working on this book, and I extend my heartfelt thanks to all of them. I would particularly like to mention Anuradha Roy, Arvind Krishna Mehrotra, Bikram Grewal, Divybhanusinh Chavda, Ravi Singh, Shahid Amin, Sudhir Vyas, Theodore Baskaran, Toby Sinclair, and Vivek Khadpekar for their comments and suggestions. Sangita Gupta provided valuable assistance by not only locating inaccessible sources, but also by scanning material and making it available in no time.

I acknowledge the cooperation I received from the editorial team of the Oxford University Press India during the preparation of this anthology. A Fulbright Fellowship awarded to me during this period enabled me to tap library resources in some American universities and fish out worthwhile material, and I thank the United States Educational Foundation, India for providing me with this opportunity.

To my friends—Abid Shah and Zaheer Babar—I am grateful for support at crucial junctures. Siddiq Ahmed Siddiqi, my uncle, helped me to stay focused, and I thank him for countless, inspiring discussions. The reader will agree that for undertaking a project such as this, an interest in reading must be a minimum prerequisite. In my case, I owe this to my parents, who, ever since my childhood days, encouraged me to read. My mother, Mehjabeen Jalil, opened a whole new world for me—the world of literature and books—by bringing books and magazines from the school where she worked as a librarian. It is to her that I dedicate this book.

New Delhi Abdul Jamil Urfi
November 2007

Introduction

The interest in birds at all levels—from the popular to the scientific—is immense, and continues to grow by the day as birdwatching and nature study find more converts. Birds enjoy greater popularity than any other group of living organisms—butterflies, mammals, and plants included. This is not surprising considering the diurnal habits of birds, their engaging behaviour patterns, calls, plumage, and historically, their association with the conservation movement. But why an anthology of writings on Indian birds, and what could possibly be the best writings on this subject?

Birds are excellent model organisms for understanding key issues in ecology, animal behaviour, evolutionary biology, and conservation. Writers, poets, travellers, and painters down the ages have taken note of these winged creatures and some truly remarkable works on birds are on offer. Over the last century, during the course of the development of ornithology as a scientific discipline, professionals have derived support from an army of nature lovers and birdwatchers. Our times are characterized by an active popular interest in the conservation of biological diversity and the preservation of the natural environment. Given the special popularity of birds and the diverse types of writings on them, a compilation of some of these remarkable works is in order. The present anthology hopes to be a step in this direction.

Richard Mabey, well-known nature writer and anthologist, observes that 'deliberate attempts to portray the life of nature in prose face a huge philosophical barrier'. This is because,

[h]ere, after all, is language, one of the most exquisite human inventions, resonant with the structures of human consciousness, being used to describe a world about whose inner states and meanings we can know virtually nothing. It forces us to rely on external clues,

on empathy, and most notoriously on anthropomorphism, the assumption that nature shares human motives and feelings. Yet attempts to sidestep this by, for example, denying any inner lives to species other than our own, or attempting to contain their behaviour within apparently objective description (that is, description based on our definitions and categories), can also suffer from a kind of backdoor human-centeredness.[1]

Therefore, when I embarked upon the task of compiling an anthology of bird writings I did so with some trepidation. Also, being professionally involved with the study of birds, I wanted to ensure that the biases of my profession did not cloud my judgement while selecting material for this book. If any such biases had indeed managed to creep in then it would have been singularly unfortunate, since much of the technical literature on birds is too specialized to be of interest to the general reader.

II

My task was a daunting one. Principally, it was to decide how widely to throw the net and how big to keep the mesh size. To begin with, one could include pieces on each bird group, or take a different approach, by incorporating articles on birds found in different habitats—Himalayan birds, grassland birds, wetland birds, etc. Alternatively, one could go by the writers, by having articles by all the well-known ornithologists till date. Thirdly, one could take a chronological approach by picking up writings on birds from different historical periods; start with the earliest recorded writings, ramble though the medieval period, move on to the arrival of the Europeans on Indian shores, and finally end in the post-Indian independence era. Such an approach, if adopted, would follow a familiar and expected trajectory. Ancient Indian texts, dating to several hundred years BC, are replete with descriptions of birds.[2] Therefore, beginning with a somewhat

---

[1] Mabey, Richard, *The Oxford Book of Nature Writing*, Oxford: Oxford University Press, 1999.
[2] Dave, K.N., *Birds in Sanskrit Literature*, New Delhi: Motilal Banarsidass, 1985.

pastoral view of nature and the anthropomorphism of birds and other living creatures in ancient Indian texts, one then encounters meticulous descriptions of birds and other wildlife in the writings from the medieval period, generally stemming from the interest in *shikar* and hunting. Next comes the age of exploration and discovery of the mysterious Indian subcontinent, whether accidental or planned, in the larger interests of European imperialism. And finally, we come to the so-called 'modern times', post-independence India, with a focus on conservation, science, and also love of the outdoors—probably a manifestation of the values of the burgeoning Indian middle class.

Our bird and natural history heritage is rich, but equally so are the tapestry of writings on these subjects. As I turned to fellow birdwatchers, and wildlife experts for suggestions and ideas, I discovered that an anthology of writings on Indian birds could be prepared in as many ways as the number of people whose help I solicited. Encountering dizzying paradigm shifts, as I met one expert after another, I finally decided that the best way of doing the job would be to go by instinct and I tried to recollect all that I had read during the course of my evolution as a birdwatcher and ornithologist. I was always enamoured by writings that were in some sense 'alive', that is, either they had a story in them, or in some undefined way, were unique and transmitted the joy and thrill of seeing and studying birds, whether in their natural habitat or in residential premises and gardens. So those writings, which in my opinion ranked high on the basis of their reading pleasure, and not solely on the basis of their information content or ornithological value, were given priority. I especially tried to locate pieces which I felt would appeal to a broad spectrum of readership, and touched upon the experiences of our species in relation to the natural world, both from within and from outside; the passage of time and seasons and in a sense a wider perspective on life, with birds as a medium.

The writings included here are those which were written originally in English by writers for whom English was their first language or an adopted language. This naturally restricts the writings to one particular time period (writings from the British period up to the present day), and to a large extent, to one particular social class. In doing so, I have virtually excluded the wealth of writings in Indian languages. But given that this book is in English, such a step had to be taken, although I have tried to rectify this anomaly by including a few translations as well. The bulk of the writings included here are drawn from works of prose, though I have been unable to resist the temptation of including a poem as well. This compilation is by no means comprehensive, and suffers from the omission of several deserving articles. Some excellent pieces could not make it to the final cut, because of permission and copyright-related issues, which is indeed regrettable.

III

In the English language, a kaleidoscope of writings on birds is available. Many readers would be familiar with the writings of the legendary hunter and conservationist, Jim Corbett, whose stories on hunting man-eating tigers are admirable not just for their detail of the natural and cultural landscape, but also for their sheer beauty of expression. Corbett's writings usually mention birds in passing, but there is a galaxy of several other British writers whose writings on Indian birds are well regarded for their style and presentation. One name which comes to mind first is that of Edward Hamilton Aitken, also known simply as 'EHA'. Whether it was due to his long stay in India or his temperament, EHA developed an endearing writing style which seldom failed to establish a direct link with the reader. His writings are liberally peppered with amusing anecdotes, and I suspect that in spite of the stiff upper lip which most British residents in India at the time assiduously cultivated, EHA remained a boy at heart. EHA's most famous book, *A Naturalist on the Prowl*, begins thus:

I have always felt a strange pleasure in seeing without being seen. Even when I was an indolent little man of six it gave me rare delight to hide under a sofa and peep at the feet of everybody who passed through the room. 'Ha! He does not know that I am here,' I said to myself and 'chortled'. I cannot quite satisfactorily analyze this kind of enjoyment and am not sure it is very respectable, but it is very human. Stolen waters are sweet, and bread eaten in secret is pleasant. I have long since given up the pastime of prying into the secret ways of my kind, and to crawl under furniture would now be irksome to me; but I wander into the jungle, where 'things that own not man's dominion dwell,' and there I prowl, climb into a tree, sit under a bush, or lie on the grass, and watch the ways of my fellow-creatures, seeing but unseen, or, if seen, not regarded ...

Coming to Indian writers, one of the most notable ornithological figures of the last century—Sálim Ali—has the distinction of being not only a subject expert *par excellence*, but also a master of the English language. Indeed, his first major 'discovery'— an understanding of the polygynous breeding behaviour of the Baya Weaver bird was acclaimed as much for its original observations and deductions, as for its writing style, winning an award for the best writing on nature in English by an Indian when it was published.

It is of course impossible to completely disentangle observations on nature from the life and times of their authors. Science, which is often regarded as an embodiment of truth, is seldom free from the personalities of the scientists and the influences of their times. Writings on birds cannot be an exception and I hope that in the selections presented here the perceptive reader will also glean this 'non-ornithological leakage'. Sálim Ali's autobiography, *The Fall of a Sparrow*, is replete with vivid descriptions of the politics of his times, the conflicts of class interest, and notions of racial superiority. Interestingly, some of his writings on birds also reflect the prevalent biases of the time. Consider, for instance, the description of the mating behaviour of the Bustard-Quail as described in *The Book of Indian Birds*:

The normal condition among birds is that where the sexes differ in coloration, it is the male who is the brighter coloured and more showy. He displays his splendour before the female, courts her and if need be fights furiously with rival males for her possession. In the Bustard-Quail, however, the role of the sexes is reversed. Here it is the female who is the larger and more brightly coloured, and who takes the initiative in affairs of the heart. She decoys eligible males by a loud drumming call, courts them sedulously, displaying all her charms before them, and engages in desperate battles with rival Amazons for the ownership of the favoured one. As soon as the husband is secured, the preliminaries over, and the full complement of eggs laid, she leaves him to his own devices and wanders off in search of fresh conquests. The unfortunate husband is saddled with the entire responsibility of incubating the eggs and tending the young which, to his credit, he discharges admirably and with great solicitude. By feminine artifice the roving hen manages to inveigle another unattached cock, who is likewise landed with family cares. And she is once again in the market, for a third husband! In this manner each hen may lay several clutches of eggs during a single season which, accordingly, is much prolonged.

An excellent piece of writing, but note the uninhibited use of such expressions as 'decoys eligible males', 'courts them sedulously', 'the unfortunate husband', 'fresh conquests', and 'feminine artifice' in the quoted passage. Such expressions would be considered politically incorrect today, and would greatly annoy many readers.

IV

This anthology is somewhat loosely structured although it does make pretence of following a pattern of sorts. It is divided into six sections, each being a collection of writings on a broad, though specific theme. The first section, 'Birds and the Human Mindscape', includes human interest angle stories about birds in mythologies and, how they have figured in the writings of poets, statesmen, and travellers. It starts off with pieces which deal with birds in the writer's imagination, whether it be the *Panchatantra* tales, the Biblical story about 'How the Hoopoe got its Crest', or a more contemporary portrait of the familiar rooster ('The Cock-A-doodle-doo'). Many writers have treated animals as human characters, some of the most memorable ones being those created by Rudyard Kipling in his *Jungle Book*. In the extract from this famous book included here, although the central character is not a bird but the mongoose, Rikki-Tikki-Tavi, several bird characters, notably Darzee, the Tailor bird, figure prominently.

The other kind of writings included in this section are those in which birds provide a diversion, or some sort of a fulcrum, in the chain of human thoughts ('Delightful Distractions' – the title of the subsection borrowed from one of the pieces included in this section). Interestingly, the Sarus crane figures prominently in some of the pieces. For instance, in the *Jahangirnama*, Emperor Jahangir records the devotion in Sarus crane couples for each other. Also, in the piece by Khushwant Singh, when Sher Singh, one of the main characters in the novel, *I shall not hear the Nightingale*, shoots a crane (for 'baptism in blood') and he is immediately confronted with a jumble of conflicting emotions of guilt and pride. Besides several other fine pieces, including those by Jim Corbett, Mark Twain, D.A. Stairmand, M. Krishnan, R. Lokaranjan, Simon Barnes, and a short piece by Jawaharlal Nehru, the extracts from *Ghubar-e-Khatir* by Maulana Abul Kalam Azad stand out. A noteworthy feature of Azad's piece is the details he recorded on the behaviour of the sparrows which used to visit his prison cell in the Ahmednagar Fort Prison, where he as well as other important leaders of the freedom movement were imprisoned, immediately after the 'Quit India' resolution was adopted by the Indian National Congress in August 1942. Of the multitude of sparrows who used to visit Azad's prison cell, three sparrows – Moti, Qalandar, and Mulla – stood out from the crowd. Azad's description of Mulla is quite interesting.

One of the sparrows has a very stout build and is extremely contentious. Its tongue is ever active, its gait has a swagger and it picks up a fight with whoever it sees. No neighbourhood sparrow can set foot in its domain; several brave ones tried but were laid down in the very first encounter. Whenever a gathering of the company is held on the floor, it comes with its peculiar swagger, casts a glance all round and jumps up on a high seat ... Now tell me what other name would suit it if not Mulla?

A caged bird is a very sorry thing. The famous Urdu poet Mohammad Iqbal touched upon this sentiment very poignantly in his poem '*Parinde Ki Faryad*' (The Song of the Caged Bird). But throughout human history, birds have been popular as pets and entertainers, and for sport. Nowadays, due to the movement by animal rights activists, the idea of keeping birds as pets has lost some of its popularity, and shikar, falconry, and bird fights are now banned. From our point of view, what is interesting is the way in which writers and naturalists have touched upon this subject. Writings on these topics constitute the second section, 'Sport, Entertainment, and Falconry'. The intention here is not to glamorize either falconry or shikar, but only to present a sample of writings on a subject, which at one point of time, was a favourite human past-time.

This section starts off with a piece extracted from Jim Corbett's 'The Talla Des Man Eater', in which the author describes the scene at an organized shoot in the twilight years of the British Raj. After bagging a number of birds and other game, the hunting party pauses to observe a drama unfold in the skies as a ground owl tries to escape from the talons of a falcon. In the end, the owl manages to save his life and Corbett's observation is pithy,

The reactions of human beings to any particular event are unpredictable. Fifty-four animals had been shot that morning—and many more missed—without a qualm or the batting of an eyelid. And now, guns, spectators, and *mahouts* were unreservedly rejoicing that a ground owl had escaped the talons of a peregrine falcon.

In a similar vein, Sálim Ali describes the large scale, organized duck shoots (termed as 'holocausts') that were a common phenomenon during this period. 'Hunting with the Sheikhs' by Mary Anne Weaver is a powerful narrative about organized *houbara* hunts undertaken by the Arab Sheikhs till some decades ago. Although the setting of this story is in Sindh, Pakistan, it has a strong sub-continental flavour. The other pieces in this section cover a variety of writings on birds used for entertainment, bird fighting, pigeon flying, and falconry, as well as caged birds. The writings included here are those by the Mughal emperors Babur and Jahangir, Abdul Halim Sharar (about bird fighting and pigeon flying in Lucknow), EHA, and a vivid description of a patridge fight in the heart of old Delhi by the well known writer William Dalrymple. Also include in this section is an account of the falcons of India by T.C. Jerdon and the classic piece 'Falconry Fights' by R.S. Dharmakumarsinhji.

Following the age of exploration and discovery, many Europeans started arriving in India, and the British, who were successful in gaining control over India after defeating rival powers, went about the business of administering the country and, in the process, also exploring and documenting its natural wealth. This period saw a number of dedicated naturalists and biologists who also took serious steps towards the establishment of natural history research in India. As a result, writings by Englishmen constitute a large share of the literature on Indian birds. Many of these writings reveal a feeling of bafflement, surprise, and amazement at encountering Indian customs, manners, and birds, but the dilemmas are often laid to rest in a characteristic, humorous manner, somewhat typical of the British. A case in point is the description of the Common Hawk-Cuckoo in the story by Fletcher and Inglis. The ear splitting and nerve racking cries of this bird so irritated the Britishers that they named it the 'Brainfever Bird'.

Interestingly, the huge amount of interest in birds and natural history that civil officers, foresters, policemen, and other servants of the British Empire exhibited, as is evident from the years of publication of their books, is mostly dated post-1857. Also, a number of books on birds of this period are about garden birds and bear such titles as *Birds in My Indian Garden*, indicative of the British people's fondness for home gardens. The third section of this book 'Naturalists on the Prowl' (the title borrowed from EHA's famous book) presents a sample of these writings. Besides the inimitable EHA, this section also showcases the writings of some other well known British writers such as R.S.P Bates, Douglas Dewar, B.B. Osmaston, and C.H. Donald.

Although systematic works on Indian birds date back to ancient and medieval times, the modern 'scientific' approach to studying birds is said to have begun with the arrival of the Europeans on Indian shores. According to Sálim Ali,[3] one of the earliest 'modern' or 'scientific' accounts of Indian birds was published in 1713 by Edward Buckley, an East India Company surgeon in Madras. Buckley's work contains descriptions and drawings of twenty-two birds found in and around Fort St. George. Following

---

[3]Ali, Sálim, *Bird Study in India: Its History and its Importance*, Maulana Azad Memorial Lecture, New Delhi: Indian Council for Cultural Relations, 1978.

this, Colonel W.H. Sykes from the Bombay army published *A Catalogue of the Birds of the Deccan* in 1832. An interesting aspect of Sykes's work is that he described a number of new species, many of which he named after Hindu deities, such as *Milvus govinda* (Pariah Kite), *Hippolais rama* (Tree Warbler), *Petrocinchla pandoo* (Blue Rock Thrush), and *Hypsipetes ganeesa* (Southerly Black Bulbul).

However, for all practical purposes, Indian ornithology, as we understand it today, took off with the publication of the two-volume *Birds of India* by T.C. Jerdon, which appeared between 1862 and 1864. The other major figure in Indian ornithology is A.O. Hume, who besides writing a number of books on birds, also edited a journal, *Stray Feathers*, devoted to the study of Indian birds. A notable feature of Hume's career as an ornithologist was his extensive network of correspondents, scattered across India and abroad, including among them some of the most eminent naturalists and sportsmen of the day such as Brian Hodgson, Leith Adams, C.T. Bingham, W. Blanford, Edward Blyth, T.C. Jerdon, Eugene Oates, Ferdinand Stoliczka, Charles Swinhoe, and Samuel Tickell, among others.

The fourth section 'Natural History and Science' showcases writings that provide insights into the development of ornithology in India. It starts off with a sample of writings extracted from the memoirs of the Mughal emperors, Babur and Jahangir, followed by a selection from Hume's writings. Other pieces in this section, including 'Suggestions on How to Run a Bird Survey' by Hugh Whistler, and 'The Non-violent Scientific Study of Birds' by J.B.S. Haldane, provide a glimpse of modern approaches to bird study. Mason and Maxwell-Lefroy's introduction to 'Economic Ornithology' makes for interesting reading, coming as it does from the official agricultural establishment of British India.

But, presumably, the reason why most of us want to read about birds is simply because we love them, indulge in bird watching, and savour being in the outdoors. And generally, there is more to it than just watching birds! Although the fifth section entitled 'Birdwatching and Beyond' includes a range of articles by Indians as well as foreign visitors, my favourite one is the hilarious piece by Zai Whittaker ('Misty Binoculars and Other Survival Strategies among Birdwatchers') for its tongue-in-cheek, somewhat anti-birding slant. It takes me back to the days when, as a boy, I had taken to bird watching and used to go out on field outings with fellow birdwatchers, each invariably trying to impress the other by his bird identification abilities.

Times were when, if you looked up on a bright sunny winter afternoon, you could see plenty of vultures, floating like 'tea trays in the sky'. But now, chances are, you will mostly see empty space or, if you happen to be standing close to a garbage disposal pit, then plenty of kites hovering around. Who would have thought that one day the vultures—one of our commonest birds and also perhaps, the most abhorred—would disappear so suddenly? But this is exactly what has happened, bringing about unprecedented problems in its wake and posing new questions. Carcasses of dead animals lie strewn across the countryside as there are no vultures to dispose them. (One hears reports that the numbers of feral dogs have increased in recent years because of abundant food being available.) But it is not just the vultures that are disappearing from our skies. The sparrow too, it is feared, is declining in number throughout the country. The familiar '*chiriyas*' who used to be considered a nuisance due to their habit of building nests in the cups of ceiling fans and creating a mess are now becoming less common. The writing is clearly on the wall: something is seriously wrong, and we will all have to pay for it.

Witnessing the destruction of bird habitats and learning that many species are threatened is not pleasant and birdwatchers, since they are out in the field quite often, get a first-hand perspective about environmental degradation. Small wonder then that many articles written on birds nowadays end with pleas to save natural habitats or rally for support for their protection. In this section, at least three articles focus on the decline of bird populations and try to make a case for conservation. Prakash Gole's piece 'The Pair beside the Lake', focusing on the plight of the Sarus Crane, is noteworthy for the linkages that he draws between development and environmental harmony.

The last section, 'Personalities and Controversies', in a sense, completes a cycle. It again places our interest in birds in the context of humans and their approaches to bird study. Over the last century, Sálim Ali played a larger-than-life role in shaping the development of ornithology and conservation in India. While he was deeply rooted in the fashions and prejudices prevailing in his time, instead of being a prisoner, he was a trendsetter, acutely aware of contemporary political realities. This section includes pieces on him and by him, about other ornithological personalities of his time. The bulk of this section however deals with controversies of one sort or the other in the world of conservationists and ornithologists. For instance, is the conservation of tigers more important than protecting birds and other small creatures? This was a matter of controversy some decades ago although the question seems somewhat redundant now because the tiger itself has totally vanished from many prime areas of the country. Madhusudan Katti addresses this controversy ('Are Warblers Less Important than Tigers?') and makes a case for the preservation of tiny, nondescript warblers inhabiting the forests of Kalakad Mundanthurai Tiger Reserve in South India. Sálim Ali's piece describes a controversy that enveloped the bird ringing program of the Bombay Natural History Society in 1957. More recently, there was a flutter in the scientific world when it was discovered that due to a fraud committed by a well-known ornithologist, the rediscovery of a rare species of owlet – 'Blewitti's or Forest Spotted Owlet', believed to be extinct for almost a century, was stalled as the field survey teams were misled and wasted much time and effort barking up the wrong tree.

Finally, one issue that has cropped up in recent years and which deserves special mention concerns the plans that are afoot to standardize the English names of birds. It turns out that the business of changing bird names is not so simple. Many names have historical connotations and even ornithologists—who are supposed to be un-sentimental scientists working for the 'greater common good'—tend to feel touchy, and sometimes, even patriotic, when they see attempts being made to tamper with their 'heritage'. The issues actually run much deeper, especially in today's intellectual climate. Many naturalists talk of 'neo-imperialism', the rivalry between the Europeans and the Americans for renaming birds. Certainly, there are many who feel that in the new bird taxonomy being proposed, there is more than what meets the eye. Unfortunately, there is very little good writing on this sensitive subject and the piece included here, though slightly technical, at least provides a glimpse of the controversy.

V

Although there is a vast amount of material on Indian birds, it was possible to include only some of it in this anthology, chiefly due to limitations of space. In all, there are forty-seven different authors/sources and seventy-six separate pieces of varying lengths in this anthology. Even this selection, though limited, throws up an impressive diversity of writing styles and themes. It almost seems as if these authors have lent their voice so that I, as the anthologist, can also tell a story; a story full of action and drama, for there are pieces in this book, which in my view, can bring a lump in the throat and indeed, some are loaded with suspense. One ingredient that one might think would be missing is sex and romance. However, there is plenty of that too, especially in the piece by Malcolm MacDonald in the section 'Birdwatching and Beyond', where the amorous antics of considerable Rose-ringed parakeet in a Delhi garden are discussed in considerable detail.

There are also some pieces here whose inclusion might appear to be difficult to justify in an anthology of writings on Indian birds. For instance, I have included a wonderful piece by Simon Barnes, from his best-selling book, *How to be a (Bad) Birdwatcher*. Although the piece mentions India and Indian birds in passing (the author, being a sports correspondent visited India to cover a cricket match), the reason why I decided to include this piece was because it provided a certain perspective on human life and birds. But also, it brought back memories of the period of the 1960s, during which I was born. Although I missed the peak of the hippie revolution and the coming of the flower children (the so-called 'third generation'), later, at the university, I spent countless hours discussing this 'revolution' with friends. The following extract from Barnes' book makes a point about the attitude of those days,

Being too specific would have been a bit of a *faux pas* in those days. One of the set texts of hippie-dom—a book I never cared for—was *Jonathan Livingston Seagull*. Jonathan Livingston black-headed gull? Audouin's gull? Bonaparte's gull? Don't be uncool, man: it's a seagull, just dig it. Precision was frowned upon. "Perfect speed is being there," said the book. Or perhaps it was a glaucous gull.

VI

Most of the writings included in the book have been reproduced as such, and I have generally refrained from making any alterations in the original texts. However, in some cases, when a printed mistake appeared obvious, I have made appropriate changes. The English and scientific names of birds have not been changed in any of the pieces, except for making the first alphabet capitals to allow for uniformity in the text. In those writings that have footnotes, the numbering has been altered to suit the requirements of this anthology. Thus, footnote numbers will not match numbers in the original texts. For each section, the footnotes are serially numbered. In some cases, I have provided my own footnotes, which are separately marked.

Each piece is followed by the name(s) of the author(s), the title of the book/name of the journal from where the piece was extracted, and the year in which it was written. The last bit of information is provided in most cases, though not in all. The full details of each piece are provided in the section 'Sources' towards the end of the book. For each section, the sources are arranged in the same sequence in which they appear in the book.

Birds and the Human Mindscape

1

Sketch titled 'The Indian Hoopoe' accompanying the biblical story of King Solomon in T.B. Fletcher and C.M. Inglis's *Birds of an Indian Garden* (1936)

Birds in the Writer's Imagination

The Panchatantra *stories form an integral part of growing up for most children in India. These stories probably hark back to the origins of language and the subcontinent's earliest social groupings of hunting and fishing folk gathered around campfires. Although similar animal fables are found in many other cultures, some experts believe that the* Panchatantra *tales were the prime source for other tales, such as* Aesop's Fables.

The Heron that Liked Crab-meat

A heron ate what fish he could,
The bad, indifferent, and good;
His greed was never satisfied
Till, strangled by a crab, he died.

There was once a heron in a certain place on the edge of a pond. Being old, he sought an easy way of catching fish on which to live. He began by lingering at the edge of his pond, pretending to be quite irresolute, not eating even the fish within his reach.

Now among the fish lived a crab. He drew near and said: 'Uncle, why do you neglect today your usual meals and amusements?' And the heron replied: 'So long as I kept fat and flourishing by eating fish, I spent my time pleasantly, enjoying the taste of you. But a great disaster will soon befall you. And as I am old, this will cut short the pleasant course of my life. For this reason I feel depressed.'

'Uncle,' said the crab, 'of what nature is the disaster?' And the heron continued: 'Today I overheard the talk of a number of fishermen as they passed near the pond. "This is a big pond," they were saying, "full of fish. We will try a cast of the net tomorrow or the day after. But today we will go to the lake near the city." This being so, you are lost, my food supply is cut off, I too am lost, and in grief at the thought, I am indifferent to food today.'

Now when the water-dwellers heard the trickster's report, they all feared for their lives and implored the heron, saying: 'Uncle! Father! Brother! Friend! Thinker! Since you are informed of the calamity, you also know the remedy. Pray save us from the jaws of this death.'

Then the heron said: 'I am a bird, not competent to contend with men. This, however, I can do. I can transfer you from this pond to another, a bottomless one.' By this artful speech they were so led astray that they said: 'Uncle! Friend! Unselfish kinsman! Take me first! Me first! Did you never hear this?

> Stout hearts delight to pay the price
> Of merciful self-sacrifice,
> Count life as nothing, if it end
> In gentle service to a friend.'

Then the old rascal laughed in his heart, and took counsel with his mind, thus: 'My shrewdness has brought these fishes into my power. They ought to be eaten very comfortably.' Having thus thought it through, he promised what the thronging fish implored, lifted some in his bill, carried them a certain distance to a slab of stone, and ate them there. Day after day he made the trip with supreme delight and satisfaction, and meeting the fish, kept their confidence by ever new inventions.

One day the crab, disturbed by the fear of death, importuned him with the words: 'Uncle, pray save me, too, from the jaws of death.' And the heron reflected: 'I am quite tired of this unvarying fish diet. I should like to taste him. He is different, and choice.' So he picked up the crab and flew through the air.

But since he avoided all bodies of water and seemed planning to alight on the sun-scorched rock, the crab asked him: 'Uncle, where is that pond without any bottom?' And the heron laughed and said: 'Do you see that broad, sun-scorched rock? All the water-dwellers have found repose there. Your turn has now come to find repose.'

Then the crab looked down and saw a great rock of sacrifice, made horrible by heaps of fish-skeletons. And he thought: 'Ah me!

> Friends are foes and foes are friends
> As they mar or serve your ends;
> Few discern where profit tends.

Again:

> If you will, with serpents play;
> Dwell with foemen who betray:
> Shun your false and foolish friends,
> Fickle, seeking vicious ends.

Why, he has already eaten these fish whose skeletons are scattered in heaps. So what might be an opportune course of action for me? Yet why do I need to consider?

> Man is bidden to chastise
> Even elders who devise
> Devious courses, arrogant,
> Of their duty ignorant.

Again:

> Fear fearful things, while yet,
> No fearful thing appears;
> When danger must be met,
> Strike, and forget your fears.

So, before he drops me there, I will catch his neck with all four claws.'

When he did so, the heron tried to escape, but being a fool, he found no parry to the grip of the crab's nippers, and had his head cut off.

Then the crab painfully made his way back to the pond, dragging the heron's neck as if it had been a lotus-stalk. And when he came among the fish, they said: 'Brother, why come back?' Thereupon he showed the head as his credentials and said: 'He enticed the water-dwellers from every quarter, deceived them with his prevarications, dropped them on a slab of rock not far away, and ate them. But I—further life being predestined—perceived that he destroyed the trustful, and I have brought back his neck. Forget your worries. All the water-dwellers shall live in peace.'

The Shrewd Old Gander

Take old folks' counsel (those are old
Who have experience)
The captive wild-goose flock was freed
By one old gander's sense.

In part of a forest was a fig tree with massive branches. In it lived a flock of wild geese. At the root of this tree appeared a creeping vine of the species called *koshambi*. Thereupon the old gander said: 'This vine that is climbing our fig tree bodes ill to us. By means of it, someone might perhaps climb up here some day and kill us. Take it away while it is still slender and readily cut.' But the geese despised his counsel and did not cut the vine, so that in course of time it wound its way up the tree.

Now one day when the geese were out foraging, a hunter climbed the fig tree by following the spiral vine, laid a snare among the nests, and went home. When the geese, after food and recreation, returned at nightfall, they were caught to the last one. Whereupon the old gander said: 'Well, the disaster has taken place. You are caught, having brought it on yourselves by not heeding my advice. We are all lost now.'

Then the geese said to him: 'Sir, the thing having come to pass, what ought we to do now?' And the old fellow replied: 'If you will take my advice, play dead when that hateful hunter comes. And when the hunter inferring that we are dead, throws the last one to the ground, we then must all rise simultaneously, flying over his head.'

At early dawn the hunter arrived, and when he looked them over, everyone seemed as good as dead. He therefore freed them from the snare with perfect assurance, and threw them all to the ground, one after the other. But when they saw him preparing to descend, they all followed the shrewd plan of the old gander and flew up simultaneously.

Panchatantra

Joseph Rudyard Kipling's (1865–1936) The Jungle Book, The Second Jungle Book, *and* Kim *are regarded as classics in children's literature. In the extract, Kipling paints an intimate portrait of 'Darzee', the Tailorbird. Kipling's other literary works include* Just So Stories, Puck of Pook's Hill, Mandalay, Gunga Din, 'If—', Life's Handicap, The Day's Work, *and* Plain Tales from the Hills.

Rikki-Tikki-Tavi and Darzee

Without waiting for breakfast, Rikki-tikki ran to the thorn-bush where Darzee was singing a song of triumph at the top of his voice. The news of Nag's death was all over the garden, for the sweeper had thrown the body on the rubbish-heap.

'Oh, you stupid tuft of feathers!' said Rikki-tikki angrily. 'Is this the time to sing?'

'Nag is dead—is dead—is dead!' sang Darzee. 'The valiant Rikki-tikki caught him by the head and held fast. The big man brought the bang-stick, and Nag fell in two pieces! He will never eat my babies again.'

'All that's true enough; but where's Nagaina?' said Rikki-tikki, looking carefully round him.

'Nagaina came to the bathroom sluice and called for Nag,' Darzee went on; 'and Nag came out on the end of a stick—the sweeper picked him up on the end of a stick—the sweeper picked him up on the end of a stick and threw him upon the rubbish-heap. Let us sing about the great, the red-eyed Rikki-tikki!' and Darzee filled his throat and sang.

'If I could get up to your nest, I'd roll all your babies out!' said Rikki-tikki. 'You don't know when to do the right thing at the right time. You're safe enough in your nest there, but it's war for me down here. Stop singing a minute, Darzee.'

'For the great, the beautiful Rikki-tikki's sake I will stop,' said Darzee. 'What is it, O Killer of the terrible Nag?'

'Where is Nagaina, for the third time?'

'On the rubbish-heap by the stables, mourning for Nag. Great is Rikki-tikki with the white teeth.'

'Bother my white teeth! Have you ever heard where she keeps her eggs?'

'In the melon-bed, on the end nearest the wall, where the sun strikes nearly all day. She hid them there weeks ago.'

'And you never thought it worthwhile to tell me? The end nearest the wall, you said?'

'Rikki-tikki, you are not going to eat her eggs?'

'Not eat exactly; no. Darzee, if you have a grain of sense you will fly off to the stables and pretend that your wing is broken, and let Nagaina chase you away to this bush. I must get to the melon-bed, and if I went there now she'd see me.'

Darzee was a feather-brained little fellow who could never hold more than one idea at a time in his head; and just because he knew that Nagaina's children were born in eggs like his own, he didn't think at first that it was fair to kill them. But his wife was a sensible bird, and she knew that cobra's eggs meant young cobras later on; so she flew off from the nest, and left Darzee to keep the babies warm, and continue his song about the death of Nag. Darzee was very like a man in some ways.

She fluttered in front of Nagaina by the rubbish-heap, and cried out: 'Oh, my wing is broken! The boy in the house threw a stone at me and broke it.' Then she fluttered more desperately than ever.

Nagaina lifted up her head and hissed: 'You warned Rikki-tikki when I would have killed him. Indeed and truly, you've chosen a bad place to be lame in.' And she moved toward Darzee's wife, slipping along over the dust.

'The boy broke it with a stone!' shrieked Darzee's wife.

'Well, it may be some consolation to you when you're dead to know that I shall settle accounts with the boy. My husband lies on the rubbish-heap this morning, but before night the boy in the house will lie very still. What is the use of running away? I am sure to catch you. Little fool, look at me!'

Darzee's wife knew better than to do *that*, for a bird who looks at a snake's eyes gets so frightened that she cannot move. Darzee's wife fluttered on, piping sorrowfully, and never leaving the ground, and Nagaina quickened her pace.

Rikki-tikki heard them going up the path from the stables, and he raced for the end of the melon-patch near the wall. There, in the warm litter about the melons, very cunningly hidden, he found twenty-five eggs, about the size of a bantam's eggs, but with whitish skin instead of shell.

'I was not a day too soon,' he said; for he could see the baby cobras curled up inside the skin, and he knew that the minute they were hatched they could each kill a man or a mongoose. He bit off the tops of the eggs as fast as he could, taking care to crush the young cobras, and turned over the litter from time to time to see whether he had missed any. At last there were only three eggs left, and Rikki-tikki began to chuckle to himself, when he heard Darzee's wife screaming: 'Rikki-tikki, I led Nagaina toward the house, and she has gone into the veranda, and—oh, come quickly—she means killing!'

Rikki-tikki smashed two eggs, and tumbled backward down the melon-bed with the third egg in his mouth, and scuttled to the veranda as hard as he could put foot to the ground. Teddy and his mother and father were there at early breakfast; but Rikki-tikki saw that they were not eating anything. They sat stone-still, and their faces were white. Nagaina was coiled up on the matting by Teddy's chair, within easy striking-distance of Teddy's bare leg, and she was swaying to and fro singing a song of triumph.

'Son of the big man that killed Nag,' she hissed, 'stay still. I am not ready yet. Wait a little. Keep very still, all you three. If you move I strike, and if you do not move I strike. Oh, foolish people, who killed my Nag!'

Teddy's eyes were fixed on his father, and all his father could do was to whisper: 'Sit still, Teddy. You mustn't move. Teddy, keep still.'

Then Rikki-tikki came up and cried: 'Turn round, Nagaina; turn and fight!'

'All in good time,' said she, without moving her eyes. 'I will settle my account with *you* presently. Look at your friends, Rikki-tikki. They are still and white; they are afraid. They dare not move, and if you come a step nearer I strike.'

'Look at your eggs,' said Rikki-tikki, 'in the melon-bed near the wall. Go and look, Nagaina.'

The big snake turned half round, and saw the egg on the veranda. 'Ah-h! Give it to me,' she said.

Rikki-tikki put his paws one on each side of the egg, and his eyes were blood-red. 'What price for a snake's egg? For a young cobra? For a young king-cobra? For the last—the very last of the brood? The ants are eating all the others down by the melon-bed.'

Nagaina spun clear round, forgetting everything for the sake of the one egg; and Rikki-tikki saw Teddy's father shoot out a big hand, catch Teddy by the shoulder, and drag him across the little table with the tea-cups, safe and out of reach of Nagaina.

'Tricked! Tricked! Tricked! *Rikk-tck-tck!*' chuckled Rikki-tikki. 'The boy is safe, and it was I—I—I that caught Nag by the hood last night in the bath-room.' Then he began to jump up and down, all four feet together, his head close to the floor. 'He threw me to and fro, but he could not shake me off. He was dead before the big man blew him in two. I did it. *Rikki-tikki-tck-tck!* Come then, Nagaina. Come and fight with me. You shall not be a widow long.'

Nagaina saw that she had lost her chance of killing Teddy, and the egg lay between Rikki-tikki's paws. 'Give me the eggs, Rikki-tikki. Give me the last of my eggs, and I will go away and never come back,' she said, lowering her hood.

'Yes, you will go away, and you will never come back; for you will go to the rubbish-heap with Nag. Fight, widow! The big man has gone for his gun! Fight!'

Rikki-tikki was bounding all round Nagaina, keeping just out of reach of her stroke, his little eyes like hot coals. Nagaina gathered herself together, and flung out at him. Rikki-tikki jumped up and backward. Again and again and again she struck, and each time her head came with a whack on the matting of the veranda, and she gathered herself together like a watch-spring. Then Rikki-tikki danced in a circle to get behind her, and Nagaina spun round to keep her head to his head, so that the rustle of her tail on the matting sounded like dry leaves blown along by the wind.

He had forgotten the egg. It still lay on the veranda, and Nagaina came nearer and nearer to it, till at last, while Rikki-tikki was drawing breath, she caught it in her mouth, turned to the veranda steps, and flew like an arrow down the path, with Rikki-tikki behind her. When the cobra runs for her life, she goes like a whiplash flicked across a horse's neck.

Rikki-tikki knew that he must catch her, or all the trouble would begin again. She headed straight for the long grass by the thorn-bush, and as he was running Rikki-tikki heard Darzee still singing his foolish little song of triumph. But Darzee's wife was wiser. She flew off her nest as Nagaina came along, and flapped her wings about Nagaina's head. If Darzee had helped they might have turned her; but Nagaina only lowered her hood and went on. Still, the instant's delay brought Rikki-tikki up to her, and as she plunged into the rat-hole where she and Nag used to live, his little white teeth were clenched on her tail, and he went down with her—and very few mongooses, however wise and old they may be, care to follow a cobra into its hole. It was dark in the hole; and Rikki-tikki never knew when it might open out and give Nagaina room to turn and strike at him. He held on savagely, and struck out his feet to act as brakes on the dark slope of the hot, moist earth.

Then the grass by the mouth of the hole stopped waving, and Darzee said: 'It is all over with Rikki-tikki! We must sing his death-song. Valiant Rikki-tikki is dead! For Nagaina will surely kill him underground.'

So he sang a very mournful song that he made up on the spur of the minute, and just as he got to the most touching part the grass quivered again, and Rikki-tikki, covered with dirt, dragged himself out of the hole leg by leg, licking his whiskers. Darzee stopped with a little shout. Rikki-tikki shook some of the dust out of his fur and sneezed. 'It is all over,' he said. 'The widow will never come out again.' And the red ants that live between the grass-stems heard him, and began to troop down one after another to see if he had spoken the truth.

Rikki-tikki curled himself up in the grass and slept where he was—slept and slept till it was late in the afternoon, for he had done a hard day's work.

'Now,' he said, when he awoke, 'I will go back to the house. Tell the Coppersmith, Darzee, and he will tell the garden that Nagaina is dead.'

The Coppersmith is a bird who makes a noise exactly like the beating of a little hammer on a copper pot; and the reason he is always making it is because he is the town-crier to every Indian garden, and tells all the news to everybody who cares to listen. As Rikki-tikki went up the path, he hard his 'attention' notes like a tiny dinner-gong; and then the steady *'Ding-dong-tock!* Nag is dead—*dong!* Nagaina is dead! *Ding-dong-tock!'* That set all the birds in the garden singing, and the frogs croaking; for Nag and Nagaina used to eat frogs as well as little birds.

JOSEPH RUDYARD KIPLING, *The Jungle Book*, 1894

Douglas Dewar (1875–1957), an amateur ornithologist in British India, was a prolific writer. His well-known works on birds include Birds of the Plains, Glimpses of Indian Birds, Birds of the Indian Hills, A Bird Calendar for Northern India, Himalayan and Kashmiri Birds, Birds at the Nest, The Common Birds of India, Indian Bird Life, Game Birds, *and* Indian Birds' Nests. *Dewar's other books include* Bombay Ducks, The Indian Crow, Birds of an Indian Village, *and* Animals of No Importance.

The Cock-a-doodle-doo

Ever since that far-off day in the prehistoric past, when some unknown Aryan *shikari* captured a pair of *Gallus ferrugineus* and domesticated them, the fowl has been the constant companion and friend of man. The utility of the hen bird soon rendered her indispensable to human beings, while the proud bearing and the valour of the cock gained for him the admiration of mankind.

Idomeneus bore on his shield at the siege of Troy a representation of the gallant chanticlere. The war-like Romans held the birds in high esteem; they were in the habit of using them as augurs. The method of ascertaining the will of the gods was to place food before the sacred birds. If the grain was consumed quickly, the omen was favourable; if, on the other hand, the fowls were slow in disposing of the victuals, the omen was evil. Since both cocks and hens have a habit of devouring their food as though they were travellers, determined to have their money's worth, eating dinner at a railway restaurant with the train waiting impatiently outside, it was not often that fowls gave an unfavourable omen. On one memorable occasion, however, they seem to have been off colour; the *pullarius* must have been trying experiments with them, for they refused the food offered them. This was too much for Claudius Pulcher, who was consulting them; he fairly lost his temper, seized the recalcitrant birds, and threw them into the sea, with the remark, 'If you won't eat, you rascals, you shall drink!'

Our mediaeval ancestors highly honoured the cock. Gerald Legh asserts that 'the Cocke is the royallest birde that is, and of himself a king, for Nature hath crowned him with a perpetual diademe, to hime and his posteritie for ever. He is the valliantest in battle of all birdes, for he will rather die than yielde to his adversarie.' The cock, moreover, was believed to be able to impart his valour.

Porta writes: 'If you would have a man become bold and impudent, let him carry about the skin or eyes of a lion or cock, and he will be fearless of his enemies—nay, he will be very terrible unto them.' Extract of cock was held to be the cure for consumption.

The prescription runs: 'Take a red cock, cut him into quarters, and put him into an earthenware pot with the rootes of fennell, parcely and succory, corans, whole mace, Anise seeds, and liqorice scraped and slyced, two or three clean dates, a few prunes and raysons.' Then add half a pint of rosewater and a quart of white wine and stew the whole gently for twelve hours. A teaspoonful of the resulting broth should be taken twice a day.

The fowl, alas! has now fallen from his high estate, especially in India. In this country, although it is the true home of gallinaceous birds, the *murghi* is a very degenerate creature. Natives do not understand the art of breeding, as their miserably undersized cattle, horses, and donkeys, and their mongrel pigeons, demonstrate. Indian poultry, however, are worse than undersized; they exhibit a strong leaning towards pachydermism—a fatal creed for a table bird. This the traveller is able to verify for himself at any *dak* bungalow, for murghi will inevitably appear on the table, and the would-be diner, after many ineffectual attempts to get his degenerate teeth into the bird sacrificed to him, obliged to console himself for his unsatisfied appetite by singing gently:

'That bird must have crowed when they built
the Tower of Babel,
'Twas fed by Cain and Abel,
And lived in Noah's stable,
All the shots that were fired on the field
of Waterloo
Couldn't penetrate or dislocate
That elongated, armour-plated,
Double-breasted, iron-chested,
Cock-a-doodle-doo.'

All the various breeds of poultry were at one time supposed to be descended from the common Indian jungle fowl. It is now, however, thought that Cochins and Brahmas have possibly arisen from other ancestors.

The Scrapers are a dimorphic family of birds—the sexes differ in appearance. The males are more showy and larger than the females. This is supposed to be due to sexual selection, that is to say, the preference of the ladies for gaily-coloured husbands. Each cock does his utmost to secure a goodly harem of hens. In order to gratify his ambition he must be of gallant appearance, of winning manners, and a good fighter. The former qualities enable him to obtain wives and the last to retain them when once secured.

The Rabbi Jochanan says: 'Had the law never been given us, we might still have learned politeness from the cock, who is fair spoken to the female in order to win her. "I will buy thee a dress," he whispers in the hen's ear, "a dress that shall reach down to the very ground." And when the victory is achieved, he shakes his head solemnly and cries, "May my comb perish if, when I have the means, I do not keep my word.'"

If the cock and hen birds differ in appearance, they exhibit still greater diversity in character. The cock is a warrior, valiant, careful of his honour, hot-tempered, albeit prudent, proud, and vain. The hen is the type of good-tempered bourgeoisie, humble, prone to cackle, subservient to her husband, foolish, and affectionate. The carefulness with which she bruises every grain of corn, lest it should hurt the soft palates of her chicks, the way in which she teaches her children to scrape the ground to make it yield up its good things, the tender manner in which she gathers her brood under her wings, and her anxiety and solicitude if one stray from her, are among the most homely and the sweetest sights in nature. But it is unnecessary to dilate upon the affection of a hen for her chickens; let it suffice that it has been made the subject of one of the most beautiful similes in the Bible.

Cruel man must cause the poor foolish bird many an anxious moment when he sets her to rear up ducklings. It is truly pitiful to watch her distress when the unruly brood betakes itself to the dreaded water.

There is a story told of a goose that saw a hen in this predicament, and swam up to her to cackle a few words of comfort. The hen seized the opportunity to jump upon the goose's back. The latter, although a little scandalized at the hen's familiarity, was too kind-hearted to shake her off, so swam with her alongside her duckling children. The hen enjoyed her trip so much that she repeated it the next day.

Then the goose, who hailed from Scotland, determined to float a company to take distressed hens for trips on the water at 2d. a—but stay! Methinks I hear the gentle reader complain of a pulling sensation in the leg. This will never do. Let us hie back to the young chicks.

It is characteristic of the Gallinae that their young are hatched in a highly developed state, and not blind, naked, and helpless, as is the case with most young birds. The downy chick is so precocious a baby that it needs no nest to protect it, consequently the hen does not build one, but lays her eggs on the hard ground. While yet inside the shell the chick calls out to let its mother know that it is prepared to face the troubles and dangers of this life; then the excited parent breaks the little bird's frail prison by pecking at it. An opening is soon formed and the young chick emerges, ready for a good solid meal as soon as its mother has taught it how to eat, a lesson that is quickly learned.

Although born in so highly developed a condition, the young bird differs greatly in appearance from either of its parents, and has thus to pass through a transitory, a hobbledehoy stage, before it assumes the adult plumage. Most birds live through this period hidden away in the nest, but the poor fowl has to do so in public. Hobbledehoys are always awkward, ugly creatures, and the pullet forms no exception; a more ungainly bird it would be difficult to find.

DOUGLAS DEWAR, *Bombay Ducks*, 1906

Thomas Bainbrigge Fletcher (1878–1950) was an English entomologist. Charles McFarlane Inglis (1870–1954) worked in North-East India as a planter and made several studies of birds, butterflies, and dragonflies. He was also the curator of the Darjeeling museum, run by the Bengal Natural History Society. The biblical tale of the Hoopoe and how it got its crest is extracted from their book Birds of an Indian Garden.

How the Hoopoe Got its Crest

The Hoopoe is looked on as a favourite because so the story goes, he was King Solomon's messenger, and he is known as the King of Birds from the legend of his crest and crown, which is related by the Hon. Robert Curzon as follows:

In the days of King Solomon, the son of David, who, by the virtue of his cabalistic seal, reigned supreme over genii as well as men, and who could speak the languages of animals of all kinds, all created beings were subservient to his will. Now, when the king wanted to travel, he made use, for his conveyance, of a carpet of a square form. This carpet had the property of extending itself to a sufficient size to carry a whole army, with the tents and baggage; but at other times it could be reduced so as to be only large enough for the support of the royal throne, and of those ministers whose duty it was to attend upon the person of the sovereign. Four genii of the air then took the four corners of the carpet, and carried it with its contents wherever King Solomon desired. Once the king was on a journey, in the air, carried upon his throne of ivory over the various nations of the earth. The rays of the sun poured down upon his head, and he had nothing to protect him from its heat. The fiery beams were beginning to scorch his neck and shoulders, when he saw a flock of vultures flying past. 'O vultures!' cried King Solomon, 'come and fly between me and the sun, and make a shadow with your wings to protect me, for its rays are scorching my neck and face.' But the vultures answered, and said, 'We are flying to the north, and your face is turned towards the south. We desire to continue on our way; and be it known unto thee, O king! that we will not turn back in our flight, neither will we fly above your throne to protect you from the sun, although its rays may be scorching your neck and face.' Then King Solomon lifted up his voice, and said, 'Cursed be ye, O vultures! and because you will not obey the commands of your lord, who rules over the whole world, the feathers of your neck shall fall off; and the heat of the sun, and the cold of the winter, and the keenness of the wind, and the beating of the rain, shall fall upon your rebellious necks, which shall not be protected with feathers, like the neck of other birds. And whereas you have hitherto fared delicately, henceforward ye shall eat carrion and feed upon offal; and your race shall be impure till the end of the world.' And it was done unto the vultures as King Solomon had said.

Now it fell out that there was a flock of hoopoes flying past; and the king cried out to them, and said, 'O hoopoes! Come and fly between me and the sun, that I may be protected from its rays by the shadow of your wings.' Whereupon the king of the hoopoes answered, and said, 'O king! we are but little fowls, and we are not able to afford much shade; but we will gather our nation together, and by our numbers we will make up for our small size.' So the hoopoes gathered together, and, flying in a cloud over the throne of the king, they sheltered him from the rays of the sun. When the journey was over, and King Solomon sat upon his golden throne, in his palace of ivory, whereof the doors were emerald, and the windows of diamonds, larger even than the diamond of Jemshea, he commanded that the king of hoopoes should stand before his feet.

'Now,' said King Solomon, 'for the service that thou and thy race have rendered, and the obedience thou hast shown to the king, thy lord and master, what shall be done unto thee, O hoopoe? and what shall be given to the hoopoes of thy race, for a memorial and a reward?'

Now the king of the hoopoes was confused with the great honour of standing before the feet of the king; and making his obeisance and laying his right claw upon his heart, he said, 'O king, live forever! Let a day be given to thy servant, to consider with his queen and his counsellors what it shall be that the king shall give unto us for a reward.' And King Solomon said, 'Be it so.'

And it was so.

But the king of the hoopoes flew away; and he went to his queen, who was a dainty hen, and he told her what had happened, and desired her advice as to what they should ask of the king for a reward; and he called together his council, and they sat upon a tree, and they each of them desired a different thing. Some wished for a long tail; some wished for blue and green feathers; some wished to be as large as ostriches; some wished for one thing, and some for another, and they debated till the going down of the sun, but they could not agree together. Then the queen took the king of the hoopoes apart and said to him, 'My dear lord and husband, listen to my words; and as we have preserved the head of King Solomon, let us ask for crowns of gold on our heads, that we may be superior to all other birds.'

And the words of the queen and the princesses, her daughters, prevailed; and the king of the hoopoes presented himself before the throne of Solomon, and desired of him that all hoopoes should wear golden crowns upon their heads. Then Solomon said, 'Hast thou considered well what it is that thou desirest?' And the hoopoe said, 'I have considered well, and we desire to have golden crowns upon our heads.' So Solomon replied, 'Crowns of gold shall ye have: but, behold, thou art a foolish bird; and when the evil days shall come upon thee, and thou seest the folly of thy heart, return here to me, and I will give thee help.' So the king of the hoopoes left the presence of King Solomon with a golden crown upon his head, and all the hoopoes had golden crowns; and they were exceedingly proud and haughty. Moreover, they went down by the lakes and the pools, and walked by the margin of the water that they might admire themselves, as it were in a glass. And the queen of the hoopoes gave herself airs, and sat upon a twig; and she refused to speak to the merops, her cousins, and the other birds who had been her friends, because they were but vulgar birds, and she wore a crown of gold upon her head.

Now there was a certain fowler who set traps for birds; and he put a piece of a broken mirror into his trap, and a hoopoe that went in to admire itself was caught. And the fowler looked at it, and saw the shining crown upon its head; so he wrung off its head, and took the crown to Issachar, the son of Jacob, the worker in metal, and he asked him what it was. So Issachar, the son of Jacob, said, 'It is a crown of brass,' and he gave the fowler a quarter of a shekel for it, and desired him, if he found any more, to bring them to him, and to tell no man thereof. So the fowler caught some more hoopoes, and sold their crowns to Issachar, the son of Jacob; until one day he met another man who was a jeweller, and he showed him several of the hoopoes' crowns. Whereupon the jeweller told him that they were of pure gold, and he gave the fowler a talent of gold for four of them.

Now when the value of these crowns was known, the fame of them got abroad, and in all the land of Israel was heard the twang of bows and the whirling of slings; bird-lime was made in every town, and the price of traps rose in the market, so that the fortunes of the trapmakers increased. Not a hoopoe could show its head but it was slain or taken captive, and the days of the hoopoes were numbered. Then their minds were filled with sorrow and dismay, and before long few were left to bewail their cruel destiny.

At last, flying by stealth through the most unfrequented places, the unhappy king of the hoopoes went to the court of King Solomon, and stood again before the steps of the golden throne, and with tears and groans related the misfortunes which had happened to his race.

So King Solomon looked kindly upon the king of the hoopoes and said unto him, 'Behold, did I not warn thee of thy folly, in desiring to have crowns of gold? Vanity and pride have been thy ruin. But now, that a memorial may remain of the service which thou didst render unto me, your crowns of gold shall be changed into crowns of feathers, that ye may walk unharmed upon the earth.' Now, when the fowlers saw that the hoopoes no longer wore crowns of gold upon their heads, they ceased from the persecution of their race; and from that time forth the family of the hoopoes have flourished and increased, and have continued in peace even to the present day.

T.B. Fletcher and C.M. Inglis, Birds of an Indian Garden, 1936

Delightful Distractions

Mughal Emperor Nuruddin Salim Jahangir (1569–1627) was a lover of nature and assiduously recorded his observations on birds and other animals. His memoirs, The Jahangirnama, *has been admirably translated into English by the American scholar Wheeler M. Thackston.*

The Quail and the Hawk

On the third [January 12], we decamped and once again, as we had before, we got into boats and went two and an eighth kos to the camp site in the village of Kawalhas. While I was hunting along the way, a quail flew into a bush. After a search had been made I ordered one of the scouts to surround the bush and get hold of the quail, and I went on. Just then another quail took off. No sooner had I sent a hawk after that one than the scout brought me the first one. I ordered the hawk fed on the latter quail, and I ordered the second one, which we had caught, kept because it was young. By the time this order arrived, the huntsmen had already let the hawk feed on the quail. After a time the scout said, 'If we don't kill the quail, it will die.' I commanded it killed if that was the case. When the blade was placed on its throat, it squirmed out from under the blade and flew away. After that I moved from the boat to horseback, when suddenly a sparrow was blown by the wind and impaled itself on an arrow shaft one of the scouts ahead of me was holding. It died instantly. I marveled at the twist of fate. Back there, it protected a quail whose time had not come, and within an instant, saved it from danger, while here it made a swallow whose time had come the prisoner of an arrow of destiny in the hand of destruction.

If the blade of the world moves, it will not cut a vein until God so wills.

The Story of a Pair of Cranes

A strange thing happened at this site. Before the imperial banners reached this spot, a eunuch went to the edge of a large lake in the village and captured two baby *saras*es, a kind of crane. That night, while we were camped there, two screeching adult cranes appeared in the vicinity of the ablutions tent, which had been played by the lake. Just like a person with a complaint, they came forward, screeching in terror. It occurred to me that they must certainly have suffered an injustice, most likely their young had been taken from them.

After an investigation, the eunuch who had taken the baby cranes brought them for me to see. When the cranes heard the cries of the babies, they hurled themselves on top of them. Thinking they hadn't been fed, each of the cranes put a morsel of food in the babies' mouths and did all sorts of things to console them. Then, picking the babies up, they spread their wings and headed off, yearning for their nest.

***

Around this time I witnessed the mating of saras cranes, something I hadn't seen before. It is often said among the people that no one has ever seen it. The saras is a bird something like a crane, but it is larger than a crane by a ratio of ten to twelve and has no feathers on its head, just skin stretched over bone. For a distance of about six fingers behind the eyes and down the neck it is red. Mostly they live in pairs in the wilderness, though occasionally they can be seen in flocks. People bring chicks from the wilderness and keep them in their houses, and they become tame to humans. Anyway, in my establishment there was a pair of saras cranes that I called Layli and Majnun.[1] One day one of the eunuchs told me they had mated in his presence. I ordered him to let me know whenever they looked like they were going to mate again. He came

---

[1] The cranes were nicknamed after the famous pair of lovers from Persian romance.

one morning at dawn and told me they were going to mate. I immediately ran out to watch. The female stretched her legs straight and then bent them slightly. First the male lifted one of his legs off the ground and put it on her back, and then the other. The instant he was seated on her back they mated. Then he got down, stretched out his neck, put his beak on the ground, and circled once around the female. It is possible that they have produced an egg and the young will be brought forth.

Many strange and wonderful tales have been heard concerning the affection and attachment sarases have for their mates. Since they have been heard so often and are so strange, they are worth recording. One of them is as follows.

Qiyam Khan, a khanazad of this court who was highly skilled in the art of scouting for the hunt, said to me, 'One day when I had gone out hunting I saw a saras sitting on the ground. As I approached it got up and went away. From the way it moved I could tell it was sick or hurt. I went to the place it had been sitting and found several bones and a handful of feathers it had been sitting on. I then laid a trap and retired to a secluded spot. As the bird was about to come back and sit down its leg got caught in the trap. I went forward and seized it. It felt very light. When I examined it, there were absolutely no feathers on its breast or belly. Its flesh and skin were falling away and it was infected. In fact, there was no trace of flesh anywhere on its body. It was just a handful of feathers and a few bones. Apparently its mate had died and it had been reduced to that state by separation.'

My body melted away in heart-rending separation. My soul-searing sigh burns like a candle./My day of gaiety has turned as black as the night of grief. Separation from you has made me like this.

Himmat Khan, one of my fine servants whose word can be trusted, related the following: 'A pair of sarases was spotted beside the tank in the pargana of Dohad. One of my musketeers shot one of them, cut its head off, and cleaned it. As it happened, there was a halt there for two or three days. The mate kept circling around that spot, crying out and lamenting. My heart ached over its distress, but there was nothing I could do but regret. After we decamped, we chanced to pass by there again twenty-five days later. I asked the inhabitants what had happened to the saras.

They said it had died within two days, but its bones and feathers were still there. I went to the spot myself and found the remains, just as they had said.' Many such stories are current among the people, but they would take too long to record.

The Jahangirnama: Memoirs of Jahangir, Emperor of India

Khushwant Singh (b. 1915) is a well-known journalist, novelist, and historian. His weekly column, 'With Malice towards One and All', published originally in The Illustrated Weekly of India, *is one of the most widely read columns in India. His well-known works include* A Train to Pakistan, I Shall Not Hear the Nightingale, The Mark of Vishnu and Other Stories, *and* A History of the Sikhs. *Khushwant Singh is also a nature lover and birdwatcher, and as the editor of* Hindustan Times *during the 1980s he started a nature section in the newspaper, 'On the Ridge', which became very popular among nature lovers. The extract from his novel* I Shall Not Hear the Nightingale *describes the feelings of Sher Singh—one of the central characters in the novel, after he shoots a Sarus Crane.*

Sher Singh Shoots a Crane

'O.K., brother, O.K.,' said Sher Singh in English and stood up. 'We must be quick. It will be dark in an hour.' He collected the empty cases lying on the ground and put them in his pocket. The boys also stood up and brushed the dust off their clothes. They put their guns in the jeep. One of them volunteered to stay back.

Sher Singh loaded his rifle and led the party down the canal bank towards the marsh. Dyer ran ahead barking excitedly.

They crossed the stretch of chalky saltpeter and got to the edge of the swamp. There were no birds on the water. On the other side was a *peepul* tree on which there was a flock of white egrets. Right on the top was a king vulture with its bald red head hunched between its black shoulders. Beneath the tree were bitterns wading in the mud. The birds were over a hundred yards away; well beyond Sher Singh's range of marksmanship.

The party surveyed the scene and considered the pros and cons of taking a shot from that distance. The vulture struck out its head and the egrets began to show signs of nervousness. Suddenly there came the loud, raucous cry of a Sarus crane followed by another from its mate. They were in a cluster of bulrushes not fifty yards away. The boys sat down on their haunches and stopped talking. The cranes continued calling alternately for a few minutes and then resumed their search for frogs. The vulture and the egrets on the opposite bank went back to sleep.

'Kill one of these. They are as big as any black buck,' whispered the small boy.

'Who kills cranes?' asked Sher Singh. 'They are no use to anyone. And I am told if one of a pair is killed, the other dies of grief.'

'If you are going to funk shooting birds, you will not do much when it comes to shooting Englishmen,' taunted Madan. 'You will say, "Why kill this poor chap, his widow and children will weep," or "His mother will be sad." Sher Singhji, this is what is meant by baptism in blood; get used to the idea of shedding it. Steel your heart against sentiments of kindness and pity. They have been the undoing of our nation. We are too soft.'

That was enough to provoke Sher Singh—particularly as it came from Madan. 'Oh no! nothing soft about me,' he answered defiantly. 'If it is a Sarus crane you want, a Sarus crane you will have. Come along Dyer—and if you bark, I'll shoot you too.'

Sher Singh got down on his knees and crawled up behind the cover of the pampas grass, his dog following warily behind. He stopped after a few yards and parted the stalks with the muzzle of his rifle. One of the birds was busy digging in the mud with his long beak; the other was on guard turning its head in all directions looking out for signs of dangers. Sher Singh decided to be patient. He wanted to get a little closer and also get enough time to take aim. Missing a bird of that size would be bad for his reputation.

After a few minutes, he looked through the stalks again. Both the cranes were now busy rummaging in the reeds. He crept up another ten yards, Dyer behind him. He paused for breath and once again parted the pampas stalks with the muzzle of his rifle. One of the birds was again on the lookout. Sher Singh drew the bead on the other—at the easiest spot to hit: the heavy, feathered middle of its body. The sentry crane spotted Sher Singh. It let out a warning cry and rose heavily into the air. Its mate looked up. Before it could move, Sher Singh fired. The bullet hit its mark. A cloud of feathers flew up and the bird fell in the mud.

Dyer ran across to seize it. The boys came up from behind, clapping and shouting.

Sher Singh clicked open the catch; the metal case of the bullet flew out and fell on the ground. He picked it up and put it in his pocket. He blew into the barrel and saw the smoke shoot out of the other end. He was a jumble of conflicting emotions of guilt and pride. He had mortally wounded a harmless, inedible bird. But this was his first attempt to take life and it had succeeded. Then his friends came up, slapped him on the back and shook his hand by turn. The feeling of remorse was temporarily smothered.

The shot had not killed the crane. It flapped its wings and dragged itself out of the pool of blood a few feet farther towards the water. When Dyer came up, it turned towards him and pecked away fiercely with its long, powerful beak. The snarling and snapping Alsatian kept a discreet distance. Then the other crane flew back and began to circle overhead, crying loudly. It dived down low over the dog to frighten it away.

'Leader, give the other one its salvation too. Let them be together in heaven or hell.'

'Yes, let's see you take a flying shot,' added Madan.

The argument appealed to Sher Singh. The anguished cry of the flying crane was almost human. If he did not silence it, it would continue to haunt him for a long time. If both of the pair were dead, perhaps they would be together wherever cranes went after death. Sher Singh took out the magazine of his rifle and pressed six bullets in it. He followed the crane's flight with his barrel and fired when the bird was almost above him. The bullet went through one of the wings. The bird wavered badly in its flight and some feathers came floating down. Sher Singh fired the second shot. Then the third and the fourth and emptied the magazine. The crane flew away across the swamp, ducking nervously as the bullets whistled by in quick succession.

Sher Singh blew the smoke out of the barrel once more. In his excitement he forgot to pick up the empty cases.

'Its time is not up yet,' said Madan to console him. 'Put this one out of its agony.'

Once having embarked on the bloody business, Sher Singh could not stop half way. He walked up to the injured bird and put his right foot on its neck. The crane began to kick violently and gasp for breath. Its beak opened wide showing its thin, long tongue. Sher Singh took out his revolver and fired two shots into its body. The bird's dying gurgle was stifled in its throat. Its legs clawed the air and then slowly came to a stop in an attitude of prayer. Blood started trickling from its beak and a film covered its small black eyes.

'This one is finished. Let us take it to the jeep and baptize our weapons in its blood.'

Two of the boys caught the crane by the wings from either end and dragged it out of the swamp. Dyer sniffed at the dead bird's head dangling between its trailing legs and began to run round in circles yapping deliriously. Sher Singh saw his handiwork and a lump came up in his throat. He did not respond to the backslapping and hilarity of his companions.

Before they got clear of the swamp the other crane flew back and started circling over them. They saw it high above in the deep blue sky catch the light of the setting sun; then heard its cries piercing the stillness of the dusk. Sher Singh ignored requests to have another go at the flying bird; in any case it was too high and the light was failing fast. When they got to the canal bank, it became dark. The crane flew lower and lower till they could see its gray form with its long legs almost above their heads. They shoo'd it off. The bird disappeared in the dark only to comeback again and again. Its crying told them it was there all the time, trying to reclaim its dead mate. Sher Singh wanted to get away from the place as fast as his jeep could take him. That was not to be.

KHUSHWANT SINGH, *I Shall Not Hear the Nightingale*, 1959

The famous American humorist, satirist, writer, and lecturer Mark Twain (1835–1910), well remembered for his novels Adventures of Huckleberry Finn *and* The Adventures of Tom Sawyer *and his numerous quotes and sayings, visited India in 1895. In* Following the Equator, *the account of his travels in India, Twain unmasks and criticizes racism, imperialism, and missionary zeal through his observations, woven into the narrative with classical 'Twain wit'. His observations on the Indian crow, reprinted below, illustrate this point very well.*

In Praise of the Indian Crow

I suppose he is the hardest lot that wears feathers. Yes, and the cheerfulest, and the best satisfied with himself. He never arrived at what he is by any careless process, or any sudden one; he is a work of art, and 'art is long'; he is the product of immemorial ages, and of deep calculation; one can't make a bird like that in a day. He has been re-incarnated more times than Shiva, and he has kept a sample of each incarnation, and fused it into his constitution. In the course of his evolutionary promotions, his sublime march toward ultimate perfection, he has been a gambler, a low comedian, a dissolute priest, a fussy woman, a blackguard, a scoffer, a liar, a thief, a spy, an informer, a trading politician, a swindler, a professional hypocrite, a patriot for cash, a reformer, a lecturer, a lawyer, a conspirator, a rebel, a royalist, a democrat, a practicer and propagator of irreverence, a meddler, an intruder, a busybody, an infidel, and a wallower in sin for the mere love of it. The strange result, the incredible result, of this patient accumulation of all damnable traits is, that he does not know what care is, he does not know what sorrow is, he does not know what remorse is, his life is one long thundering ecstasy of happiness, and he will go to his death untroubled, knowing that he will soon turn up again as an author or something, and be even more intolerably capable and comfortable than ever he was before.

In his straddling wide forward step, and his springy sidewise series of hops, and his impudent air, and his cunning way of canting his head to one side upon occasion, he reminds one of the American blackbird. But the sharp resemblances stop there.

He is much bigger than the blackbird; and he lacks the blackbird's trim and slender and beautiful build and shapely beak; and of course his sober garb of gray and rusty black is a poor and humble thing compared with the splendid lustre of the blackbird's metallic sables and shifting and flashing bronze glories. The blackbird is a perfect gentleman, in deportment and attire, and is not noisy, I believe, except when holding religious services and political conventions in a tree; but this Indian sham Quaker is just a rowdy, and is always noisy when awake—always chaffing, scolding, scoffing, laughing, ripping, and cursing, and carrying on about something or other. I never saw such a bird for delivering opinions. Nothing escapes him; he notices everything that happens, and brings out his opinion about it, particularly if it is a matter that is none of his business. And it is never a mild opinion, but always violent—violent and profane—the presence of ladies does not affect him. His opinions are not the outcome of reflection, for he never thinks about anything, but heaves out the opinion that is on top in his mind, and which is often an opinion about some quite different thing and does not fit the case. But that is his way; his main idea is to get out an opinion, and if he stopped to think he would lose chances.

    I suppose he has no enemies among men. The whites and Mohammedan never seemed to molest him; and the Hindoos, because of their religion never take the life of any creature, but spare even the snakes and tigers and fleas and rats. If I sat on one end of the balcony, the crows would gather on the railing at the other end and talk about me; and edge closer, little by little, till I could almost reach them and they would sit there, in the most unabashed way, and talk about my clothes, and my hair, and my complexion, and probable character, and vocation and politics, and how I came to be in India, and what I had been doing, and how many days I had got for it, and how I had happened to go unhanged so long, and when would it probably come off, and might there be more of my sort where I came from, and when would they be hanged—and so on, until I could no longer endure the embarassment of it; then I would shoo them away, and they would circle around in the air a little while, laughing and deriding and mocking, and presently settle on the rail and do it over again.

They were very sociable when there was anything to eat—oppressively so. With a little encouragement they would come in and alight on the table and help me eat my breakfast; and once when I was in the other room and they found themselves alone, they carried off everything they could lift; and they were particular to choose things which they could make no use of after they got them. In India their number is beyond estimate, and their noise is in proportion. I suppose they cost the country more than the government does; yet that is not a light matter. Still; they pay; their company pays; it would sadden the land to take their cheerful voice out of it.

MARK TWAIN, *Following the Equator*, 1897

Abul Kalam Muhiyuddin Ahmed (1888–1958), popularly known as 'Maulana Azad', was an important leader of the Indian independence movement and the first Minister for Education in independent India. A well-known writer, journalist, and scholar, the pieces included here are taken from his collection of prison letters, Ghubar-e-Khatir.

Sparrows of Ahmad Nagar Fort Prison

Ahmad Nagar Fort
17 March 1943

Dear Friend,
Many a story have I created in life. Indeed, life itself has been a story:

> *Hai aaj jo sarguzasht apni; Kal uski kahaniyan banengi*[2]
> (What we are passing through today, shall become a story tomorrow.)

[2] Amir Meenai's couplet (*Sanam Khana-e-Ishq*: 334).

Let me today tell you the story of a sparrow couple:

> *Digarha shuneed-sti, een ham shunau*[3]
> (Others you have heard, listen this too.)

The living rooms allotted to us here are specimens of the architecture of last century. Ceiling of wood beams supported by arches provide ample scope for building nests; so, whole colonies of sparrows have come up and their noisy activity continues the whole day. The Ballyganj[4] locality of Calcutta, being open and full of trees, the verandahs and cornices of houses there are always under attack by gangs of sparrows. The desolation of this place has brought to mind the ruin at home:

> *Ug raha hai dar-o-deewar peh sabza, Ghalib!*
> *Ham byabaan mein hain, aur ghar mein bahar aayee hai*[5]
> (Green shrubs have grown all around, O Ghalib!
> We are in ruins and the house is experiencing spring.)

Last year when we arrived here I was terribly upset by the nesting activities of these sparrows. On the eastern corner of the room there is a washbasin. Right above it was an old nest that had been there, no one knows since when. The sparrows collected straw all day and tried to spread it in the nest; pieces of straw fell on the wash basin and filled it with that rubbish. No sooner would I get a jug full of water that from above would start a rain of straw. On the western side the bed was laid against the wall. Above it, new construction, was on and the consequent tumult was even more annoying. These sparrows have been gifted with a tiny beak and the body is just a handful but their zeal and ambition was tremendous so that they cleared up a spanful of mortar on the wall in a matter of minutes. There is a well known saying of Archimedes[6]: 'Give me a place to stand in the atmosphere

---

[3]First line of Firdausi Toosi's *Dastan-e-Sohrab*.
[4]Ballyganj: Maulana's residence was at 19-A, Ballyganj Circular Road, Calcutta.
[5]*Dewan-e-Ghalib*: 177.
[6]Archimedes, famous mathematician of Syracuse. His inventions are world famous. A glass instrument invented by him could exactly show the

and I will shift the earth from its place.' The activities of these sparrows prove the truth of his claim. First they peck at the wall to create space for a claw-hold and then, with claws set there, strike the axe of their beak with such force that the whole body trembles. A few minutes later you find that several inches of mortar has fallen off. The building is old and one doesn't know how many layers of lime and sand have overlaid its walls so that there is a thick paint of mortar which, when broken, fills the room with a cloud of dust and the clothes are covered with layers of that dust.

The easy remedy was to renovate the building and remove all the nests. But that was not possible without calling construction labour from outside and no outsider can set foot here. Immediately as we arrived the water pipes had gone out of order. It was an ordinary mason's job. But the repair could not be undertaken till a British officer brought a permit from the Army Engineering Commanding Officer, an Englishman.

I exercised patience for a few days but then my patience was exhausted and I decided on a confrontation:

Man va gurz-o-maidan va Afrasiab
(I, with my mace came into the field against Afrasiab.)

An umbrella has come here with my luggage. I picked it up and declared war but, in a moment, I realized that this was too inadequate a weapon against these denizens of ceiling and arches. I looked with wonder now at the inaccessibility of the umbrella and then at the lofty reach of the rivals. Suddenly a couplet of Hafiz[7] came to mind:

Khyal-e-qadd-e-tu mee kunad dil-e-man
Tu dast-e-kotah-e-man been va aasteen-e-draz
(When my heart thinks of your height
You look at my tiny hand and its long reach.)

---

movements of the heavenly bodies. He also discovered the test of the purity of gold. His famous sayings: If I get a proper perch I can move and shake the globe. He died during the Roman invasion of Syracuse.
[7]*Dewan-e-Hafiz*: 193.

Now I looked for another weapon. In the verandah there was a bamboo stick for cleaning cobwebs. I rushed to pick it up and, then, what an unprecedented war ensued! The enemy was going round and round in the room and I was madly pursuing it. That brought on lips the heroic couplets of Firdausi and Nizami:

Beh khanjar zamin ra maistan kunam
Beh neza hawa ra naistan kunam[8]
(With dagger I reduce the earth to pulp
And with spear clear up the air.)

Ultimately victory was mine; in a short while the room was clear of these enemies in the ceiling and arch:

B'yak takhtan ta kuja ta khatam
Cheh garden kashan ra sar andakhtam
(In one sweep I cleared the whole place,
How dare the high-necked raise their head?)

I cast a victorious glance at every corner of the ceiling and, fully satisfied, resumed my writing. But, even fifteen minutes would not have passed that enemy's war songs and flights filled the air. As I looked at the ceiling I found that every corner was under their occupation. I got up and resumed the war with the bamboo stick:

Bar aaram dayar az hama lashkarash Beh aatish b'sozam hama kishwarash[9]
(The entire home force was mobilized to scorch the entire territory.)

This time the enemy displayed great fortitude; compelled to leave one corner they would take stand in another. Yet, they had to beat a retreat; fleeing from the room they came into the verandah and began lining up their forces again. I followed them there

---

[8]*Shahnama*, 1: 68.
[9]Refers to incidents in *Shahnama* when Alexander goes to the court of Qaidana in Andulus.

too and did not drop the weapon till the field was cleared up to a long distance beyond the boundary. The enemy force had been scattered but the fear still lurked that it may regroup and be back in the battlefield. Experience showed that the bamboo spear had put the fear of God in the enemy so that at the very sight of it they would take flight. Therefore I decided to keep it in the room for a while. If any of the enemy camp dared come hither its very sight, of this sky high spear, would put it to flight. Following that strategy I put the bamboo in the corner where there was the wash basin, in such a position that its top edge stood right at the gate of the old nest. Now, although the future was not free of apprehension, at least the satisfaction of taking all possible strategic precautions was there. The situation called up the oft repeated and oft profaned couplet of Meer:

Shikast-o-fateh naseebon se hai, vale ai Meer
Muqabila to dil-e-natuwan ne khoob kiya![10]
(Victory or defeat is a matter of luck but, O Meer,
The feeble heart faced the enemy with courage!)

By now it was 11 o'clock. I went out for lunch. After a while, as I came back and set foot in the room I was astonished to see that the entire room was under enemy occupation. They were busy in their work without any care, as if nothing untoward had happened. The worst of it was that the weapon whose fear was my trust had now become the instrument of their work. The end of the bamboo that touched the nest was now the stepping stone for entry into the nest. They brought straw from outside and, perched on that, spread it in the nest. The activity was accompanied by their chirping, as if humming the line of poetry:

Adoo shavad sabab-e-khair gar khuda khwahad[11]
(If God so wills, the enemy becomes the instrument of good.)

The tragic end of the illusion of victory broke my resolve. It was clear that it was possible to deter those enemies but it was not

---

[10] The couplet is wrongly attributed to Mir, it is in fact by Mohammad Yar Khan Amir (*Tabqat-ul-Sho'ra-e-Shauq*).
[11] *Behtreen Ash'ar*: 908; author not known.

easy to meet the challenge of their fortitude and determination! The best course in the circumstances was to admit defeat:

> *Baya keh sipar andakhtem agar jang-st!*[12]
> (Come and takeover, if it is war, I surrender!)

Now the problem was to evolve a strategy of coexistence with these unlettered guests. First was the problem of bed that lay right under the impact of new constructions. All the dirt from the scrapping of old building and the remains of the new construction fell on it. So, it was moved far away from the wall to avoid the dust and dirt. This disturbed the setting of the room but there was no other way? When the house itself is not in possession what can one do about its setting and decoration! However, the problem of washbasin was not easy of solution because the corner where it was placed was the only suitable place for it; it was not possible to move it at all! Perforce, I secured a bunch of dusters from the market and covered everything on the wash-table with a duster each. Every few minutes I dusted them up one by one. One of the dusters was kept apart for cleaning the table itself. The most problematic was the cleaning of the floor and that too was solved somehow. It was decided that in addition to the morning cleaning the room should be swept twice or thrice or more. A new broom was purchased from the market and kept hidden behind a cupboard to be used during the day twice, thrice or even more frequently. Here one sweeper prisoner is provided for every two rooms. Obviously he can't stand the whole day with a broom in hand; even if it were possible, it would not be fair to burden him so much. So I started sweeping the floor myself two-three times a day, avoiding the sight of neighbours. You will note that one has to perform menial jobs to entertain these illiterate guests:

> *Ishq azeen bisyaar kard-st va kunad!*[13]
> (Passion of love leads to great excesses and has already done.)

One day it occurred to me that if compromise has to be made why not go whole hog; it doesn't look nice that you live in the

---

[12] Line of Saadi Shirazi (*Kuliyat-e-Saadi*: 373).
[13] Line of Khwaja Fariduddin Attar (*Mantaq-al-Tayyar*: 94).

same house and live like strangers. I called for some rice from the kitchen, sprinkled some of it on the carpet in front of my sofa and positioned myself like a hunter waiting for the hunt after casting the net. See how apposite (*sic*) is the couplet of Urfi:[14]

> *Fatadam daam bar kunjashk va shadam, yaad-e-aan himmat*
> *Keh gar seemurgh mee aamad b'daam, azad mee kardam*
> (Spread the net for the tiny sparrow and was happy with the
>   confidence it gave;
> Even if the legendary phoenix is caught, I will release it.)

For a while the guests paid no attention and, if at all, no more than a sideway glance. Soon however it was clear that the neglect was no more than the ogling of the beloved. Otherwise how could the white grains on the blue carpet fail to attract attention?

> *Hoor-o-jannat jalwa bar zahid dihad, dar raah-e-dost*
> *Andak andak ishq dar kar aavurd begana ra*[15]
> (Lure of the houries in paradise lead the ascetic on the path
>   of love,
> Bit by bit the stranger is made to fall in line)

First a sparrow came and began to jump about. Apparently busy chirping, its eye was on the rice. How wonderful is the couplet of Vehshi Yazdi:[16]

> *Cheh lutfha keh dareen sheva-e-nahani ne-st*
> *Inayate keh tu daari b'man, bayani ne-st*
> (What a pleasure that there is no secret in this coquetry?
> The favour you show is beyond description!)

Then came another and went round on the carpet in the company of the first. Then third and fourth also arrived. Their glances fell now on the rice and then on the one who spread it. Sometimes it appeared as if some consultations were on and sometimes each

---

[14]*Kuliyat-e-Urfi*: 29.
[15]*Dewan-e-Nazeeri*: 39.
[16]*Dewan-e-Vehshi Bafqi*: 36.

one seemed to be deep in thought. You must have noticed when a sparrow looks around to make inquiries, its face takes on a strangely sanguine aspect. First it looks straight ahead, and then turns its neck sideways. And the face gives such impression of inquiry and research as if it is looking around in astonishment over what is happening. Every face at that moment casts such searching glances around:

> *Paayam beh pesh az sar-e een koo n'mee ravad*
> *Yaraan khabar dihed keh een jalwagah-e-kee-st*[17]
> (Never before have I set foot in this street;
> Go and tell all friends whose splendour is manifest here.)

After a while they began coming forward but not towards the grains of rice. Every time they passed it by as if to show they were not going towards the grain. Seeing this falsehood masquerading as truth brought to mind the couplet of Zahuri:[18]

> *B'go hadis-e-wafa, az tu bavar-st, b'go*
> *Shoom fida-e-droghe keh raast manind-st*
> (Tell the story of this fidelity, for you can be trusted;
> The unfortunate wretch is given to mistaking falsehood as truth.)

You are well aware that the hunter has to be more circumspect than the hunt. The moment they turned towards rice I held my breath, turned away my face and sat immobile like a rock as if not a man but a stone statute was lying there. For, I was conscious that if, out of eagerness, I showed the slightest haste the hunt would skirt the net. It was the first stage in the interplay of the beloved's coquetry and the lover's humility:

> *Nihaan az-oo beh rukhash daashtam tamashae*
> *Nazar beh jaanib-e-maa kard va sharmsaar shudam*[19]
> (I watch the fun with my face turned away from him;
> He turns the eye towards me and I pretend to be coy.)

---

[17]*Dewan-e-Nazeeri*: 66.
[18]*Dewan Mulla Nooruddin Zahoori*: 68.
[19]Couplet of Sharf Jahan Qazveeni (*Khazana-e-Amra*: 26, *she'r-ul-Ajam* 3: 18).

However, thank God that the early phase of this blandishment, masked as indifference, ended and a bold one among them advanced towards the rice. But what a wonderful move it was; I, in my immobile pose, was muttering to myself:

> *Beh har kuja naaz sar bar aarad, nyaz ham paae kam n'darad*
> *Tu va kharame va sad taghaful, man va nigahe va sad tamanna*[20]
> (Whenever faced with the coquet's tricks, humility succeeds;
> You come with your elegant gait and indifference, I bring my eager desire.)

It moved one step forward and took two steps back. I, in my heart of hearts felt, what a wonderful manifestation of the admixture of response and indifference. I wish it could be slightly changed and it could take two steps forward and one step back. How well has Ghalib portrayed the position:

> *Vida'-o-vasl judagana lazzate darad*
> *Hazar baar birau, va sad hazar baar baya*[21]
> (Departure and meeting, each has its own pleasure;
> Depart a thousand time and come back hundred thousand times.)

This coquetry of mixed response and indifference was on when, all of a sudden, one well-built cock-sparrow, distinguished among them for boldness and undaunted courage, got bored and took a bold step forward; it fell upon the rice with the carefree slogan:

> *Zadem bar saf-e-rindaan va har cheh baadabaad!*[22]
> (I attack the line of the intoxicated ones, whatever may happen!)

That initiative broke all the inhibitions, all reservations and confusion collapsed and all of them fell headlong on the rice. To borrow the English idiom, it can be said that the ice of coy hesitation suddenly broke. To think of it, all the action in this world awaits this first step forward, this initiative. Till it is taken, all steps remain stuck; once it is taken the whole world is on the move:

[20]*Kuliyat-e-Bedil*, 1: 12.
[21]*Kuliyat-e-Ghalib*: 363.
[22]*Dewan-e-Hafiz*: 144.

Na Mardi-o-Mardi qadme fasla darad[23]
(From unmanliness to manliness there is just one step.)

In this world, this company of profit and loss, success has never come to the timid ones; the cup is his who dares pick it up. The late Shaad Azimabadi wrote a wonderful couplet:[24]

Yeh bazm-e-mae hai, yaan kotah dasti mein hai mehroomi
Jo badh ke khud utha le haath mein meena usi ka hai
(This is the tavern, coy ones stay deprived;
Here the cup in his who makes bold to pick it up.)

I was so impressed with this boldness that, there and then I decided to develop friendship with it. I gave it the name Qalandar because in all his carelessness and bold initiative there was a touch of gallantry that imparted a luster to his activity:

Rahe ek baankpan bhi bedimaghi mein to zeba hai
Badha do cheen-e-abroo par ada-e-kajkulahi ko
(A touch of gallantry adds dignity to carefree disposition;
Add foppery to the peaceful brow.)

For two-three days, I entertained them this way. Twice or thrice in a day I would spread rice on the carpet. They came one by one and picked up their share. If there was a delay then Qalandar would come and start chirping as if telling me that the agreed time was passing. That confirmed that the veil of coyness was gone and the time was not far when all the hesitation would evaporate:

Aur khul jayenge do chaar mulaqaton mein[25]
(A few meetings will dispel all hesitation.)

After a few days I took the next step. I took up an empty cigarette tin, put rice into it and put it in a corner of the carpet. The guests at once took note. Some came to peck at the corner of the tin

---

[23]Author not known.
[24]Couplet of Sayyid Ali Mohammad Shaad Azimabadi (*Maikhana-e-Ilham*: 377; *Kuliyat-e-Shad*, 2: 184).
[25]*Aaftab-e-Daagh*: 44.

and some took position on the edge and started non-chalantly picking the rice. Occasionally there was a spectacle of rivalry and some competition too. When I found that they had got used to that form of entertainment, I moved the tin a bit away from the carpet. The third day it was moved further and kept right in front of me. Thus there was a move from distance to closeness. Look, this matter of distance and closeness has brought to mind the opening couplet of Bannatul Mehdi.

Love, for love begets love; how many, away from home, are worth keeping close to heart!

This closeness at first surprised the guests. They came close to the carpet but with hesitating steps and confusion in eyes. Meanwhile Qalandar came with his bold slogans so that his dauntless courage removed everyone's hesitation. It means they were all followers of Qalandar; his step forward was a signal for all to move. It pecked at rice and then, raising its head and thrusting forward its chest sang out:

The whole world sings my songs; whenever I compose one it appears on the lips of all, the world over.[26]

When matters came to this stage, another step was taken. The rice bowl was shifted from the carpet to the teapoy that was lying on the left, adjacent to my sofa, and was within my reach. They took some time to get used to the change; they came, went round the teapoy and flew away. Ultimately, here also Qalandar had to take the initiative. The moment he stepped forward the way was open to all, even as in the past. The teapoy now became at times their pleasure palace and at other times a battlefield. As they got used to coming close I felt that the process could be taken further ahead. One morning I kept the rice bowl right on my side and got busy with my writing as if totally unconcerned with the matter:

Dil-o-janam b'tau mashghool va nazar bar chap-o-raast
Taa na danand raqeeban keh tu manzoor-e-manee[27]

[26]*Dewan Abi al-Tayyab Mutnabbi*: 361.
[27]*Kuliyat-e-Saadi*: 614.

(Heart and soul engaged with you but the eyes elsewhere,
Lest the rival should know you are my sweetheart.)

A while later I heard sounds of strong pecking. As I cast a furtive glance in that direction I found that old friend Qalandar had arrived and was unhesitatingly picking the rice. As the tin was lying close at hand, his tail touched my knee. A little later others also reached. Thereafter the position was that all the time two-three of these friends played around me freely. Sometimes one of them would get on to the back of the sofa; another one would jump on to the books, and then come down and fly away chirping. At times it so happened that they mistook my shoulder for a bent branch of a tree where they could jump about. But then went back surprised or flew over just touching it with their claws. The matters had not reached the situation described by Vehshi Yazdi:

Hunooz aashiqi va dilrubaiye na shuda ast
Hunooz zori va mard aazmaiye na shuda ast
Hameen tawaza' aam-st husn ra ba ishq
Myan-e-naaz-o-nyaz aashnai na shuda ast[28]
(The meeting point of love and beloved is not yet reached,
Measuring of strength and manliness is not yet reached,
For the time being it is the usual relation of beauty and love,
Between blandishments and gallantry the meeting point is
  far away.)

However, these flying deer gradually felt convinced that this figure on the sofa is not dangerous like human beings, though a human being. See, the magic of love that cannot tame human beings tames these wild birds:

Dars-e-wafa agar bood zamzama-e-muhabbate
Jum'aa b'maktab aavurd tifl-e-gurezpae ra[29]
(If lesson in fidelity were a song of love,
It would bring the reluctant child to school every Friday.)

---

[28]*Dewan-e-Vehshi Bafqi*: 25.
[29]*Dewan-e-Nazeeri Neshapuri*: 26.

Several times it happened that, while writing, some impressive idea came to be penned or the narrative brought to mind some exhilarating couplet and I began tossing my head and shoulders, or uttered 'ha' involuntarily and then I heard the whirr of flight. What had happened? I found that a couple of these informal friends were busy in their game undisturbed till, suddenly, they realized that the rock was moving and they flew away. No surprise if they be feeling that there is a rock on the sofa which sometimes becomes a man!

***

Ahmad Nagar Fort
18 March 1943

Dear Friend,
The story that was started yesterday has not yet reached its conclusion. Today I will narrate to you the next chapter of this *Mantaq-al-Tayyar*[30] ('Bird-logic'). I don't know if you would have enjoyed it or got bored with it if you were present, but I find that my mind has forgotten to stop story-telling. The more a story gets prolonged the greater is the desire for story-telling:

> *Farkhanda shabe bayad va khush mahtabe*
> *Taa baa too hikayat kunam az har babe*
> (Let the night smile and the sun laugh
> Till I read you the story chapter by chapter.)

Between me and these friends in the arches of the ceiling there was left only a slight distance and that too disappeared in a few days.

To come down from their nests to the sofa they required only a few steps. For that they used the fan as first step and my head and shoulders as the second. From outside they would come straight to their nests and, perched there, look around to survey the scene in the room, then fly straight on to the fan. Jumping from there some times my head was made a playground or my shoulders

---

[30]*Mantaq-al-Tayyar.* Famous work of Khwaja Fareeduddin Attar in which temporal and spiritual wisdom is presented through birds.

were accorded the honour for staging their procession. See, these sparrows have brought to mind a poetic construct of Momin Khan:

> *Jaulaan ko hai uski qasad-e-pamal*
> *Ae khak! Naveed-e-sarfarazi*[31]
> (His fun and frolic are out to destroy all,
> For the dust it's a message of great honour!)

This unforeseen glorification, in the first instance surprised me and I have to confess, to my shame, that I was shaken. This rude lack of appreciation must naturally have offended those easy-going ones! It was however a temporary error and I soon realized it. As I got used to it my shoulders became so immune to shock that they could well be used as the upper storey rather than the tower top. From the fan they would descend directly on my shoulder and from there jump on to the sofa, all the time merrily chirping. Quite often they jumped from the shoulder and sat on my head. You know once Aatishi Qandhari had made a boat of his eyes in a couplet which Badayuni has reproduced.[32]

> *Sarishkam rafta rafta be too darya shud, tamaasha kun*
> *Baya, dar kashti-e-chashman nasheen va sair-e-darya kun*
> (See, in your absence my tears have become a river,
> Now come and sit in the boat of my eye and enjoy the river scene.)

And, our Sauda was hesitant:[33]

> *Aankhon mein doon us aaina-roo ko jagah, vale*
> *Tapka kare hai baskeh yeh ghar, nam bahut hai yaan*
> (I would like to accommodate that radiant beauty in my eyes;
> What to do, the place is always leaking and it is very damp.)

But in my present situation I had to follow the humble entreaty of the Sheikh of Shiraz:[34]

---

[31] *Kuliyat-e-Momin*, 1: 383.
[32] *Muntakhib-ul-Tawareekh*, 3: 180.
[33] *Kuliyat-e-Sauda, Dewan-e-Awwal*: 102.
[34] From Chapter 1 of *Gulistan* (*Kuliyat-e-Saadi*: 25).

> *Gar b'sar-o-chashm-e-man nasheeni*
> *Nazat b'kasham keh naazneeni*
> (If you sit on my head and in my eyes
> I enjoy the frolic, for you are my sweetheart.)

As things came to this I thought of making another experiment; one morning I delayed putting out the rice bowl. The virtuous guests paid repeated visits and not seeing the dinner spread, whirled around raising a racket. Now I took out the bowl and put it on the palm of my hand. The Qalandar jumped at it the moment he got sight of the bowl, went round it and, perched on my thumb, began to peck at the rice at great speed. The speed was partly an expression of its assertive dandyish nature and partly because of the wait for the feast. As a result of the fast blows of its beak the grain spilled out and one fell at the root of my finger. It gave a blow there too; so deep was the consequent hurt that I can't just explain; had I not got used to the tyrannies of these tyrants I would have given out a shriek:

> *Man kushta-e-karishma-e-mizgaan keh bar jigar*
> *Khanjar zad aan chunaan keh nigah ra khabar na shud!*[35]
> (So smitten am I with the magic of those eyes
> That my eyes don't take notice where their dagger hits.)

Now I raised my hand with the bowl on it high in the air. Another hen-sparrow came soon thereafter. You will shortly know that its name was Moti. It took a few rounds above my hand and flew away, as if trying to ascertain what would be a safe spot to land on this island. A little later it came again and perched straight on the wrist via the elbow. From the wrist it descended on my palm and started its beak-assault. See, I have changed the 'hand-assault' to 'beak-assault'. I know that idioms do not admit of such interference but I am helpless because I have to deal with those who instead of using hands for assault use the beak:

> *Draz-dasti-e-een kotah-aasteenaan been.*[36]
> (Look at the aggression of these frail beings!)

---

[35]*Dewan-e-Vehshi Bafqi*: 58.
[36]Line of Hafiz Shirazi (*Dewan-e-Hafiz*: 28).

But this last experiment set my inquisitive mind on another course; I felt ashamed of my shortcoming in expression of love that, despite having a hand, I was wasting their blows on a poor piece of tin. Next day I dispensed with the tin-cover, put the rice on the palm of my hand and extended it on to the sofa. First of all came Moti and raised its neck to look around why the tin-cover was not visible! This is the most beautiful hen-sparrow here. These days in a beauty contest, the charmer that proves most bewitchingly attractive, is named after the country she belongs to *viz.*, Miss England, Mademoiselle France. In other words the radiance of one beautiful face brightens up the face of an entire country and nation:

> *Kunand khwesh-o-tabaar az too naaz va mee zebad*
> *B'husn-e-yak tan agar sad qabeela naaz kunad!*
> (Everybody, irrespective of relationship feels proud of you;
> And rightly too; the beauty of one should be pride of hundreds of clans.)

To use this practice on Moti it can be christened Madam Ahmad Nagar Fort.

> *Een nigaheest keh shaista-e-deedare hast*[37]
> (The eye has the ability to appreciate your beauty.)

Slim body, long neck, tapering tail and round eyes reflecting a vocal innocence; when it comes to pick grain it casts a glance at me at every step. We communicate silently with our glances. It can understand the language of my eyes and I, its. Haa! what depth of perception has Vehshi Yazdi displayed in this matter?[38]

> *Karishma garm-e sawal-st lab m'kun ranja*
> *Keh ahtiyaj beh purseedan-e zabani ne-st*
> (The magic raises questions but the lips are silent;
> It is not proper to verbally inquire about the need.)

In any case this time also its artless eyes conveyed something, then unhesitatingly jumped on to the root of my thumb and started

---

[37] Urfi's line (*Kuliyat-e-Urfi*: 289).
[38] *Dewan-e-Vehshi Bafqi*: 37.

pecking for the grain. Its beak was like a sharp knife that could pierce through my palm but it stopped at superficial wounds:

> *Yak naavak-e-kaari z'kamaan-e-too n'khurdam*
> *Har zakhm-e-too mohtaaj beh zakhm-e-digram kard*[39]
> (You have not given me a single deep wound;
> Each wound you give spurs desire for another.)

Each time it turned its head towards me, as if asking if it caused pain. How could I, addicted to the pleasure of agony, reply to that:

> *Een sukhan ra cheh jawab ast too ham meedani*[40]
> (The answer to this query you too know quite well.)

You must have seen this couplet of Mirza Saib:

> *Khwesh ra bar nok-e-mizgaan-e-sitam keshaan zadam*
> *Aan qadar zakhme keh dil me khwast, dar khanjar na bood*[41]
> (The tyrannical beauties give blows with the edge of eyelashes,
> But the wound looked for does not seem to be in that knife.)

In that I had to make only one change, from 'mizgaan' (eyelashes) to 'minqaar' (beak):

> *Khwesh ra bar nok-e-minqaar-e-sitam keshan zadam*
> *Aan qadar zakhme keh dil mee khwast, dar khanjar na bood*

I can't say anything about pain but every blow of the beak left a wound on the surface of the palm:

> *Raseedan hai minqaar-e-huma bar ustkhwaan Ghalib*
> *Pas az umre b'yadam daad-e-rasm-o-rah-e-paikaan ra*[42]
> (The wounds received from the beak of the phoenix, Ghalib,
> I recall after ages as the compliment to love's blandishments.)

---

[39] Zaki Hamdani's couplet (*Khareeta-e-Jawahar*: 112).
[40] *Dewan-e-Hasan Sajvi Dehlvi*: 352.
[41] This couplet was not found in *Kuliyat-e-Saib* but in *Khareeta-e-Jawahar*: 138 and *Shama'-e-Anjuman*: 373. It has been attributed to Faseehi Harvi; Maulana is mistaken.
[42] *Kuliyat-e-Ghalib*: 372.

If the commoners of this place be ignored, there are a few personalities that deserve notice. You have already been introduced to Qalandar and Moti. Here is a brief account of Mulla and Sufi. One of the sparrows has a very stout build and is extremely contentious. Its tongue is ever active; its gait has a swagger and it picks up fight with whoever it sees. No neighbourhood sparrow can set foot in his domain; several brave ones tried but were laid down in the very first encounter. Whenever a gathering of the company is held on the floor it comes with its peculiar swagger, casts glances all around and jumps upon a high seat. Then starts its unending 'choon-chaan' that calls to mind the figure of Qa-Aani's little preacher at the main mosque:[43]

> *Vee vaizke aamad dar masjid-e-jami' Choon barf hama jama sped az paa taa sar*
> *Chashmash b'soo-e-chap va chashmash b'soo-e-rast Taa khud salaame kunad az mun'im-o-muztir*
> *Z'insan keh khramad beh rasan mard-e-rasan baaz Aahista khrameedi va mauzoon-o-muaqqar*
> *Faarigh na shuda khalq z' tasleem-o-tashahud Barjast choo boozina va b'nashist beh mimbar*
> *Vangah beh saru gardan va reesh va lab va beeni Bas ishwa b'aavurda, sukhan kard chuneen sar*
> (That little preacher comes into the main mosque, Dressed in white, like snow, from head to foot;
> One glance right, another towards the left, Himself salutes all and sundry in the assembly;
> The man is a rope-dancer and performs rope-tricks, Walks slowly to look proper and dignified;
> Before the people have responded to his salutations, He jumps like a monkey on to the platform
> See there his straight high neck, his beard and lips, He talks like an accomplished coquet.)

Now tell me what other name would suit it if not Mulla? There is another sparrow, just the opposite of this one. Whenever you look at it, it is lost in itself;

---

[43]*Dewan-e-Qa'ani*: 322.

Kaan ra keh khabar shud, khabrash baaz nayamad[44]
(You talk to it, there would be no response.)

At best you hear a squeak and that too so feeble that you can hardly catch a word. It is like a man ever absorbed in himself and occasionally raising the head to utter a 'ha':

*Taa too bedari, nala kasheedam, varna
Ishq kar-eest keh be aah-o-fughaan neez kunand*[45]
(I cry only while you are asleep;
Passion of love forbids wailing and crying.)

Other sparrows pursue it around as if tired of its silence but it does not stir its tongue. But, if you are able to comprehend the language of eyes, its silence does communicate:

*Too nazar-baaz na-i, varna taghaful nigah-st
Too zabaan fehm na-i, varna khamoshi sukhan-st*[46]
(You are not a fop to know that neglect means ogling
Nor do you know the tricks of language to comprehend silence.)

Seeing its condition I named it Sufi and the fact is—

Jama-e bood keh bar qamat-e-oo dokhta![47]
(It was a coat as if specially stitched to fit him.)

In the morning when all in the colony come out there is a unique hustle and bustle in the yard and verandah. Some of them jump about on the flower pots and some use Crotine as the swing. One couple decided on a wash and waited for watering of the plots; the moment water was poured the couple jumped down into the tank and started fluttering the wings. Another couple, unable to find water around, started a dust-wash, following the *Qur'anic* injunction 'if you don't get water for ablutions, make

[44]Line from Introduction of *Gulistan* (*Kuliyat-e-Saadi*: 2).
[45]Iqbal's couplet (*Zaboor-e-Ajam*: 101).
[46]Couplet of Zahoori Tarshezi (*Dewan*: 46).
[47]Line of Hafiz Shirazi (*Dewan-e-Hafiz*: 112).

use of the dust on earth'.⁴⁸ First so much earth was dug with its beak that its body may sink in dust up to the chest and then, sitting in the pit, started flitting about ferociously so that it raised a veritable dust-storm. At a little distance Mulla is busy in a combat that is so absorbingly strange:

> *Ladte hain aur haath mein talwar bhi nahheh*⁴⁹
> (Strange combat that combatants have no weapon in hand.)

If you look for hand, there is no weapon; indeed the hand itself is missing:

> *Dahan ka zikr kya yaan sar hi ghayab hai girebaan se*⁵⁰
> (What to speak of the mouth, the head itself is missing.)

But look at their beaks that work for all kinds of weapons; so fiercely do they merge into each other that it is difficult to distinguish one from the other and you can well witness *jadaal-e-Saadi ba muddai dar bayaan-e-tuwangri-o-darveshi*⁵¹ (Saadi's combat between the powerful and the saintly).

> *Oo dar man va man dar oo futada.*⁵²
> (He and I fell in to each other's embrace!)

When the combat takes place in the air, they do not know where they fall. Several times they fall on my head. Once they landed right in my lap. I caught one in one hand and the other, in the other:

> *Mere dono haath nikle kaam ke*⁵³
> (Both my hands proved useful.)

---

⁴⁸*Qur'an, Surat-al-Nisa'* 4: 43; it means: 'If you don't get water for ablutions, make do with dust.'
⁴⁹*Dewan-e-Ghalib*: 139.
⁵⁰*Dewan-e-Zauq* (Compiled by Azad): 239.
⁵¹Title of the last story in *Gulistan* (*Kuliyat-e-Saadi*: 117).
⁵²*Kuliyat-e-Saadi*: 121.
⁵³Slightly modified line of Daagh (*Yaadgar-e-Daagh*: 112).

The body of each was in my grip with only the neck out; yet, they were eager to peck at one another. When I released the grip both flew up on to the fan, chirping for a long while as if saying to each other:

Raseeda bood balae, vale b'khair guzasht[54]
(But for the evil one, you were done for.)

For quite some time the sound of an infant was coming out of the hen-sparrow's nest. It would come and not pick up more than one or two grains of rice and rush to the nest. The moment it reached there the infant's noise would rise. She repeated this every one second or two. Once I counted it made seven trips in one minute.

The zoologists who have studied the specific qualities of birds say that a hen-sparrow feeds its infant 250 to 300 times in a day; if the total quantity of feed be compared with the weight of the infant its mass would not be less than that of the infant's body. But, the digestive power of the infants is such that the moment one grain is fed its begins to get dissolved. That is why the growth rate of the birds is much faster than that of the four-legged animals and they reach adulthood quickly. This view was confirmed by the speed of Moti's activity.

As the infant's wings grow the angel of intuition comes and whispers into the ears of the mother that it is time to give them flying lessons. It seems such whisperings were on in the ears of Moti. One morning as it descended from its nest I saw a tiny infant falling with half open wings. Moti would approach it and ask it to fly. But there was no sign of any effect on the infant; it lay still with eyes closed and wings spread. I picked it up and found that its wings were not yet fully developed and the fresh trauma of fall had unnerved it. That brought to mind the following couplet of Nazeeri:[55]

Beh vaslash taa rasm, sad baar bar khaak afgand shouqam
Keh nau-parwazam va shaakh-e-buland-e-aashiyan daram

---

[54]Line of Asifi Harvi (*Imsaal-o-Hukum*: 2: 868).
[55]*Dewan-e-Nazeeri*: 293.

(The passion to meet you threw me into dust a hundred times;
Am new to flight and have my nest on a high branch.)

Anyway, I put it on the carpet. Moti picked up grains of rice and fed him; as the infant opened its mouth it would start chirping a bit feebly and then resume its still posture. The whole day passed like that. The same position continued the next day too. The mother would incite it to fly but it was so inert that there was no response; I thought it won't survive. But, the third day there was a strange development; it went and stood on a line of sun that had spread out in the room. Its wings were down, the feet turned and eyes closed as usual. Then I saw that it suddenly opened its eyes and shook itself and looked up into the stratosphere. Thereafter it opened and closed the wings a few times and then it jumped up, went into the field and disappeared into the skies like a rocket. So strange and unexpected was this spectacle that for once I doubted my eyes that I might have mistook it for the flight of another sparrow. The happening was however so manifest that there was no scope for doubt. Strange that for two days all the efforts of the mother failed to slightly raise it above the ground and now it displayed such revolutionary zest that in the very first flight broke all shackles and boundaries to disappear in the unbounded expanse of the spheres. How to express in words; the spectacle sent me into such rapture that the following couplet struck me so forcefully that the neighbours were startled:

Neeroo-e-ishq been keh dareen dasht-e-be-karaan
Gaame na rafta em va b'payaan raseeda em
(See the force of love that in this desert;
It has never been traversed but now reached on foot.)

It was nothing but a simple miracle of life that is ever before our eyes and we don't want to understand it. The infant sparrow had developed the flying faculty and had come out of the corner of its nest to stand face to face with the heavens but what had not awakened yet was its 'self-awareness', it was ignorant of its real nature. The mother's repeated hints, the repeated touch of the flowing breeze and the instigations of the life's tumult, nothing warmed up its cold inner self:

Kleem, shikwa z' taufeeq-e-chand, sharmat baad!
Too choon b'rah na nahi pae, rehnuma cheh kunad[56]
(Kleem, your complaint of the guide is a shame;
If your foot is not rightly placed what can the guide do?)

The moment its 'self-awareness' awoke and it realized that it is 'a flying bird', every dead faculty came alive. The lifeless body that could not stand, stood erect like a cypress; the knees that could not carry the weight of the body, were straight; the fallen wings that showed no movement, began to contract and balanced themselves; the whole body was shaken by the lighting zeal to fly. It could be seen that all the shackles of helplessness were broken and the enterprising bird was measuring the boundless extremes of skies like the eagle:

Baal b'kusha va safeer az shajr-e-toba zan
Haif bashad choo too murghe keh aseer-e-qafasi![57]
(With open wings the bird flies from the heavenly tree;
It's a shame on the bird that remains caught in the cave!)

The entire revolutionary transformation of frailty to strength, from ignorance to awakening, from helplessness to lofty flight and from death to life took place in the twinkling of the eye. If you think over it, this is the essence of the entire story of life:

Tae neshawad een rah b'drakhshsheedan-e-barqe
Maa bekhabaraan muntazir-e-sham'a va charaghem[58]
(This path is covered in a lightning interval;
The ignorant ones wait for the flame and lamp.)

What requirements of flight were not available to this one newly caught in the life's cage? Nature had equipped it with all necessary faculties and the mother was constantly spurring it to flight, but as long as its 'self-awareness' had not awakened and the realization had not dawned that it was a bird meant to fly

---

[56]*Dewan-e-Kleem*: 241.
[57]*Dewan-e-Hafiz*: LL 341.
[58]Couplet of Ali Quli Beg Aneesi Shamloo (*Shama'-e-Anjuman*: 26).

high, all its equipment remained inert. In the same manner, no external stimulus can awaken a person till his 'self-awareness' does not wake up. The moment he discerns his inherent nature and comes to realize his truth, the transformation takes place in a moment and, in one jump, he rises from the dust of the earth and reaches the heavenly heights. Khwaja of Shiraz had hinted at the same truth to say:

> *Cheh goyamat keh b'mae-khana dosh mast-e-kharab*
> *Sarosh-e-aalm-e-ghaibam cheh muzdaha dad ast*
> *Keh ae buland-e-nazar, shahbaz-e-sadd-e-rah nasheen!*
> *Nisheman-e-too na een kunj-e-mehnatabad ast*
> *Tura az kingra-e-arsh meezanand safeer*
> *N'danamat keh dareen damgah cheh uftad ast*[59]
> (They say that the tavern has paralysed the wine drinker,
> What good tiding does the prophet's voice bring?
> O high minded, the eagle is your model against obstructions;
> This corner of daily grind is not your place;
> You have been thrown down from the corner of paradise
> Don't know the cause of this fall into this trap.)

MAULANA AZAD, *Ghubar-e-Khatir* (*Sallies of Mind*), 1946

Legendary hunter, conservationist, and naturalist Jim Corbett (1875–1955), famous for his writings on the hunting of man-eating tigers and leopards, had an inimitable writing style by which he gradually built up suspense and tension. The extract from the story 'The Chowgarh Tigers', about how he shot the man-eater while holding a clutch of nightjar eggs in his hand could be shikari yarn, as has been pointed out by experts, but is a good sample of Corbett's writing style. Corbett's most well-known books are Man-eaters of Kumaon, The Man-eating Leopard of Rudraprayag, *and* The Temple Tiger and More Man-eaters of

[59]*Dewan-e-Hafiz*: 37.

Kumaon. Corbett also established India's first national park, now named as Corbett National Park. One of the five remaining subspecies of tigers is named after him—Panthera tigris corbetti, more commonly called Corbett's tiger.

Nightjar Eggs

The ravine was about ten yards wide and four or five feet deep, and as I stepped down into it a nightjar fluttered off a rock on which I had put my hand. On looking at the spot from which the bird had risen, I saw two eggs. These eggs, straw-coloured, with rich brown markings, were of a most unusual shape, one being long and very pointed, while the other was as round as a marble; and as my collection lacked nightjar eggs I decided to add this odd clutch to it. I had no receptacle of any kind in which to carry the eggs, so cupping my left hand I placed the eggs in it and packed them round with a little moss.

As I went down the ravine the banks became higher, and sixty yards from where I had entered it I came on a deep drop of some twelve to fourteen feet. The water that rushes down all these hill ravines in the rains had worn the rock as smooth as glass, and as it was too steep to offer a foothold I handed the rifle to the men and, sitting on the edge, proceeded to slide down. My feet had hardly touched the sandy bottom when the two men, with a flying leap, landed one on either side of me, and thrusting the rifle into my hand asked in a very agitated manner if I had heard the tiger. As a matter of fact I had heard nothing, possibly due to the scraping of my clothes on the rocks, and when questioned, the men said that what they had heard was a deep-throated growl from somewhere close at hand, but exactly from which direction the sound had come, they were unable to say. Tigers do not betray their presence by growling when looking for their dinner and the only, and very unsatisfactory, explanation I can offer is that the tigress followed us after we left the open ground, and on seeing that we were going down the ravine had gone ahead and taken up a position where the ravine narrowed to half its width; and that when she was on the point of springing out on me, I had disappeared out of sight down the slide and she had involuntarily given vent to her disappointment with a low growl. Not a satisfactory reason, unless one assumes—without any reason—that she had selected me for her dinner, and therefore had no interest in the two men.

Where the three of us now stood in a bunch we had the smooth steep rock behind us, to our right a wall of rock slightly leaning over the ravine and fifteen feet high, and to our left a tumbled bank of big rocks thirty or forty feet high. The sandy bed of the ravine, on which we were standing, was roughly forty feet long and ten feet wide. At the lower end of this sandy bed a great pine tree had fallen across, damming the ravine, and the collection of the sand was due to this dam. The wall of overhanging rock came to an end twelve or fifteen feet from the fallen tree, and as I approached the end of the rock, my feet making no sound on the sand, I very fortunately noticed that the sandy bed continued round to the back of the rock.

This rock about which I have said so much I can best describe as a giant school slate, two feet thick at its lower end, and standing up—not quite perpendicularly—on one of its long sides.

As I stepped clear of the giant slate, I looked behind me over my right shoulder and—looked straight into the tigress's face.

I would like you to have a clear picture of the situation.

The sandy bed behind the rock was quite flat. To the right of it was the smooth slate fifteen feet high and leaning slightly outwards, to the left of it was a scoured-out steep bank also some fifteen feet high overhung by a dense tangle of thorn bushes, while at the far end was a slide similar to, but a little higher than, the one I had glissaded down. The sandy bed, enclosed by these three natural walls, was about twenty feet long and half as wide, and lying on it, with her fore-paws stretched out and her hind legs well tucked under her, was the tigress. Her head, which was raised a few inches off her paws, was eight feet (measured later) from me, and on her face was a smile, similar to that one sees on the face of a dog welcoming his master home after a long absence.

Two thoughts flashed through my mind, one, that it was up to me to make the first move, and the other, that the move would have to be made in such a manner as not to alarm the tigress or make her nervous.

The rifle was in my right hand held diagonally across my chest, with the safety-catch off, and in order to get it to bear on the tigress the muzzle would have to be swung round three-quarters of a circle.

The movement of swinging round the rifle, with one hand, was begun very slowly, and hardly perceptibly, and when a quarter of a circle had been made, the stock came in contact with my right side. It was now necessary to extend my arm, and as the stock cleared my side, the swing was very slowly continued. My arm was now at full stretch and the weight of the rifle was beginning to tell. Only a little further now for the muzzle to go, and the tigress—who had not once taken her eyes off mine—was still looking up at me, with the pleased expression still on her face.

How long it took the rifle to make the three-quarter circle, I am not in position to say. To me, looking into the tigress's eyes and unable therefore to follow the movement of the barrel, it appeared that my arm was paralysed, and that the swing could never be completed. However, the movement was completed at last, and so soon as the rifle was pointing at the tigress's body, I pressed the trigger.

I heard the report, exaggerated in that restricted space, and felt the jar of the recoil, and but for these tangible proofs that the rifle had gone off, I might, for all the immediate result the shot produced, have been in the grip of one of those awful nightmares in which triggers are vainly pulled of rifles that refuse to be discharged at the critical moment.

For a perceptible fraction of time the tigress remained perfectly still, and then, very slowly, her head sank on to her outstretched paws, while at the same time a jet of blood issued from the bullet-hole. The bullet had injured her spine and shattered the upper portion of her heart.

The two men who were following a few yards behind me, and who were separated from the tigress by the thickness of the rock, came to a halt when they saw me stop and turn my head. They knew instinctively that I had seen the tigress and judged from my behaviour that she was close at hand, and Madho Singh said afterwards that he wanted to call out and tell me to drop the eggs and get both hands on the rifle. When I had fired my shot and lowered the point of the rifle on to my toes, Madho Singh, at a sign, came forward to relieve me of it, for very suddenly my legs appeared to be unable to support me, so I made for the fallen tree and sat down. Even before looking at the pads of her feet I knew it was the Chowgarh tigress I had sent to the Happy Hunting Grounds, and that the shears that had assisted her to cut the threads of sixty-four human lives—the people of the district put the number at twice that figure—had, while the game was in her hands, turned, and cut the thread of her own life.

Three things, each of which would appear to you to have been to my disadvantage, were actually in my favour. These were (a) the eggs in my left hand, (b) the light rifle I was carrying, and (c) the tiger being a man-eater. If I had not had the eggs in my hand I should have had both hands on the rifle, and when I looked back and saw the tiger at such close quarters I should instinctively have tried to swing round to face her, and the spring that was arrested by my lack of movement would inevitably have been launched. Again, if the rifle had not been a light one it would not have been possible for me to have moved it in the way it was imperative I should move it, and then discharge it at the full extent of my arm. And lastly, if the tiger had been just an ordinary tiger, and not a man-eater, it would, on finding itself cornered, have made for the opening and wiped me out of the way; and to be wiped out of the way by a tiger usually has fatal results.

While the men made a detour and went up the hill to free the buffalo and secure the rope, which was needed for another and more pleasant purpose, I climbed over the rocks and went up the ravine to restore the eggs to their rightful owner. I plead guilty of being as superstitious as my brother sportsmen. For three long periods, extending over a whole year, I had tried—and tried hard—to get a shot at the tigress, and had failed; and now within a few minutes of having picked up the eggs my luck had changed.

The eggs, which all this time had remained safely in the hollow of my left hand, were still warm when I replaced them in the little depression in the rock that did duty as a nest, and when I again passed that way half an hour later, they had vanished under the brooding mother whose colouring so exactly matched the mottled rock that it was difficult for me, who knew the exact spot where the nest was situated, to distinguish her from her surroundings.

JIM CORBETT, *Man-eaters of Kumaon*, 1944

A whole generation of Indian naturalists and wildlifers has grown up reading M. Krishnan's (1912–95) nature columns. Though his mainstay was The Statesman, *he also wrote for* The Hindu, The Indian Express, The Illustrated Weekly of India, *and* Shankar's Weekly *among others. A member of the Indian Board for Wildlife, he was awarded the Padma Shree in 1970.*

Rescuing a Fledgeling

After a solid breakfast I smoked my favourite pipe and, while my table was being cleared and dusted, had a nice, cold wash. Then I had a cup of strong, hot coffee. I was preparing to work. By two o'clock I had decided on the plan of work—before tackling hard jobs it is wise to spend a moment in planning the attack. And as I sat down at last to the hateful, necessary thing, a commotion broke out in my backyard, a series of high, thin squeaks and quivers, like the 'ghosts that did squeak and gibber in the Roman streets'. White-headed Babblers are excitable creatures, and there is a clan of them living in and around my compound, sounding the alarm at each passing cat and human. However, there was a sustained hysteria in their alarm now, something in the way their *chee-chee-chees* and tremulous *chrrrrrrs* rose in outraged pitch till they were choked altogether, that called for immediate investigation, and I stepped out to the corner of my backyard.

On the clean-swept ground beneath the mango tree was a fledgeling babbler, just out of the nest, fluttering weakly against the corner of the compound wall and the bathroom, falling to the ground after each futile attempt to gain the top of the wall. Two adult babblers were on the ground beside it and three more in the boughs above, all encouraging the premature adventurer with frantic voices and quick flirts of their round, half-spread wings and loose-feathered tails. And strolling along the top of the wall towards this domestic group was a small, grey cat.

The cat was the first to see me. It froze in its tracks, gave me one intense, green-eyed look, and disappeared down the other side of the wall into my neighbour's territory. The birds flew into the tree at my approach, but when I was right under the tree and just a step from the fledgeling, they flew away in a loud body to a drumstick tree some twenty yards away, and there continued their alarm even more agitatedly than before—I noticed that not one of them was facing me and that some hopped down to the grass beneath the drumstick tree, as if what excited them lay there.

The moment its elders left, the infant babbler crouched low and was instantly turned into a small, grey, shapeless, immobile lump; it did not move or bat an eyelid even when I touched it. Here was an intriguing situation. The fledgeling on the bare ground was exposed to every passing enemy, and the older birds would not come to its rescue so long as I was near; they would persist with their instinctive demonstration at the grass beneath the drumstick tree. Should I move the youngster beyond the drumstick tree, where it would be safe in the innumerable bolt-holes offered by a pile of broken brick and scrap, or should I leave it where it was, keeping an eye on it from a distance and watching further proceedings? Finally I retreated to a point equidistant from both trees, after taking a photograph of the fledgeling, and sat down to watch.

Till five o'clock I sat stolidly on, observing happenings. First all except two of the babblers (the parents of the grounded youngster?) left ostentatiously, whirring and skimming on weak, blunt wings over and beyond my roof. The birds that stayed behind struck to the drumstick, tree, twittering feebly from time to time. Next, a pair of ill-assorted baby-snatchers arrived on the scene, a jungle-crow and a house-crow; they perched on the compound wall and cocked their heads from side to side, looking at me and at the ground beneath them with sly, sidelong glances. After a while they hopped towards the mango tree.

This was the signal for the waiting pair of babblers to fly headlong into the mango tree, yelling blue murder and the rest of the clan was there at once, as if by magic. Routed by the pack of yelling, gibbering babblers, the crows fled to a coconut top some distance away. The performance was repeated several times, the babblers leaving the mango tree and even my compound, the crows approaching furtively, the babblers returning in screaming force at once to mob and drive away the enemy. The grey cat, which appeared on the wall again, was also mobbed and chased away, but the passage of a shikra low overhead was marked by silence.

All the time the little one stayed put. I doubt if it as much as lowered its bill by a fraction of an inch in all those three hours.

But it closed its eyes and did not open them except when the crows, whose proximity was proclaimed by the furious babblers, were near. It was evident that no attempt to induce the youngster to move to safer quarters would be made so long as I was there. The sky was darkening, and rain imminent. I decided I had watched long enough, and taking the fledgeling gently in my hand, deposited it on the scrap heap, and it promptly disappeared into a crevice.

In a moment the adult babblers had joined it and, the new ground being sufficiently far from me, vociferously encouraged the fledgeling to essay flight. However, in the further fifteen minutes that I watched, it did not succeed in getting out of my compound—the babblers have no nest here, but probably have one in my neighbour's compound. Next morning they were more successful, and the youngster cleared the wall after a few tries. Apparently, a day and a night make all the difference in development to a fledgeling learning to fly.

M. Krishnan, 1956

In this memorable piece describing the simple incident of a Paradise Flycatcher nesting in an army camp in Udhampur, Kashmir, for several years in succession, Brigadier R. Lokaranjan reflects upon the passage of time and seasons. An avid birdwatcher, Brigadier Lokaranjan frequently contributed to the Newsletter for Birdwatchers.

Delightful Distractions

The letter from Shivrajkumar Khacher, in the correspondence column of the October 1971 issue of the *Newsletter* regarding the nest of the Paradise Flycatcher seen at the same place in July 1957 and '58, then again in 1971, reminded me vividly of a somewhat similar experience of mine.

Perhaps the first time I consciously observed these fascinating birds was during a short visit to Udhampur (J&K) in 1963. Not wanting to attract undue attention watching the birds, I merely loitered outside the bosses's office, where I was summoned, long enough to notice the nimble twisting and turning of the male birds with their streaming, waving, long tails. It was all very fascinating to me. I also noticed a nest, which I presumed must be the Paradise Flycatcher's. On my return to my rather bleak station (it happened to be Leh) reference to Sálim Ali's *Book of Indian Birds*, describing the bird's nest which was so distinctly a woven cup in a forked portion of a branch, with a nicely plastered look on the outside, confirmed my views. I forgot this incident of the identification of the nest, but the memories of seeing these birds—rather their tails that morning remained quite vivid right up to the time six years later, when it so happened that I moved there, and occupied the same office. The month was April, in 1969. There were mulberry trees amongst the others in the area and these were full of fruit, and the commotion of various birds was to say the least very noisy; but what a delightful distraction from work!

My constantly peeping out of the window while files marked 'urgent' kept coming in, and my readjusting my chair and office table to a better-suited position in the midst of all the confusion of files and visitors, caused some problems to my PA particularly, but he gallantly refrained from giving me the slightest indication of what appeared to him no doubt a marked eccentricity in my office seating arrangements! I later however shared the secret of watching the birds with him, as my binoculars were carefully concealed in the 'pending' tray and he had to be told! An unforgetable experience of the nesting of a white-eye, that fluffy little greenish yellow fellow, just a few feet from my window was another experience shared with a chosen few, but that I will write about some other day.

To get back to the Paradise Flycatchers of Udhampur, I was time and again distracted from important work, but always rewarded with the most impressionable and striking antics of these birds. Above all, 'the agile fairy-like movements of the male as he twists and turns in the air after flies, with his tail ribbons looping or trailing behind' (I'm quoting Sálim Ali) 'a spectacle of exquisite charm'!

It was such a pleasant surprise when a few days later I suddenly noticed a nest! 'Ah' I thought as I now recollected the forgotten incident of six years earlier, 'could it possibly be the same nest?' I racked my brain and paced up and down to construe the earlier loitering outside this office. Where could I have stood? Which was the tree? And so on! I was soon quite sure that it was the same nest.

Then a year went by. In April/May 1970, I had the same delightful experience as winter ended and it was spring. Dry twigs so rapidly bore millions of green leaves—birds' voices created a din—mulberries, appeared as if from nowhere—the 'fairies' of paradise reappeared—and then THE NEST. The same one was occupied. Again I had the same joyful experience of observing all this and particularly the posterior of the bird just visible, seated on its eggs! I saw the whole thing repeated in 1971 also.

I have now left that place on transfer. The sad thing is that the old office may not be there in 1972. A new office is being built, 'pakka' this and 'pakka' that, a storied building a hundred yards away, with a car park outside, and so on! Give me back the old tent in an orchard or garden, back in 'the field'. But it cannot be so, I know. It may be appropriate here to quote Thomas Moore:

> When time who steals our years away,
> Shall steal our pleasures, too
> The memory of the past will stay
> and half our joys renew.

R. LOKARANJAN, *Newsletter for Birdwatchers*, 1971

D.A. Stairmand *worked for a business house in Bombay, and indulged in birding, by either driving up to Khandala or the Borivali National Park, as well as other areas, every weekend. He regularly contributed to the* Newsletter for Birdwatchers.

Castaway with Birds

For many years now the BBC radio has run a programme called 'Desert Island Discs' in which a person is asked to assume that he or she is cast away alone on a desert island with rescue only a remote possibility. The 'castaway' is allowed ten records to take onto the Desert Island and required to give reasons for his or her choice. As a slight—and appropriate—variation I would like to choose ten species of birds I would have with me—assuming the almost impossible and completely horrifying possibility that no birds already existed on or visited the island—and give my reasons. I would, however, make one or two pre-conditions about the island; it must have a good lake, rivers and a varied type of habitat. And the birds would, at minimum, be pairs; in some species small parties.

1. Pitta: This delightful bird would be my first choice. Once the bird became trusting it would be a great joy to watch hopping around on the ground and digging violently into the mulch for insects all the while, with leaves flying and the pitta keeping a dead-pan face. In its breeding season the pitta would call from high up in the trees and at all times its colours would be flamboyant.

2. Dipper: There is no more exciting ground bird to watch than the dipper as it plunges into deafening, foaming, high velocity water in mid-torrent. In more peaceful moods of evening I would watch and listen to the dipper singing happily—if somewhat hoarsely—from a stone in mid-steam while a female would be bathing and preening herself on a nearby rock in readiness for the night.

3. Osprey: To see an osprey hover over water about 60 feet up, then drop and strike the water with a most tremendous splash before disappearing below the surface and then emerge a second or two later with its prey held under foot, is a great experience. A really thrilling spectacle. A pair of ospreys used to be at Vihar Lake, Bombay, around about March and I hope they still are. If any readers in the Bombay area have never watched an osprey fishing I suggest that they could do no better than try their luck at Vihar.

4. Spotted Owlet: I visited the Ghana, Bharatpur, last July and was very fortunate to stay at the Rest House and have Mr Panday

for company. Adjacent to the Rest House balcony are rings on a stand—originally intended to hold potted plants. However, these rings are put to a far better use. Just on dusk electric lights are switched on around the Rest House and these attract night flying insects and, with them, a party of spotted owlets. The owlets use the rings as perches and these absolutely delightful birds sally to and fro from the stand all night (I know, because I got up from bed at mid-night at 4 a.m. to continue my watching which had started at 7 p.m.). Sometimes the birds—up to 7—descended to the ground and moved quickly to pick up insects but mainly they flew and returned to a perch with the victim held tightly in their claws. Whatever they did, they were delightful (sorry, but I just must use this word time and again) and form one of my most lasting and vivid impression of Indian birds.

5. Yellow Wagtail: I had to have a wagtail—they are so beautiful in summer dress, but it was a hard decision to discard the grey wagtail and the yellowheaded wagtail. All wagtails are really lovely birds but the few yellow wagtails I have seen in India in summer plumage in late winter or early summer just tipped the scale in favour of this species. I make no excuses, but realize I shall probably incur the wrath of many grey wagtail enthusiasts!

6. Common Green Bee-eater: A strange choice, you may say. Why not the chestnut-headed or, more to the point, the blue-tailed? Perhaps it is strange, but then although I know Beethoven is the *greatest* composer of Western Classical music my *favourite* composer is Delius. So my favourite bee-eater is the common green—you all know it—and I have a party of 35 on my Desert Island. Thank you. What fun they will be!

7. Malabar Whistling Thrush: It may be a bit shy but when the rainy, cloudy, stormy weather comes it will mate and build its nest and still find time to sing a lot of the day. And I can watch those marvellous flashes of cobalt against its blue-back background! I found the Himalayan whistling thrush a little disappointing—it's a fraction too big and sturdy and its song inferior. The whistling schoolboy is never disappointing.

8. Pied Kingfisher: A very handsome bird and really the most exciting kingfisher that one can watch for long period of time. It perches on vantage points over water with its huge bill pointed downwards like a rifle, then takes off to hover over water before

plunging excitingly for its prey. It does this for hour after hour and I could watch it for year after year. Sad though, that I cannot take the three-toed forest kingfisher, too.

9. Little Egret: Elegant, active and, above all, the most beautiful of all birds in its breeding plumage. Need I say more?

10. Racket-tailed Drongo: A fascinating, intelligent and beautiful bird. With some luck I could tame it to visit my hut and rant on as drongos do. Always it would be at the very centre of birdlife.

So my list of ten has ended and I have had to omit personal favourites and wonderful birds such as the flamingos, falcons, redstarts, tits, sunbirds, swallows, sarus crane, flycatchers (Paradise, black and orange, fantail), minivets, pelicans and cormorants (which could have helped me with fishing), woodpeckers, rollers, orioles, hornbills, etc.

The 'Desert Islander' is usually allowed a 'bonus' and for this I'll take a bird I have never seen in the wild—the blackheaded sibia. The one I remember was in a large mixed aviary at Ahmedabad Zoo and its colours and calls were pure delight.

D.A. STAIRMAND, Newsletter for Birdwatchers, 1972

Simon Barnes is the award winning chief sports writer for The Times of London. *Author of a number of books on wildlife, conservation, and travel, and of three novels, he is also the well-loved and controversial columnist of* Birds (*a publication of the Royal Society for the Protection of Birds*).

Falling in Love Again

That's right, the Mascara Snake—fast and bulbous.
—Captain Beefheart

It is remarkable how much of our lives we spend doing things we really don't want to do. More remarkable still is how much time we spend doing stuff we think we are enjoying, only to realize later on that it wasn't enjoyable at all, and that we'd far sooner have been doing something else.

I feel these things very strongly when I look back at the wasted years: the years when I wasn't bird-watching, or even getting *the Observer's Book of Birds* by heart. I read books at school, all the time, and that wasn't a waste of time at all. But I did things like playing soccer, which secretly bored me, and getting involved in drama, which I secretly hated. When I reached the sixth form I was interested in revolutionary politics, for which I secretly had no taste whatsoever, still less talent or understanding.

Perhaps if the right opportunity, the right person had come along, I'd have been off a-birding. But perhaps not: I wished to be—needed to be—both successively and simultaneously a good mate, a fancy-Dan intellectual and a dangerous radical. Birdwatching can cater to all those desires; I will demonstrate later on how hanging out peanuts for your chickadees is an act of revolution. But I didn't realize that at the time.

Then there was traveling and going to university and that sort of thing. I was involved in the traditional pursuits of youth, of course. But again, looking back, I secretly hated an awful lot of rock and roll and I was secretly terrified of drugs. The other part was all right, though no doubt an understanding of the way birds do it—in song and dance and finery—would have helped me on my way with my callow courtships.

This was the end of the sixties, the start of the seventies, and it was considered a fine, even enviable thing to be 'into nature.' Or, better still, 'heavily into nature.' A fine thing, so long as you didn't get too specific. 'Wow, man! Far out!' That was the OK response to a sunset, a tree in the acid green of spring, or the flight of a seagull. I think it's a pretty good response, on the whole, and I still think it when I see a barn owl, say, or I hear the first willow warbler of spring. Sometimes I even say it; it slips out in moments of unexpected delight. But then no birdwatcher, however good, however bad, ever really grows out of the wow response. You just get to add other things to it.

Being too specific would have been a bit of a faux pas in those days. One of the set texts of hippiedom—a book I never cared for—was *Jonathan Livingston Seagull*. Jonathan Livingston black-headed gull? Audouin's gull? Bonaparte's gull? Don't be uncool, man: it's a seagull, just dig it. Precision was frowned upon. 'Perfect speed is being there,' said the book. Or perhaps it was a glaucous gull.

There was a period in which I listened to the dawn chorus every day, though not entirely on purpose. I did it by staying up for it, rather than getting up for it. I well remember the daily horror of drawing back the curtains and finding instead of night a pale blue-gray morning alight with the cacophony of birds. 'Hey, wow, man, it's the tweeties! That's too heavy.'

We compared the sound unfavorably to that of Captain Beefheart playing the saxophone. If you have ever heard Captain Beefheart playing the saxophone, you will understand that the tweeties were not soothing. It was a violent invasion, the new day interrupting our night, a great shock to the senses and to the mind. Odd, now to think that I can listen to the dawn chorus—blindfolded if you wish—and know most of the birds that sing. I can tell a thrush from a blackbird, and even, when the wind is southerly, a blackcap from a garden warbler. Had I been able to do such a thing then, I would have done better to keep quiet about it. Jut listen and say *wow*. Not such bad advice, as I say. But you can add other things to it, which is what happened to me and why I am writing this book. The more you know, the more you wow.

After university, I started work on local papers and my principal memory of the first couple of years is of anxiety and stress and the feeling that I was never going to get anywhere. It was a horribly uncomfortable time, with no birds to help me through. These days, out on assignment, in the stresses and the self-pity that come from high-octane work, I almost always manage to slip away and spend a couple of hours birdwatching. I am a sportswriter by profession, but I have never been monomaniacal about sports, or anything else for that matter. A healthy biodiversity of interest is something to cultivate, I think. I remember, for example, a roseate spoonbill seen in front of a garbage incinerator in Tampa, when I was there covering the Super Bowl. It sent me back to the football with a spring in my step and a little calmness in my brain.

Birds are great removers of stress—so long as you are not a twitcher. A twitcher might be defined as someone who actively seeks stress in birdwatching. The very name came about because of the neurotic behavior on view when these people are close to a rarity and believe they might miss it. But, for most of us, birdwatching works the other way: making life both richer and calmer, a pretty good double, I think you will agree.

So, I had no birds to help me with the stresses of the first years in a professional life. Just the memory of the pleasures I had once found in birds—or at least in the idea of birds. That and the occasional more-than-casual glance at such chance-countered delights as a kestrel hovering by the motorway, or a bunch of jackdaws riding a Ferris wheel of air in a winter wind. I had a hankering for birds, a nostalgia for birds. I just needed the right moment, the right excuse, the right place, the right person, the right bird.

Meanwhile, my father continued his high-profile, high-stress stuff at the BBC. He didn't do any birdwatching either. He looked at them all right, especially when he was on holiday in Cornwall. He, too, was waiting for the right place, the right person, the right bird.

I was unable to supply such things for him, nor he to me. The years of radical politics hadn't brought out the best in either of us. My father, for some reason, saw it all as a rejection of himself and everything he had worked for. He was unable to take things with a detached, boys-will-be-anarcho-syndicalists sort of air. And I had a wicked way of winding him up. The word *bourgeois* would do it every time. Never failed. Television was bourgeois, wearing a suit was bourgeois, drinking wine was bourgeois—not that this ever stopped me from enjoying my fair share.

We were not altogether reconciled during the university years. If I had dropped the political labeling and posturing, I was still intolerant of what we called 'straight society.' This too he took as a rejection. Partly it was, but the sixties were more than a Freudian rebellion-for-the-sake-of-rebellion. No. We all felt we were on the verge of creating a new way of living: more tolerant, less stressful, more spiritual, more meaningful—richer and more amusing. The accepted routes to this desirable state were smoking plenty of dope, wearing a great deal of fancy clothes, and listening to an awful lot of music. The aims had a lot going for them; the routes were flawed. I have since discovered a far more effective way of getting closer to all those desired things. If you're still reading, you might guess what it is.

But please remember that there was, in the full outpouring of hippiedom, a deep and heartfelt desire to put things right. There was a feeling everywhere that things had gone badly amiss in modern Western life, and that it was our job—our duty, our destiny—to make it better. Or at least to do our best. Perhaps you think I should have grown out of all that sort of thing. I have not. I still feel that there are one or two areas in which Western society has got it a bit wrong; I still feel that we should do our best to make it better. I don't think I am unique in this, either.

And quite a lot of all this comes down to bird-watching; it really does. Birds are not only a delight; they are a cause, a battle, a purpose, a meaning—and no trivial one either. I will be talking more about that later on. But before the meaning comes the joy. Marriage, for many, is what gives life its meaning. And in marriage, before the meaning comes the joy. You don't find the meaningfulness of marriage without first falling in love. I want every reader who likes birds—who, as it were, *fancies* birds—to move on. Stop ogling them from afar and make your big move. Stop admiring birds; start falling in love with them.

The sixties had substance behind the poses, you see, though the posing seemed pretty important at the time. Naturally, this involved a fair amount of perversity. Take Tim. I was to meet him a few years later, and he was to teach me a huge number of birdwatching skills and, more importantly, pleasures. He was so perverse that he not only refused to wear a tie; he refused to wear a shirt as well. But birdwatching with Shirtless Tim opened my eyes.

My father obviously approved of my getting a job after university—he had been half-expecting me to decamp to India and never be seen again. Amazing to think that had I gone, I would have done so without the guidance of Sálim Ali, the doyen of Indian ornithology and creator of India's first field guides. I once saw a shikra—a jet-propelled Asian hawk—when covering a cricket match in Bangalore, but that is by the way.

But I wanted to write for my living, and journalism offered the only obvious option. So with a suit and a tie, and hair at last a little trimmed, I set off to work for the *Surrey Mirror* in Redhill.

One of the first stories I did was about somebody who had a pet owl—a tawny, as I recall. It shat on my suit. As if to say: 'Bourgeois!'

I remember talking to my father during the worst of my time on local papers, bullied, oppressed, and persecuted by an editor who admitted later, 'Of all the journalists I have had under me, he was the one I hated the most.' I groaned and told my father about the horrors of having to go to work the following morning. He said: 'I have never not enjoyed going to work.' There was still a gap in understanding, then. I felt oppressed by the scale of his achievement and his total absorption with his professional life; he felt bewildered and disappointed by my poor showing and my lack of relish for the great world of jobs.

Well, I did go to India. I went as part of my annual leave, not as a dropout. Though I was, naturally, seeking the meaning of life. And I loved India from the first moment. How could anybody not? My heart was filled with the East after that, and I had to go and live in Asia. A couple of years later, I was living on Lamma Island, a forty-five-minute ferry ride from Central District, Hong Kong, working as a freelance journalist, traveling all the time to various thrilling places around the region. And seeing the occasional thrilling bird. I was never quite sure whether my ignorance was part of the pleasure. I loved the mystery of the great birds I knew nothing about, but there was something inside me that wanted to put a name to it. Obscurely, I felt that the name mattered.

Lamma has or had a good few nice birds—it's been built up hugely since my time. In particular, there is the bird that nobody ever sees. In spring, it sings a four-note call, again and again: the first and third notes the same, the second note a semitone lower, the last with a drop of a minor third. So a musician told me, anyway. When the spring ran hot and strong in the blood, and the moon was high and the wooded slopes and the maniacally manicured market gardens were washed in silver light, the Chinese cuckoo sang all night. 'One more bot-tle! One more bot-tle!' If ever a bird sang the national anthem of a place, it was the Chinese cuckoo. In those boozy days at the far end of the empire's tether, it was a recommendation we seldom rejected.

I didn't know it was a Chinese cuckoo then, of course. But when a friend of mine—another lost birdwatcher who never watched birds—told me it was a Chinese cuckoo, I was quietly satisfied. Obscurely, I felt it mattered. Obscurely, I felt that the meaning of things, the ability to put a name to things, was important. And obscurely I wanted to put a name on even more things. I didn't know how, of course, but I had a feeling that it was important.

SIMON BARNES, How to be a (Bad) Birdwatcher, 2004

India's first Prime Minister, Jawaharlal Nehru (1889–1964), besides being a towering leader of the Indian freedom movement, a writer, scholar, and historian was also a lover of nature. Later, his daughter, Indira Gandhi, as India's fourth Prime Minister, took a giant stride for wildlife conservation by initiating Project Tiger. The piece reproduced here is noteworthy for its straight talk on the conservation of birds and animals.

Foreword
[to The Wildlife of India]

Wild life? That is how we refer to the magnificent animals of our jungles and to the beautiful birds that brighten our lives. I wonder sometimes what these animals and birds think of man and how they would describe him if they had the capacity to do so. I rather doubt if their description would be very complimentary to man. In spite of our culture and civilisation, in many ways man continues to be not only wild but more dangerous than any of the so-called wild animals.

Nature is said to be red in tooth and claw, and life is precarious in the forest. The strong prey on the weak and the weak develop subterfuges and camouflages to protect themselves. But this eternal way of the forest is due principally to the quest for food. Man does not eat man, but he kills him for other purposes; and even where he does not kill the body, he kills the spirit. We are strange mixtures of good and evil, of civilisation and barbarism, of the divine and the base. We talk in one language and act in another way. We hold aloft noble ideals and shout many slogans, but in our behaviour we belie them. We talk of peace and our manner of doing so is often aggressive and warlike.

In India, perhaps even more than in other countries, there is this difference between precept and practice. In no country is life valued in theory so much as in India, and many people would even hesitate to destroy the meanest or the most harmful of animals. But in practice we ignore the animal world. We grow excited about the protection of the cow. The cow is one of the treasures of India and should be protected. But we imagine that we have done our duty by passing some legislation. This results not in the protection of the cow but in much harm to it as well as to human beings. Cattle are let loose and become wild and become a danger not only to crops but to human beings. They deteriorate and the very purpose for which we value the cow is defeated.

In many other countries, even children take great interest in animals and birds. There are innumerable books on the animal world, and many people take arduous journeys to see some rare bird. Societies of Bird Watchers are formed, not to kill them but to see and study them. How many of our people know even the names of the less common birds? How few books we have about birds and animals.

I welcome this new interest in India in the preservation of wild life. I cannot say that we should preserve that form of wild life which is a danger in our civilised haunts or which destroys our crops. But life would become very dull and colourless if we did not have these magnificent animals and birds to look at and to play with. We should, therefore, encourage as many sanctuaries as possible for the preservation of what yet remains of our wild life. Our forests are essential for us from many points of view. Let us preserve them. As it is, we have destroyed them far too much. It is true that as population grows, the need for greater food production becomes necessary. But this should be by more intensive cultivation and not by the destruction of the forests which play a vital part in the nation's economy.

In this context I welcome this excellent book by Mr. E.P. Gee, who is one of the best known authorities in India on the subject. Although a citizen of the United Kingdom, he has spent half his life time in observing and photographing wild life in India. He has been a member of the Indian Board for Wild Life since its inception in 1952. I had the pleasure of meeting him in 1956 when I visited Kaziranga Wild Life sanctuary in Assam.

I hope that this book will help in furthering interest in this fascinating subject among our young people. I agree with the author that it is much more exciting and difficult to 'shoot' with a camera than with a gun and wish that more and more adventurous young men would give up the gun in favour of the camera. We must try to preserve whatever is left of our forests and the wild life that inhabits them.

JAWAHARLAL NEHRU, 'Foreword' to E.P. Gee's
The Wildlife of India, 1964

Sport, Entertainment, and Falconry

2

An advertisement of arms, ammunition, and fishing tackle in one of the pre-Independence English dailies

HUNTING AND SPORT

The following extract is taken from Jim Corbett's famous story 'The Talla Des Man-Eater'. It provides a vivid description of shooting birds and other game from the back of elephants in the foothills of the Himalayas, during the twilight years of the British Raj.

Hunting at Bindukhera

Nowhere along the foothills of the Himalayas is there a more beautiful setting for a camp than under the Flame of the Forest trees at Bindukhera, when they are in full bloom. If you can picture white tents under a canopy of orange-coloured bloom; a multitude of brilliantly plumaged red and gold minivets, golden orioles, rose-headed parakeets, golden-backed woodpeckers, and wire-crested drongos flitting from tree to tree and shaking down the bloom until the ground round the tents resembled a rich orange-coloured carpet; densely wooded foothills in the background topped by ridge upon rising ridge of the Himalayas, and they in turn topped by the eternal snows, then, and only then, will you have some idea of our camp at Bindukhera one February morning in the year 1929.

Shooting from the back of a well-trained elephant on the grasslands of the Terai is one of the most pleasant forms of sport I know of. No matter how long the day may be, every moment of it is packed with excitement and interest, for in addition to the variety of game to be shot—on a good day I have seen eighteen varieties brought to bag ranging from quail and snipe to leopard and swamp deer—there is a great wealth of bird life not ordinarily seen when walking through grass on foot.

There were nine guns and five spectators in camp on the first day of our shoot that February morning, and after an early breakfast we mounted our elephants and formed a line, with a pad elephant between each two guns. Taking my position in the centre of the line, with four guns and four pad elephants on either side of me, we set off due south with the flanking gun on the right—fifty yards in advance of the line—to cut off birds that rose out of range of the other guns and were making for the forest on the right. It you are ever given choice of position in a line of elephants on a mixed-game shoot select a flank, but only if you are good with both gun and rifle. Game put up by a line of elephants invariably try to break out at a flank, and one of the most difficult objects to hit is a bird or an animal that has been missed by others.

When the air is crisp and laden with all the sweet scents that are to be smelt in an Indian jungle in the early morning, it goes to the head like champagne, and has the same effect on birds, with the result that both guns and birds tend to be too quick off the mark. A too eager gun and a wild bird do not produce a heavy bag, and the first few minutes of all glorious days are usually as unproductive as the last few minutes when muscles are tired and eyes strained. Birds were plentiful that morning, and, after the guns had settled down, shooting improved and in our first beat along the edge of the forest we picked up five peafowl, three red jungle fowl, ten black partridge, four grey partridge, two bush quail, and three hare. A good *sambhar* had been put up but he escaped before rifles could bear on him.

Where a tongue of forest extended out on to the plain for a few hundred yards, I halted the line. This forest was famous for the number of peafowl and jungle fowl that were always to be found in it, but as the ground was cut up by a number of deep nullahs that made it difficult to maintain a straight line, I decided not to take the elephants through it, for one of the guns was inexperienced and was shooting from the back of an elephant that morning for the first time. It was in this forest—when Wyndham and I some years previously were looking for a tiger—that I saw for the first time a cardinal bat. These beautiful bats, which look like gorgeous butterflies as they flit from cover to cover, are, as far as I know, only to be found in heavy elephant-grass.

After halting the line I made the elephants turn their head to the east and move off in single file. When the last elephant had cleared the ground over which we had just beaten, I again halted them and made them turn their heads to the north. We were now facing the Himalayas, and hanging in the sky directly in front of us was a brilliantly lit white cloud that looked solid enough for angels to dance on.

The length of a line of seventeen elephants depends on the ground that is being beaten. Where the grass was heavy I shortened the line to a hundred yards, and where it was light I extended it to twice that length. We had beaten up to the north for a mile or so, collecting thirty birds and leopard, when a ground owl got up in front of the line. Several guns were raised and lowered when it was realized what the bird was. These ground owls, which live in abandoned pangolin and porcupine burrows, are about twice the size of a partridge, look white on the wing, and have longer legs than the ordinary runs of owls. When flushed by a line of elephants they fly low for fifty to a hundred yards before alighting. This I believe they do to allow the line to clear their burrows, for when flushed a second time they invariably fly over the line and back on the spot from where they originally rose. The owl we flushed that morning, however, did not behave as these birds usually do, for after flying fifty to sixty yards in a straight line it suddenly started to gain height by going round and round in short circles. The reason for this was apparent a moment later when a peregrine falcon, flying at great speed, came out of the forest on the left. Unable to regain the shelter of its burrow the owl was now making a desperate effort to keep above the falcon. With rapid wing beats he was spiralling upwards while the falcon on widespread wings was circling up and up to get above his quarry. All eyes, including those of the *mahouts*, were now on the exciting flight, so I halted the line.

It is difficult to judge heights when there is nothing to make a comparison with. At a rough guess the two birds had reached a height of 1,000 feet, and when the owl—still moving in circles—started to edge away towards the big white cloud, and one could imagine the angels suspending their dance and urging it to make one last effort to reach the shelter of their cloud. The falcon was not slow to see the object of the manoeuvre, and he too was now beating the air with his wings and spiralling up in ever shortening circles. Would the owl make it or would he now, as the falcon approached nearer to him, lose his nerve and plummet down in a vain effort to reach mother earth and the sanctuary of his burrow? Field glasses were now out for those who needed them, and up and down the line excited exclamations—in two languages—were running.

'Oh! He can't make it.'

'Yes he can, he can.'

'Only a little way to go now.'

'But look, look, the falcon is gaining on him.' And then, suddenly only one bird was to be seen against the cloud. Well done! Well done! *Shabash*! *Shabash*! The owl had made it, and while hats were being waved and hands were being clapped, the falcon in a long graceful glide came back to the *semul* tree from which he had started.

The reactions of human beings to any particular event are unpredictable. Fifty-four animals had been shot that morning—and many more missed—without a qualm or the batting of an eyelid. And now, guns, spectators, and *mahouts* were unreservedly rejoicing that a ground owl had escaped the talons of a peregrine falcon.

JIM CORBETT, The Temple Tiger and More Man-eaters of Kumaon, 1954

The following extract from Sálim Ali's autobiography The Fall of a Sparrow *depicts a scene at the large-scale duck shoots at Bharatpur ('holocausts' as the author terms them) that were common in the late nineteenth and early twentieth centuries.*

Bird Holocausts at Bharatpur

Before it became a waterbird sanctuary and then a National Park, Keoladeo Ghana was the fantastic private duck-shooting preserve of the Bharatpur rulers. As long as the present Maharaja retained his powers there was no question or thought on his part of voluntarily giving up his shooting rights and converting it into a sanctuary. Apart from his casual shooting of a few ducks for sport and the table, he traditionally used to lay on three or four big shoots every season to which all sorts of VIPs and some less than V were invited—viceroys, governors, top civil and military brass, and brother princes from other states. Numbered butts were allotted to the guns, distributed in strategic spots all over the lake. The whole operation worked with the mechanical precision of an army manoeuvre, with men of the state forces drilled as beaters to keep the birds moving over the guns and not letting them settle. Enormous holocausts were 'accomplished' at some of these gargantuan shoots, and there are several records of two to three thousand birds killed in a single day, and three records even of over four thousand. The all-time record bag of 4,273 ducks and geese to 38 guns for Keoladeo Ghana was made in November 1938, with Lord Linlithgow, the ruling viceroy, as the presiding slayer. Although the lord sahib's own contribution to the bag was not impressive, he did distinguish himself by creating what must surely be a world record, of firing 1,900 rounds of 12-bore ammunition from his own shoulder on that day. Only one who has let off even a paltry hundred shots in a morning's snipe shoot and got his shoulder black and blue will appreciate the magnitude of this almost superhuman performance, even after making due allowance for his lordship's considerable weight and substance.

SÁLIM ALI, *The Fall of a Sparrow*, 1985

As a senior correspondent for The New Yorker, *Mary Anne Weaver covered the political events from the Indian subcontinent and her books and dispatches during the Zia-ul Haq regime in Pakistan are well known. During the same period, the plight of the bustards throughout the subcontinent was in the news. One of the chief reasons for their decline was attributed to falconry practised by the Arab sheikhs, who went out to hunt the Houbara in Pakistan. Weaver's superb narrative paints a lucid picture of these hunts.*

Hunting with the Sheikhs

Abrar Mirza, the conservator of wildlife for the Province of Sind, in Pakistan, is by nature a rather doleful man, and he always appears to be in a state of crisis. When I first met him, late last November, he was especially anxious, because he had been waiting for weeks—waiting for the full moon, and waiting for the rains, and waiting for the houbara bustard, an endangered species of a fast-flying and cursorial desert bird that migrates to Pakistan each autumn from the former Soviet Union and from the Central Asian steppes. A good many Arab sheikhs and princes were also waiting—discreetly—in opulent Karachi palaces.

Mirza dreads November, he has often said, because his entire life is put on hold. The responsibilities of his position include the delicate job of monitoring the Arab royal hunts. He is a bit puzzled by them, and can't really explain why, with the arrival of the houbara, scores of Middle Eastern potentates—Presidents, ambassadors, ministers, generals, governors—descend upon Pakistan in fleets of private planes. They come armed with computers and radars, hundreds of servants and other staff, customized weapons, and priceless falcons, which are used to hunt the bird. Mirza considers it all a little excessive. But then the houbara bustard has been a fascination to the great sheikhs of the desert for hundreds of years. Poets have written about it. Old men of the desert have sung of it in tiny tea stalls. Even today, Arab diplomats, in well-appointed embassies abroad, discuss the advent of the season, and discuss it endlessly.

'The bird is a month late!' Mirza announced one morning when I stopped by his office, in Karachi, and found him at his desk, which was covered with mounds of papers and with half-finished cups of tea. He would make a fine Inspector Clouseau: middle-aged, wiry, although with a bit of a paunch. 'Only a handful have arrived. And I am being held responsible, as though it's all my fault. Look at these telegrams!' He threw a mass of papers into the air. They were urgent messages from the Pakistan government—the majority of them from the Ministry of Foreign Affairs, which was most stressed.

It is Foreign Ministry that awards the visiting Arab dignitaries special permits to hunt. Pakistanis themselves have been prohibited from killing the houbara since 1972. Yet each season, which lasts from November until March, their countryside is carved up, like a giant salami, into ever smaller parts. Some sheikhs—among them Zayed al-Nahayan, the President of the United Arab Emirates and the chief shareholder of the Bank of Credit & Commerce International, or B.C.C.I.—receive permits that cover thousands of square miles. No other hunters may cross the invisible line that separates Sheikh Zayed's personal hunting grounds from those of, for example, the Saudi Princes Naif and Sultan, or the Dubai leader, Sheikh Maktoum. At least, that is so in principle.

'Look at this!' Mirza nearly shouted, flailing a piece of paper before my eyes. Across the top was stamped 'CONFIDENTIAL MOST IMMEDIATE'; it was a message from Colonel S.K. Tressler, the chief of protocol. Sheikh Maktoum would soon be arriving from Dubai, and a party of royal Bahrainis was hunting on his turf—not even Dubaians but *Bahrainis*. Mirza was instructed to sort the muddle out. Then, there was a party of hunters from the royal family of Qatar 'sneaking around,' Mirza said, on Saudi Arabia's turf. And a member of the Dubai royal family was reported to have bagged two hundred birds in a protected national park, in the company of the honorary game warden, who was a member of the Pakistani parliament.

'None of this would have happened if it hadn't been for Abedi,' Mirza said. He meant Aga Hassan Abedi, the Pakistani who had founded B.C.C.I. 'He was the one who first arranged hunting outings in Pakistan for the sheikhs. He set up everything for them—from doing their shopping to providing bribes and geisha girls. The more he provided, the more their deposits filled his bank.'

I had my first inkling of the royal houbara hunts during a visit to Pakistan a few years ago when, late one evening, I entered the elevator of my Karachi hotel and, to my astonishment, found myself in the company of two Arabs with falcons on their arms. After a bit of research, I sought out a friend of a friend from the Ministry of Foreign Affairs, a man I will call Ahmed, and he agreed to let me accompany him to the Karachi airport on a night when he was to receive an advance party of one of the sheikhs. He made me promise not to reveal that I was a journalist.

The sheikhs are obsessive about their privacy. Some have built personal airfields to protect themselves from public view. Some have constructed huge desert palaces, surrounded by fortress-like walls. Some live in elaborate tent cities, guarded by legions of Bedouin troops. They have their own communications equipment, road networks, security forces, and police. Totally closed off to outsiders, their hunting fiefdoms are, in effect, Arab principalities. They sprinkle the vast deserts of Balochistan, Punjab and Sind, covering hundreds of miles. The sheikhs move in and out of them like phantoms, giving rise to any number of outlandish stories, many of which turn out to be true. There is, for example, the story that the late King Khalid of Saudi Arabia transported dancing camels in a C-130 to join him on his hunt. There is the story that Prince Sultan, who is the Saudi Defense Minister, slaughtered seventy sheep and lambs every day to feed his royal entourage.

It was well past midnight when Ahmed and I reached the airport. (The sheikhs of the desert have always preferred to travel in the middle of the night.) A Pakistani Army major met us in the V.I.P. lounge, where a small group of Arab diplomats, in tailored silk suits, sat in a corner, sipping cups of sugary tea. They shook hands with Ahmed and nodded politely to me. Then black stretch limousines whisked us to a remote section of the airfield, which had been cordoned off by Pakistani troops to assure the sheikh's entourage of total privacy.

As we waited on the tarmac, the arriving planes lit up the night sky. Flying in formation—observing protocol, apparently—an executive Learjet was followed by two customized Boeings and a fleet of reconfigured C-130s, which flew too abreast. They had all been designated 'special V.V.I.P. flights' by the Pakistani government. There would be no customs clearance, no passport control—the royal entourage enjoyed extraterritorial status in Pakistan. The lead planes touched down, and a red carpet was hastily unrolled. We all hurried to it, and stood in a slightly dishevelled line.

'This is the sixth flight this week,' one of the Arab diplomats told me, exhaustion in his voice.

'Do you accompany them on the hunts?' I asked.

'Good heavens, no,' he said, smoothing one of his silk lapels, 'I'm basically a fisherman myself.'

Two military officers in dress uniform got off the executive jet and walked briskly toward us, carrying attache case and swagger sticks. They were followed by other members of the sheikh's personal staff—a purser, a physician, a royal chamberlain—all in kaffiyehs and flowing camel-coloured robes. Security men in khaki uniforms hurried from one of the Boeings and fanned out across the field. The doors of the C-130s opened, and immense vehicles began rolling down the ramps.

From a distance, the vehicles were merely dots of colour—canary yellow, bright red, black, and white. Then they lumbered by us: two-thousand-gallon water tankers and eight-thousand-litre fuel tankers—dozens and dozens of them—in militarily precise lines. Now planes were landing all around us, ramps were quickly dropped, and jeeps, Range Rovers, and Land Cruisers raced down. They had all been customized for the royal houbara hunts, to make areas once inaccessible easily accessible now. They had open backs and convertible tops, and were equipped with special gauges, special shock absorbers, and special tyres. Their drivers were dressed in Bedouin robes, and wore exceedingly dark glasses, even though the night itself was exceedingly dark.

There was a din, deafening at times, as camp managers shouted instructions in Arabic, as gears ground and brakes slammed, as more and more heavy equipment was disgorged. Security men dashed back and forth. Cranes labored across the runway and carefully unloaded satellite dishes and communications equipment. From time to time, I glimpsed generators, air-conditioners, mobilebars, VCRs. 'They're totally self-sufficient in the desert,' the diplomat who preferred fishing said. 'Some of them even drill their own water holes. Providing water for an entourage of three hundred people is a problem.' He shook his head.

During all the commotion on the runway, I had become separated from Ahmed, and now I went in search of him. I found him among a group of agitated officials, standing in a tight circle beneath a wing of one of the planes. 'The mobile palaces are new,' he told me, 'and they don't know how to get it down.' Looking up, I saw an unwieldy dark-blue structure, about fifty feet long and perhaps thirty feet wide, stuck at the top of the ramp. It was a customized Mercedes, and prominent on its hood was a now slightly askew gold-plated royal crest. 'When they first began coming,' Ahmed said, 'even King Khalid and Sheikh Zayed slept in a tent.'

In 1929, H.R.P. Dickson, a British colonial officer who had served in Kuwait, described the houbara's yearly arrival on the Arabian Peninsula as 'a season for rejoicing.' He wrote, 'The rains are close at hand and ... the houbara have arrived. They are verily, like the manna of old, Allah's reward to those who have endured the summer heat.'

By the nineteen-sixties, the houbara had been hunted almost to extinction in the Middle East. 'There was near hysteria when the bird disappeared,' an Arab ambassador told me. The kings, sheikhs, and princes hurriedly dispatched scouting parties abroad. They recruited British and French scientists to attempt to breed the houbara in captivity. They called upon Japanese technicians to develop special tracking devices and customized vehicles for the hunt. It was the beginning of what would become a multimillion-dollar industry. But none of the endeavors solved their most pressing problem: Where could they hunt the houbara bustard *now*?

Pakistan was believed to have one of the largest migratory populations of houbara in the world, but no one was quite certain, then or later, how large it actually was. For although the houbara was declared an endangered species in 1975, largely as a result of the high-tech hunting of the sheikhs, no international conservation group had ever done a comprehensive study on the bird's distribution worldwide. After a good deal of debate, experts at an international symposium in Peshawar, Pakistan, in 1983 finally agreed that Pakistan's houbara population probably numbered somewhere between twenty and twenty-five thousand birds. In retrospect, the figure seems extremely low. The houbara reproduces at a rate that increases its numbers by only about five per cent a year, and the conservation officials I spoke with on this trip told me that the Arab hunting parties were bagging at least six thousand birds a year, and even that figure was considered a very conservative government estimate. (Sheikh Zayed alone brings a hundred and fifty falcons with him.)

Although General Muhammad Zia ul-Haq, Pakistan's President in 1983, supported the symposium, he ignored its unanimous appeal that houbara hunting be banned in Pakistan altogether for at least five years. For while the number of Arab royal falconers was small—perhaps two or three dozen men—they were all immensely powerful, and immensely rich, and they put millions of dollars into their hunts. They also provided Pakistan—whose per-capita G.N.P. was only three hundred and fifty dollars a year—with some three and a half billion dollars annually in military and economic aid and in remittances of two million Pakistanis working in the Gulf. So, despite appeals from Prince Philip, who is the president of the World Wildlife Fund, and from other conservation groups, the sheikhs and princes continue to hunt.

One evening in early December, I was invited to a dinner for one of the visiting sheikhs. It was held at the elegant Karachi home of the Talpurs, one of the great feudal families of Sind. They were the ruling family of the district of Mirpur Khas and controlled vast tracts of land, where members of the royal families of Dubai and Qatar had begun to hunt, including Sheikh Muhammad, a Dubaian prince for whom the dinner was being held. None of the guests seemed certain of precisely who he was, although they all assured me that he was definitely a very influential sheikh.

A billowing *chamiana* tent of red, white, blue, and yellow had been set up in the middle of the Talpurs' lawn. It was filled with imitation-Louix XV wing chairs and upholstered settees arranged in a large rectangle. Bearers in starched white jackets served whiskey and gin in tall glasses that had been wrapped delicately in paper napkins. (Alcohol is forbidden in the Islamic Republic of Pakistan.) As everyone waited for the Sheikh to arrive, I greeted a number of Pakistani ministers and former ministers. It was an impressive gathering of Karachi's feudal, political, and financial elite, for, if the Arab sheikhs and princes were attaching greater urgency to the houbara hunt this year, so in a sense, was the government of Pakistan. It had vacillated during the Gulf War, agonizing over what it could do that would be acceptable at home and yet would not displease its Arab patrons or the United States. In the end, ten thousand troops were sent to the Gulf, with orders not to fight. Officially, they were sent to Saudi Arabia to guard its religious shrines. But no sooner had the troops been dispatched than Pakistan's zealous mullahs—whom Saudi Arabia had been funding for years—announced with some flourish, that they had recruited thirty thousand volunteers to fight on the side of Iraq. Nobody knew for certain where Pakistan stood, and no government was more irritated than the government of Saudi Arabia. The houbara bustard was now a pawn on the geopolitical chessboard.

'We must find a proper seat for you,' my host, Nawab Abdul Ghani Talpur, said to me. 'You must not be so close to the Sheikh as to be conspicuous, but you must not be so far away that he can't see you and invite you to join him on his settee.' It was finally decided that I should sit between a feudal landlord and a member of parliament. The landlord was a short, plump man with betel-stained teeth who was wearing a reddish-orange toupee. He said that the sheikhs had been hunting on his private lands for nearly a decade. We all hurried to sit in our assigned places as the Sheikh's arrival was heralded by screeching sirens and by guards scurrying to take up positions along the perimeter of the tent, their Kalashnikovs at the ready.

'His Majesty,' the Nawab announced, and we all jumped to our feet.

'He's not "His Majesty",' the landlord whispered dismissively. 'He's merely the brother-in-law and the cousin of the ruler of Dubai, and he's not a very good hunter, either. When he didn't find any houbara in my desert tracts, he moved his entire camp— servants, vehicles, falcons—into Kirthar National Park. He killed more than two hundred houbara in ten days, and he killed gazelles and ibex, too.'

'Why was that permitted?' I asked.

'No one has ever written, either Jesus or the Prophet Muhammad, that Pakistan must be poor.'

That was the way many of my conversations in Pakistan went. I met game wardens wearing jewelled watches that were gifts from the sheikhs. Politicians, chief ministers, and former chief ministers received lavish residences or customized cars. Some of them shopped frequently in London—flying back and forth in one or another of the sheikhs' private planes.

Sheikh Muhammad bin-Khalifa al-Maktoum swept into the tent. His face showed no emotion as he went from guest to guest. A slight man with a Vandyke beard, he was dressed in a black robe trimmed with gold, and a white kaffiyeh. For some reason, he carried a shepherd's wooden crook in his hand.

After I introduced myself, he asked me if I live in Pakistan.

'No, Your Excellency, I've come for the houbara hunt.'

'We're not hunting,' he said, rather tartly. 'We're only training falcons.' And he moved on.

I asked the landlord how much a typical royal hunt cost.

'Well, when you take everything into account—The hunting vehicles, minus their electronic fittings, cost at least twenty thousand dollars each; then add the costs of their falcons and private planes; and, of course, there are the out-of-pocket expenses.' He laughed a guttural laugh. 'The controller of Sheikh Muhammad's household told me that he paid about two hundred thousand dollars out of pocket for this particular trip. He's spent a total of about nine million dollars thus far, and he bagged about six hundred birds. That works out to about fifteen thousand dollars a bird.'

He then quickly added that that figure was low. The sheikhs normally spent between ten and twenty million dollars for a typical royal hunt.

I glanced at Sheikh Muhammad, now sitting on a gilded sofa at the head of the tent. He sat rigid, seemingly bored, with the shepherd's crook held upright in his hand. The etiquette of the evening was that one was not permitted to leave one's seat unless summoned by the Sheikh. We sat for over two hours, and only three of the sixty-odd guests were invited to the royal settee.

'Have you ever been with a sheikh on a bustard hunt?' the landlord asked me as the evening dragged on. 'It's the craziest thing I've ever seen, but it's like a religion to them. They're out in the desert from dawn to dusk, covered with dirt and dust. The driver is submerged in one of those jeeps, as if he were in an A.P.C.—armored personnel carrier. 'The sheikh sits next to him in an elevated seat that swivels at a hundred and eighty degrees, I guess it's a good hobby, if you're into that kind of thing.'

'What kind of thing?' I asked him.

He looked somewhat startled, then said, 'My lady, these Arabs eat the houbara for sexual purpose—it's full of vitamins.'

A Falcon trainer told me that if I really wanted to see a hunt I should go to Balochistan, to Chagai district. It was a Saudi hunting area—a place called Yak Much. Bordering Afghanistan and Iran, Chagai is on one of two migratory paths by which the houbara enters the country, and is thus one of the most preferred hunting areas in Pakistan.

A sixteen-hour car ride, through desert and mountains and tribal lands, brought me to Quetta, the provincial capital of Balochistan—the only province in the country where the houbara is known to breed. And, in increasing numbers each year, eggs, chicks, and birds are being smuggled out, primarily to Taif, Saudi Arabia, where French scientists—in a multimillion-dollar effort of limited success—are currently attempting to breed the houbara in captivity. In the fall of 1991, Mirza had confiscated five such consignments—some five hundred birds in all—in just six weeks.

Shortly after my arrival in Quetta, I called on the provincial wildlife minister, Jam Ali Akbar. He told me that he really wasn't much of a wildlife person himself. He wrote pop music, and was the president of Balochistan's Roger Moore Fan Club. I asked him if the provincial government was doing anything to protect the houbara.

'It's impossible—it's a federal-government matter,' he said. 'And these sheikhs are extremely attached to this little bird. It's not a simple matter.' He shook his head. 'The wildlife people say this shouldn't be permitted. But then the sheikh's agents come, bringing, priceless gifts, like diamond-studded gold Rolex watches. And sometimes, I've heard, they dispense briefcases containing a couple of thousand dollars—and you can keep the briefcase, too. The sheikhs say that these are migratory birds, so we lose nothing. And if we don't permit it they'll simply go somewhere else.'

Quite by accident, I met Balochistan's largest falcon dealer, Mir Baz Khetran, one afternoon in my hotel. His presence there shouldn't have surprised me. Royal hunters had begun arriving, en route to their hunting grounds, and falcons had become a familiar sight throughout the hotel. Mir Baz and his brother, Lal Muhammad, dealt in falcons together, which was largely illegal in Pakistan, and Lal Muhammad also served as one of the chief minister's key advisers—on wildlife.

It was Lal Muhammad who trapped the falcons, Mir Baz explained when we chatted in my room at the hotel; their servants trained them, and then he himself sold them to the sheikhs. His falcon empire had ensured him a seat in Parliament, and he had been a Cabinet minister in Benazir Bhutto's short-lived government. Mr Baz was in his early forties, and had a round, puffy face and dark hair. He wore sparkling rings and a good deal of cologne; gold chains covered his chest, which was half exposed.

'Such hectic times,' he said, slumping in his chair. 'The falcon season lasts for only four months.' (The most expensive falcons migrate with the houbara from Siberia.) 'But fortunes, Madam!—fortunes can be made. There is a huge competition between these Arab sheikhs. And if a sheikh sees a falcon that he judges to be *hurr*, "or noble and free," and if that bird is nearly white or totally black—both are extremely rare—that sheikh, Madam, nearly has a heart attack. He simply must buy it, and he will pay *such* money for beauty.'

'How much?' I asked.

'Nothing less than the equivalent of eight thousand dollars. The record price for Balochistan this year was twenty-five lakhs'—a hundred and twenty thousand dollars—'for a shahin, which was caught in the northern border area, near Zhob. By the time it reaches the Middle East, it will bring much more.'

Mir Baz then said, 'you know, Madam, these Arabs consider the houbara an aphrodisiac.'

'So I've head,' I replied.

'But some of them, Madam, eat one houbara a day—sometimes two, if it's a special occasion. That means they may eat as many as *five hundred* birds a year!'

Several nights later, I was invited to dinner by one of Balochistan's tribal chiefs—Nawab Akbar Khan Bugti, the leader of the second largest of the province's seventeen major tribes. An aristocratic Anglophile who had spent the last forty-five years in and out of government, in and out of favour, and in and out of jail, Bugti was now the leader of opposition in the Provincial Assembly and one of Balochistan's most powerful men.

When I arrived at his home, I found his receiving room crowded with other tribal chiefs. Sardars, mirs, and maliks sat cross-legged on a Bukhara rug and lounged against pillows piled against a wall. The Nawab greeted me warmly; I had known him for some time. He then went from guest to guest, and each reported on the site of one or another of thirty or so royal parties hunting in his tribal lands.

'Where is Yak Much?' I asked the Nawab, after he had spoken to his guests.

'In the middle of God's country,' he replied. 'It's miles and miles from nowhere—nothing but tons and tons of sand. And it's totally off limits to everyone except the Saudis. Ask *them*.' He pointed out two men on the other side of the room, and then introduced me to Ali Ahmed Notezai and Sakhi Dost Jan. They were the kingmakers of Chagai district, of which Yak Much was a part.

Notezai was a member of the Provincial Assembly, and was allegedly involved in the smuggling trade. He reminded me of a penguin type: a large man, he had a broad, menacing face, and his teeth were betel-stained. He wore a brown waist-coat over his *shalwar kameez*—the Pakistani national dress—and a white turban, cockaded and lofty, that tied from behind, so its folds of soiled cloth streamed down his back. His wealth, which was considerable, was also said to be grounded in the smuggling trade, and he had the reputation of being a bit of a Robin Hood. When I asked him about that, he said there was no point in robbing the poor.

Both men had known the Saudi Defense Minister, Prince Sultan, for years—ever since he began hunting in Chagai district, where he held exclusive sway over nearly twenty thousand square miles. They told me that the Saudi hunters would be led by one of the Prince's sons: Prince Bandar, the Ambassador in Washington, or Prince Khalid, who had commanded Saudi forces during the Gulf War, or Prince Fahd, the governor of Tabouk

Province. But it would definitely be a son. 'In this wild mavericking, they don't trust even their brothers,' Notezai explained.

'What is so fascinating about the houbara?' I asked.

'The sheikhs tell me it is the ultimate challenge for the falcon,' the Nawab replied. 'Much of the fascination is in the flight; it can go on for miles. The falcon is the fastest bird on earth, and the houbara is also fast, both on the ground and in the air. It is also a clever, wary bird, with a number of tricks. Part of the lure is in *finding* it. You can spend half a day following its tracks. It's a contest—your wits against its. Then, there's the contest between the two birds. The houbara tries to stay on the ground, where it is difficult, sometimes impossible, for the falcon to strike. The falcon tries to coerce it, cajole it, frighten it into the air. There the falcon reaches for the sun, and then comes down on the houbara—but it must stay above. Otherwise, the houbara, whether as part of its defensive armor or in its reaction to fear, emits a dark-green slime violently from its vent. Its force is so strong that it can spread for three feet, and it can temporarily blind the falcon, or glue its feathers together, making it unable to fly. The sheikhs have told me that, once that happens, many falcons will never hunt the houbara again.'

The Nawab called for a servant and gave him instruction in Balochi. The servant left the room, and he returned carrying a custom-built leather case. He placed it at my feet.

'Open it,' the Nawab said.

I did. Nestled inside, protected by a fur lining, was a 24k.-gold-plated Kalashnikov. It was a gift to the Nawab from the Minister of Defense of the United Arab Emirates, who hunted in Balochistan each year. It was the size of the normal Kalashnikov but was perhaps three pounds heavier, because of the gold. It was engraved with the royal coat of arms, and its two magazines were also plated in 24k. gold. The Nawab handed it to me. I had held a Kalashnikov before, but I had never held three pounds of gold.

'In the old days, we would hunt the houbara on foot or camelback,' the Nawab said. 'We would try to outsmart it, using the camel as a shield. The houbara knows the camel, since the camel grazes in the areas where the houbara feeds. You couldn't go directly for the bird, or it would flee. So you circle it on camelback, making the circle ever smaller—the houbara would watch, mesmerized, confused. But now customized vehicles have replaced camels, palaces have replaced tents. They use radar, computers, infrared spotlights to find the bird at night. What is the challenge? What is the thrill? The odds have changed immensely for the houbara. The poor bird doesn't stand a chance anymore.'

YAK MUCH ('One Date Palm') is a desert village of about a hundred people, one gas station, and a few little food stands and shops. And, on close inspection, I found that it now has five date palms. Its most distinctive feature is a large green board at the village line, which in bold lettering announces 'NO HUNTING PERMITTED.' Since the houbara breeds here, Yak Much is, in principle, a protected sanctuary.

A mile or so beyond the sign was the Saudi royal camp. My driver was the first to spot it. There was nothing around us except desolate miles of sand, but then, stretched along the horizon, we saw lines and lines of tents. If we hadn't been looking for them, we could easily have passed by. The camp was deep in the desert, five miles off the road, and as we continued along the highway we could see the tents one moment, and the next moment they would disappear.

We left the highway at an unmarked point—there was no road—and careered across the desert, lurching around bushes and shrubs. Then the camp came into focus—scores and scores of black, brown, and white pyramidal forms. Against the flat emptiness of the desert, the tents suggested a gathering of giant dinosaurs. The camp sprawled over some ten acres, in two concentric circles, bringing a medieval city to mind. The inner tent city, of forty-four *chamianas*, was surrounded by perhaps sixty smaller tents. They stood like a wall, as if to keep all outsiders out. The perimeter was guarded by Pakistani levies and border militiamen, dressed in blue or gray sweaters and berets. Some were swathed in blankets against the desert chill. The inner city was guarded by security men in the retinue of Prince Fahd, who would lead the Saudi royal hunt.

Vehicles were lined up in neat patterns on the perimeter of the camp; water tankers, oil tankers, petrol tankers, and a fleet of customized hunting jeeps. There were immense yellow cranes, to pull the vehicles out of the sand if the need should arise; a mobile workshop, which was fitted with everything necessary to overhaul a car; and huge refrigerator trucks, to carry the hunting bag out. Silver satellite dishes were anchored in the desert rock. From inside the camp, you could make a phone call to any place in the world. I spotted two royal falcon trainers whom I had met at my Quetta hotel. They carried mobile telephones, and their falcons were perched upright on their arms.

There were now about a hundred falcons inside the camp for the seventy or eighty royal hunters who would accompany Prince Fahd. Only the Prince's favourite falcons would arrive with him. I asked the chief of the Pakistani security detachment how long it had taken to assemble the camp, and he said only four days. The hunting vehicles—there were sixty—and the heavy equipment, tents, generators, and fuel had all been transported from Jidda by C-130s to the airport in Dalbandin, which was the closest town to Yak Much, thirty-five miles away.

Officials in Dalbandin had told me that the Saudi royal parties which usually hunted two to three thousand birds during their month long stay had no beneficial impact on the local economy; they'd given residents only two generators (which didn't work), a mosque (which they didn't need), and the airport (which was used almost exclusively by the hunters themselves).

At the camp the following evening, after Prince Fahd himself had arrived, I sat in a Land Cruiser next to the dining tent, whose vast brown folds, with intricate gold stitching, billowed in the wind. The tent was surrounded by some twenty-five security men, who stood at smart attention with their Kalashnikovs.

I sat in darkness, my head covered with the hood of my cape. It was bitterly cold. The wind was ferocious. Land Cruisers and Range Rovers began to arrive. As I waited for Prince Fahd's personal physician, whom I'd met earlier in the day, I watched Dalbandin's notables saunter toward the dinning tent, where they had been invited to dine with the Prince. The visiting wildlife minister, Jam Ali Akbar, was flanked by servants and guards carrying two carpets, which were gifts for Prince Fahd. Ali Ahmed Notezai strutted like a peacock as he entered the tent. Sakhi Dost Jan, wearing his brown waistcoat and flowing white turban, shouted instructions here and there. Earlier that day, I had spoken to both men about the possibility of my meeting Prince Fahd.

'Impossible,' Notezai said. 'The Prince doesn't want to meet any women this time.'

'I'm not a woman. I'm a journalist.'

He shrugged. 'It's all the same,' he said.

The Prince's personal servants ferried bottles of mineral water and huge trays of food between the tents; roast lamb with dates and rice; hot nan bread; hummus; tahini; baskets of fruit. I watched two trainers open a large wicker basket near my jeep and pull out two baby houbara with clipped wings, to be used for training falcons. Carrying the birds in their left hands, they walked off, each with a falcon perched on his right wrist.

I left the jeep and stood in darkness near the entrance of the dining tent. Inside, Prince Fahd, dressed in a camel-coloured woollen robe embroidered with gold thread, sat cross-legged on an Oriental carpet, receiving his guests. The floor of the *chamiana* was covered with exquisite Kashan and Persian antique carpets and rugs; bolster pillows, in silk cases sewn with gold thread, lined with walls. In a far corner, there was a network of cellular phones, and other communication equipments hooked to a satellite dish. Behind the prince, like a ceremonial guard, thirty-five hooded falcons stood at attention. They perched on specially designed, hand-carved *mashrabiyya* stools, etched with ivory and gold. The falcons were of three different kinds—different in colour, age, and size. Despite their magnificence, however, all were dwarfed by a peregrine that stood at the Prince's side, on the arm of his chief falcon trainer. She had travelled with Prince Fahd on the royal flight, and during the entire evening she never left his side.

Sakhi Dost Jan was the last of the V.V.I.P. guests to depart. He stood outside the dinning tent, flanked by bodyguards and aides. He gesticulated, then shouted. A Saudi intelligence officer flailed his arms. Other Saudis came up and encircled the two men. 'What is happening?' I asked one of the guards.

'Rupees, Madam,' he said. 'Lakhs of rupees.' He rolled his eyes.

After some ten minutes of negotiations, an aide of Prince Fahd's appeared, and presented Dalbandin's godfather with two bulging leather saddlebags. Sakhi Dost smiled his toothy smile. He then got into his Range Rover and roared away.

One of the guards brought me a plate of food and a cup of tea. I looked down at the dark meat, which was surrounded by rice. 'Is this the houbara?' I asked.

'Yes,' he replied.

I hesitated momentarily, and then took a few bites. The meat was tough and stringy—it reminded me a bit of goat—and left a bitter aftertaste. Far from arousing amatory impulses, it had an irritating tendency to stick in my teeth. How could anyone eat five hundred of these birds a year? As I pondered the mysterious ways of the desert, Prince Fahd's physician came over to chat.

'Is it true that the houbara is an aphrodisiac?' I asked.

He looked amused, and shrugged his shoulders. 'No,' he replied. 'It's basically a diuretic. But they *think* it's an aphrodisiac.'

The howling of dogs and the chanting of mullahs woke me at dawn. No sooner had I started a fire in my tiny fireplace, in Dalbandin's government guesthouse, then one of the royal trainers whom I'd met in Quetta the previous week—I'll call him Farouq—pounded on my door. 'We're taking the falcons out!' he said. I was to accompany him back to the Saudi camp.

We left the highway before we reached the main turnoff to the camp, and drove into the desert for perhaps a mile, to a spot where another trainer and a driver waited in a customized, carpeted Range Rover. Both men carried hooded falcons—one a shahin and one a saker—on their gauntleted right arms.

I was instructed to sit in the back seat of the open jeep, with the other trainer and the hooded saker, which seemed dangerously close to my left knee. Farouq—with the hooded shahin now perched on his back-gloved wrist—took the revolving bucket seat in the front. He adjusted it to its maximum height, and towered some three feet above us, in midair.

The sun was just beginning to rise, and the sky was violet-pink. All around us, the flat emptiness of the desert stretched endlessly. The silence was broken only by the wind and the grinding of the Rover's gears. From time to time, we passed black slate formations that resembled giant marshmallows burned in a bonfire.

The trainer next to me, I'll call Mahmud, wore sandals and bright Argyle socks. 'Her name is Ashgar,' he said of the hooded saker on his arm. 'And she's just a year old. That is the perfect age for this particular bird.'

Ashgar was extremely light in colour, almost blond, and measured perhaps thirty inches from her head to the tip of her tail. White spots on the tips of her feathers, which resembled polka dots, blended quite smartly with the red leather hood and jesses she wore.

'Her talons are like steel if she grabs you. That's why we wear gloves,' Mahmud said, stroking Ashgar and giving me a pleasant smile. He then told me that Ashgar was from Iran, and had been a particularly sought-after bird, not just for her colour but for her 'soul'.

I studied the falcon more closely. A tiny solar cell, covered by glass, was attached to her tail feathers, and a thin metal aerial affixed to it rose from her feathers up the bottom of her back. It was a French-made radio transmitter, a tracking and homing device slightly larger than a watch cell; it had an especially sensitive receiver that had been devised purely for the houbara hunts. Mahmud said that the transmitter weighed about five grams and had a radius of some eight miles. It gave off a constant beep once the bird was on the wing. 'If she is lost during the hunt, we can retrieve her by the next day, maximum,' he said. 'Even when she parks for the night, we get a constant signal in our jeeps.'

'Can the transmitter be used to track a houbara?' I asked.

'Only indirectly,' he replied. 'If the falcon catches a houbara, the beeper tells us where they are. But, basically, we track the houbara by radar or two-way radio.'

The wind became fierce as we raced across the desert at eighty miles per hour, searching for houbara tracks, and knocking down everything in our path: shrubs, bushes, even tiny trees. I glanced ahead at the driver, who was wearing goggles and a crash helmet and was bent over the wheel intently. I suspected that at one time or another he had driven a tank.

A friend had told me earlier that the Yak Much desert was more like the Middle East than anywhere else in Pakistan was; you could travel for days without seeing another human being. We had travelled for more than forty miles, and although I'd seen no human beings, I had certainly seen their traces: plastic bags, abandoned jerricans, and discarded tyres. There were some areas where the hunting vehicles had so flattened everything in sight that a plane could have landed with ease.

Then Farouq shouted, 'There are the tracks!'

They were unmistakably those of the houbara—three-toed footprints dotting the sand.

Farouq stroked the shahin's underbreast, whistled softly in her ear, then raised his gauntleted arm above his head. 'A–hoh, a–hoh, a–hoh,' he chanted, above the noise of the wind, as he removed the shahin's jesses and hood with a single quick movement of his free hand 'Strike! Strike! Strike!' The shahin cast her piercing eyes incessantly around, bobbed her head, and then lurched forward, leaving Farouq's arm. She soared into the air, her radio transmitter and aerial visible in the feather of her tail. She flew low—barely off the ground—to conceal herself, and was often out of our sight as we raced across the desert, following her path. We were guided by her radio beeps.

'It should be four or five minutes,' Mahmud said, and he explained that the shahin has extraordinary vision: she could sight for over a mile. But we raced along for twenty minutes before we spotted the shahin and a houbara, on the ground. At first, they were tiny, indistinct forms in a mustard field. Then, as we surged ahead, I lost sight of the houbara.

'There she is!' Farouq shouted.

'Where?'

Even with high-powered binoculars, I couldn't find the houbara, and it was perhaps only ten yards away, concealed and camouflaged—its contours and buff-and-sandy-grey colouring blended perfectly with the desert and the bushes and shrubs. When I finally did spot it, it was frozen behind an absurdly small bush, and uttered no sound. It was a baby, weighing perhaps two pounds. The shahin circled overhead, then swooped down, attempting to frighten the houbara off the ground. The houbara tried to enlarge itself by spreading its wings, and watched our every movement with unblinking yellow eyes. Then, in an instant, it had taken off. It darted across the desert like a roadrunner, its long legs seemed not even to touch the ground. Its tail was spread like a peacock's, and its chest was thrust out.

We raced, dashing, lurching, and jolting, in huge zigzag circles, following the two birds. Then both took to the air—an absolutely cloudless blue sky. You could distinguish the houbara by its white undersurface and wings. The shahin soared and dipped, her vast wingspan spread majestically. The houbara eluded her, and tried to gain attitude. From time to time, the birds almost disappeared, becoming tiny, inky webs, but they were never completely cut of sight—we had our high-powered binoculars in addition to our radio beeps. This hunt was a far cry from the romantic image of the lone Arab walking across the desert in his flowering robe with his pet falcon perched nobly on his arm.

The shahin soared for the sun, and came down on the houbara, attempting to break its neck. The houbara flew on furiously, and the shahin struck again. The two birds spiralled downward. We found them near a tamarisk bush, struggling on the ground. The baby houbara lay exhausted but was still trying to kick. The first thing that the shahin had done was blind its yellow eyes, so that it could not run or fly away. Farouq cut open the houbara's stomach, retrieved its liver, and fed it to the shahin. He then hooded the falcon and ritually slit the baby houbara's throat, to conform to dietary laws.

'Now it's *halal*,' he said—'permitted in Islam.'

There was a time, Wahajuddin Ahmed Kermani, Pakistan's retired Inspector General of Forests, told me, when the houbara had been so plentiful in Pakistan that you could count them from the roadways 'like butterflies in a field.' But that was in the nineteen-sixties, before the great sheikhs and falconers began hunting in Pakistan.

I called on Kermani, one of his country's most respected environmentalists, at his Karachi bungalow. If any Pakistani had attempted to save the houbara, he was that man. As we sat in his drawing room one morning, sipping cups of tea, he described, his efforts to save it as 'the only failure of my life.' He went on to say, 'For a quarter of a century, the hunting has been intensive and sustained. They go through the desert like an invading army. It's slaughter, mass slaughter. They kill everything in sight.'

When I asked him why the government of Pakistan had done so little to deal with the situation, he replied, 'Because we lack the moral fibre and the moral courage.'

Kermani applauds the efforts of Tanveer Arif, the president of the Society for Conservation and Protection of the Environment, or SCOPE, a Karachi-based group that challenged the houbara hunt's legality in the Sind High Court. 'The hunts are sheer hypocrisy, and totally contrary to our laws,' Arif told me one afternoon. 'Since 1912, in the days of the Raj, the houbara has been a protected species. Yet, while Pakistanis are being arrested and prosecuted if they're found to be hunting the bird, Arab dignitaries are given diplomatic immunity.' Although in September the Sind High Court ruled in SCOPE's favour, its decision had had little impact on the Pakistani government.

Like Kermani, Arif is deeply upset that international pressure to ban the royal hunts is not being brought to bear on the government of Pakistan. Twenty-three countries, including India, Iran, and the former Soviet Union, have legislation that protects the houbara, or bustards generally, and in the vast majority of these countries there is a ban on all hunting.

After making my trip to Pakistan, I asked Paul Goriup, the leading houbara expert at the International Council for Bird Preservation, in Cambridge, England, whether he thought the international community was doing enough.

'International efforts are exceedingly scant,' he replied. 'The houbara is merely a distraction, not a priority. There's no doubt that in the Pakistani provinces of Sind and Punjab the population, which was once sizable, is now terribly diminished. Balochistan is thus the only area left that is worth hunting in—and the problem there could be severe. There's a breeding population, and if the sheikhs hunt after February'—they always do—'then it's a disaster, for they impinge on the breeding population for the next year.'

'It's a stalemate in Pakistan,' Goriup went on. 'The Pakistanis see the Arabs breaking Pakistan's own laws, yet huge sums of money are involved. As for the Arabs, they realize that the houbara is declining outrageously, yet they continue to hunt. Still, they're worried, and I'm absolutely convinced that they would accept regulations if the regulations were there.' He thought a moment, and then said, 'I've maintained consistently that the houbara should be protected by the United Nations' Bonn Convention on Migratory Species, because such protection would elevate the problem to an international level. We could set up protected areas. Money would flow the right way. We must restore habitats and breeding grounds. This is the only way the houbara can be saved.'

MARY ANNE WEAVER, *The New Yorker*, December 1992

Entertainment

Zahir-ud-din Mohammad Babur (1483–1530), founder of the Mughal Empire in India, recorded his impressions of the land he conquered in the Baburnama.

Parrot

Parrots occur in Bajaur and lower. In the summer only, when berries are ripe, they come to Nangarhar and Laghman. Parrots are of many varieties. One is the sort that is taken to our country and taught to talk. Another sort is smaller than that parrot, and it too can be taught to talk. This sort is called 'jungle parrot,' and there are many of them in Bajaur, Swat, and that region. When they fly in flocks of five or six thousand, a difference in the size of the bodies of these two variations is evident, although their colouration is the same.

Another sort is even smaller than the jungle parrot. Its head is red, and tops of its wings are also red. An area of two fingers at the tip of its tail is white. Some of this kind have iridescent heads. This kind cannot be taught to speak. They call it a Kashmir parrot. Yet another sort is like the jungle parrot but smaller. Its beak is black and it has a wide black ring around its neck. Beneath its wings is red. It learns to speak well. We used to think that parrots and myna birds said whatever they were taught, not that they could think on their own. Recently, however, Abu'l-Qasim Jalayir, a member of my close retinue, told me something strange. He had covered the cage of a parrot of this kind, and the bird said, 'Uncover me. I'm stifling.' Another time the porters who were carrying it sat down to rest as passersby were coming and going. The parrot said, 'The people have gone. Aren't you going?' The responsibility for the veracity of this report lies with the one who told it. Without hearing it with one's own ear it is difficult to believe.

Another kind of parrot is a beautiful bright red. There are other colours too. Since I do not remember exactly what they are, I haven't written them in detail. The red one is nicely shaped. It can be taught to talk, but unfortunately its voice is as unpleasant and shrill as a piece of broken china dragged across a brass tray.[1]

The Baburnama: Memoirs of Babur, Prince and Emperor

The following extract from The Jahangirnama *by Nuruddin Salim Jahangir provides an account of the use of trained pigeons in carrying messages.*

Carrier Pigeons

Much has been heard of the Baghdad pigeons that were called carrier pigeons during the time of the Abbasid caliphs.[2] Carrier pigeons are actually a third larger than wild pigeons. I ordered the pigeon raisers to teach them, and they trained several pairs so that when we let them fly from Mandu at the beginning of the day, if there was a lot of rain, they reached Burhanpur in a maximum of two and a half watches, or even one watch and a half. If the weather was very clear, most arrived in a watch and others arrived in four gharis.

The Jahangirnama: Memoirs of Jahangir, Emperor of India

[1] May be the Indian loriquet (*Loriculus vernalis*).
[2] The Abbasid caliphs ruled, mostly from Baghdad, from AD 749 until the house was extinguished by the Mongols in 1258.

Abdul Halim Sharar (1860–1926) was a well-known essayist, historian, and novelist. He wrote a column for the Awadh Punch, *and in 1887 started his famous monthly magazine* Dil Gudaz, *which he continued to edit till his death. His book* Lucknow: The Last Phase of Oriental Culture *which contains a vivid description of Awadh life at the height of its glory, also includes a chapter on bird fighting for entertainment and sport. The book, originally written in Urdu, was admirably translated by E.S. Harcourt and Fakhir Hussain. Colonel E.S. Harcourt served in the British Indian Army, and lived for many years in Lucknow. Later he taught Urdu and Persian at Oxford University. Fakhir Hussain, born and brought up in Lucknow, belongs to one of the eminent literary families of the city whose important members are mentioned in Sharar's book.*

Bird-fighting and Pigeon-flying in Lucknow

Organizing fights between beasts of prey in Lucknow was confined to the royal family and nobles of the court. Keeping the animals, training them, controlling them after the fight and protecting spectators from injury were beyond the means of the richest men; let alone the impecunious. For this reason fights between beasts of prey were only witnessed in Lucknow whilst the court existed, and when the court disappeared so did the terrifying amphitheatres.

But bird-fighting was different. Rich and poor alike could indulge in it. Any interested person, if he took the trouble, could train cocks and quails to fight. The birds that people in Lucknow used for fighting were cocks, bush quails, partridges, *lavwa*, *guldum*, *lals*, pigeons and parrots. Lucknow pigeon-flying and quail-fighting were famous throughout the country, but nowadays educated people, who make a show of modern culture, are apt to ridicule these sports. They are totally unaware of the degrees of perfection to which their devotees have raised them, having in fact made them a fine art. However, when they go to Europe and see that these 'frivolous' sports are also practised there, they will at least be sorry for their utterances in regard to the interest taken in them in their own country.

COCK-FIGHTING

Although every sort of breed of cock will fight, the best fighter is the *asil*, the thoroughbred, and it is a fact that there is no braver beast than the thoroughbred cock. Braver than tigers, they would sooner die than turn away from a fight. Experts believe that the breed came from Arabia and this appears reasonable, as thoroughbreds are found mostly in Hyderabad Deccan, the area in India where Arabs came to settle in greatest number. The breeds of cock in the mountainous regions of India originated in Persia.

A well-known Lakhnavi cock-fighter used to tell the tale that his cock was unluckily beaten in a fight. He was distressed and went to the Sacred Najaf in Iraq where he spent some months in divine worship. He prayed day and night that God, as *sadqa* [charity] for the sake of his Imams, might grant him a cock which would never be beaten in a flight. One night in a dream he received a revelation: 'Go into the wilds.' The next morning when he awoke he went into the desert taking a hen with him. Reaching a valley he heard the sound of crowing. He approached and released his hen. A cock, hearing the hen, came out of the scrub and the man managed to seize it. Its progeny was such that never again was he put to shame in a cock-fight.

Interest in cock-fighting dates from the time of Navab Shuja ud Daula, who was extremely fond of the sport. Navab Sadat Ali Khan, in spite of the fact that he was very abstinent, also enjoyed cock-fighting. His interest had a great effect on society and in addition to the Lucknow nobles, Europeans at the court also became its devotees. General Martin was an expert at cock-fighting and Navab Sadat Ali Khan used to bet his cocks against those of the General.

For fighting purposes in Lucknow, the cock's claws were tied so that they could not cause much damage, whilst their beaks were scraped with penknives and made sharp and pointed. When the two cocks were released in the cock-pit, their owners stood behind them, each trying to get his own cock to deal the first blow. When the cocks started to fight with beak and claw their owners incited and encouraged them, shouting. 'Well done my boy, bravo! Peck him, my beauty!' and 'Go in again!' On hearing the shouts of encouragement the cocks attacked each other with claw and beak and it seemed as if they understood what was being said to them.

When they had been fighting for some time and were wounded and tired out, both parties, by mutual consent, would remove their birds. The owners would wipe clean the wounds on the cock's heads and pour water on them. Sometimes they would suck the wounds with their lips and make other efforts, whereby the cocks were restored to their former vigour in the space of a few minutes. They were then once again released into the cock-pit. This method of *pani* was continued and the fights would last four to five days, sometimes even eight or nine days. When a cock was blinded or was so badly hurt that he could not stand and was unable to fight, it was understood that he had lost. It often happened that a cock's beak was broken. Even then, whenever possible, the owner would tie up the beak and set the cock to resume the fight.

In Hyderabad the sport is much more violent. There they do not tie up the claws but scrape them with penknives and make them like spearheads. As a result the fight is decided within the space of an hour or so. The practice of tying up the claws in Lucknow was probably adopted to lengthen the fight and thus to provide longer entertainment.

When preparing cocks to fight, the owners would show their skill not only in the feeding and upkeep: they also massaged the bird's limbs, sprinkled it with water, tended its beak and claws and displayed their dexterity in tying up the claws and removing any signs of fatigue. From fear that the beak might be injured by pecking food from the ground they sometimes fed grain by hand.

Great interest was taken in the sport until the time of Wajid Ali Shah. In Matiya Burj cock-fights were held in Navab Ali Naqi's residence and some English people from Calcutta would bring their birds to fight there.

In addition to kings, many nobles were interested in cock-fighting. Mirza Haidar, the brother to Bahu Begam, Navab Salar Jang Haidar Beg Khan, and Major Soirisse, who lived at the time of Nasir ud Din Haidar and used to set his cocks against the King's and Agha Burhan ud Din Haidar, were all fond of the sport. The last-named nobleman always kept, throughout his life, two hundred to two hundred and fifty birds. They were kept with scrupulous care and cleanliness and ten or eleven men were employed to look after them. Mian Darab Ali Khan was a great devotee, as was Navab Ghasita.

The respected Pathans of Malihabad were also adherents of the sport and had very good breeds of game-cock. In Lucknow there were many who were considered outstanding experts: Mir Imdad Ali, Shaikh Ghasita and Munavar Ali had acquired such skill that they could tell from the noise a cock made whether it would win its fight. Safdar Ali and Saiyyid Miran, a *vasiqa dar*, were also famous. In latter days the names of the following were well known: Fazal Ali Jamadar, Qadir Jawan Khan, Husain Ali, Nauroz Ali, Muhammad Taqi Khan, Mian Jan, Dil, Changa, Husain Ali Beg and Ahmad Husain. None of these men is now alive.

These were the people who perfected the sport of cock-fighting in Lucknow but nowadays I think that interest in the sport is greatest in Hyderabad Deccan. Many noblemen, landowners and officers are devotees. They have an unequalled stock of game-cocks and give great care to breeding.

QUAIL-FIGHTING

The interest in quail-fighting came to Lucknow from the Panjab. Some gypsies from this area, whose women were of easy virtue, went to Lucknow in the days of Sadat Ali Khan and brought bush quail with them, which they used to cause to fight. (Some well-known courtesans of today are descended from these gypsies.) There are two kinds of quail, bush quail and button quail. In the Panjab there is only the bush quail, which is bigger and stronger than the button quail. In Lucknow both the bush quail and the button quail exist. The button quail is small and delicate but it is the more powerful fighter and its fighting the more interesting to watch. In Lucknow, the button quail was considered more suitable for fighting.

Quail-fighting does not require a large arena nor does one even need to leave one's house to watch a performance. One can sit comfortably in a room on a nice clean carpet and watch the fight. For this reason the sport was very popular in Lucknow society. Dainty bamboo cages adorned with strips of ivory were made to keep the quails.

To prepare a quail for fighting it is first necessary to keep him wet with drops of water and to hold him in one's hand for hours. He then becomes quite tame and starts chirping and chirruping. After this he is starved and subsequently given a purgative containing a large amount of sugar so that his inside is thoroughly cleansed. Then late at night his trainer shouts the word 'ku' into his ear and this is known as *kukna*, winding up. By these methods the quail loses his surplus fat and any awkwardness and his body becomes very active and strong. The more diligently these details are carried out the more efficient is the quail when the bird begins to fight.

For the fight, grain is sprinkled over the floor and the quails are taken out of their cages, their beaks having been previously sharpened with penknives. When they are set against each other their fight is much the same as that of cocks. They strike with their beaks and feet. With their beaks they wound and lacerate their opponent's head and with their feet they sometimes split open his crop. The fight usually lasts fifteen to twenty minutes but it can go on longer. Eventually the vanquished quail turns and runs. When he has once run away he will never stand up to another quail again.

There are three stages in the development of a quail which are considered landmarks in the bird's career. The first is when he is originally caught, tamed and then has his initial fight. If he wins several fights and never runs away, he is put into a cage at the end of the fighting season. This is the time when he sheds his old feathers and moults, which is know as *kuriz*, the first moult period. When this period is over, he enters his second year and his second stage of progress. The bird is then known as *nau kar*, an apprentice. After this, when he has moulted a second time, known as the second kuriz period, and is trained for the third year's fighting, the bird himself is known as *kuriz*. This is the third and final stage of a quail's progress. It is generally acknowledged that a nau kar is stronger than a newly-entered quail and that a kuriz is stronger than a nau kar. A newly-entered quail could scarcely stand up to a kuriz even if the bird had two beaks. Expert quail-fighters and enthusiastic noblemen use only kuriz for fighting; they consider fights between first-year birds to be a very poor from of sport.

Many artifices and harsh practices can be employed in quail-fighting. Some people occasionally put bitter and poisonous oil or *etar* on their quail's beak so that the other quail after a few encounters moves back and abandons the fight. Should the bird go on fighting, he will die after the encounter. Some people introduce intoxicants into the sport and a few hours before a fight, administer such a strong drug to their bird that he does not realize when he is injured in the fight and never thinks of running away. He goes on fighting like one possessed until he has driven his opponent from the pit.

Interest in quail-fighting in Lucknow produced such expert fighters that they were unrivalled elsewhere. Some people seeing a famous and successful quail could produce a very ordinary quail and make it resemble the other in every detail. Then when they got the chance they changed the birds round in the course of conversation. This was of course a sharp practice amounting to theft.

Some experts acquired the ability to re-train a quail which had run away and make it fight and beat very good third-year birds. There was also an expert in intoxicants who used to prepare such efficient pills that he received one hundred rupees for ten of them and people were only too willing to buy them. These people showed their highest skill in the field of the medical treatment of quail. They cured quail which were very sick, and on the point of death. They diagnosed their ailments with accuracy and used such suitable ingredients in their medicine that physicians were astounded. They even made great efforts to breed quail from eggs which were kept in dried grass, but without success.

Quails were given high-sounding names such as 'Rustam', 'Sohrab' (Persian fairy-tale heroes), and 'Shuhra-e-Afraq' (World-Famed). Large bets were made on the encounters and I myself have seen a fight on which one thousand rupees were wagered. Some kings took an interest in the sport and Nasir ud Din Haidar was fond of watching quail-fighting on a table set before him.

The hunting of quails is also interesting. At first it was merely an interest which led to some carefree people, who never otherwise left the town, going into the country and fields to get some fresh air; but now many make their living by it. They say that quails leave the mountains at night and fly about at great altitudes. Hunters get hold of a certain kind of quail having a loud call which they then train to call throughout the night; these are known as call-birds. Nets are erected round a field of pulse and call-birds are put into it. On hearing them, the quail in the sky come down into the field, where many of them collect during the night. In the early morning they are driven into the nets where they become enmeshed and are then put into cages.

PARTRIDGE-FIGHTING

Partridges, when fighting, leap into the air more than other birds. Interest in the sport is taken only by villagers and lower-class people. Noblemen and gentlefolk never go in for it.

The birds are trained by being rolled in the dust and made to race. They are fed with termites to make them worked up and excited. But this is really no sport and was never taken by refined society, though the lower classes in Lucknow took it up quite extensively and still indulge in it.

LAVWA-FIGHTING

The *lavwa* is a variety of partridge, smaller than a quail. While other birds fight over grain, they fight over a female. A fight would begin, therefore, with a cage containing a female lavwa being placed before two male birds to make them angry. The sport was more popular in the State of Rewa and other regions of central India, but people in Lucknow followed it to a certain extent. Lavwas are usually trained by being starved and rolled in the dust and their fighting is more attractive to watch than quail-fighting. The lavwa spreads his wings, closes with his opponent, rises like a blossoming flower and then descends. Some rich men in Lucknow took an interest in the sport and there was also an expert at the late Wajid Ali Shah's court at Matiya Burj. Lavwa fighting began before quail-fighting but eventually the latter became more prevalent and the former died out. The method of catching the birds is quite interesting. Like quails, they have the habit of flying about at great altitudes. People tie an earthenware pot on top of a pole, cover the mouth of the pot with a dried animal skin, drive a stick through the skin, and fasten it inside the earthenware pot. A cord is tied to the stick and when this cord is pulled a loud humming sound is emitted from the skin. This noise attracts the lavwas so that at night they fly down to the nets around the pole. In the early morning they are caught in the nets in the same way as quails.

GULDUM-FIGHTING

Most people call the *guldum* a nightingale, but this is incorrect. The nightingale is a song-bird from Badakhshan and other parts of Persia, whereas the bird in question is called *guldum*, Rose-Tail, because it has red feathers like a rose beneath its tail. Villagers and lower-class people often make them fight but better-class people have never taken much interest in the sport.

Guldum-fighting is not unattractive to watch. During training the birds fight over grain sprinkled on the ground. When they fight, both birds fly into the air as they close with each other, become enmeshed and then descend still entangled.

LAL-FIGHTING

Lals are really only suitable as cage-birds, but in order to get some short-lived pleasure, people also used to make them fight. In the first place, it is difficult to tame the lal sufficiently to ensure that it will not fly away directly if it is released from its cage. Secondly, it is not easy to work these birds up to a pitch of excitement sufficient to make them attack and fight. However, when they do fight, they fight very well. They become entangled as they fly into the air, close with each other and go on fighting for a long time; in fact their fights last longer than those of other small birds. Lal-fighting was never popular in Lucknow. There were only one or two experts and as a general rule people were opposed to the sport which, in any case, was confined to the lower classes.

PIGEON-FLYING

Pigeons can be classed among those tame animals in which human beings everywhere have always taken interest, from ancient times until today. There are several varieties. Among the fliers are *girah baz* and *goley* which are kept merely because of their beauty and bright colours. The best-known varieties are *shirazi, guli, peshawari, gulvey, laqa, lotan, choya* or *chandan* and some others. The *yahu* never stops cooing night and day and because they utter the sound 'Yahu, O God', they were very popular with the pious Muslims and holy men.

I believe that tumbler pigeons were first brought to Lucknow from Kabul. These were the first variety to be flown. After these came the goley, a variety which was originally bred in Arabia, Persia and Turkestan. The tumbler pigeon, if it is released in the morning, will fly for hours. They rise up into the sky, vertically, and encircle the house. If water is put into a basin in the courtyard their reflection can always be seen in it. Sometimes they fly about all day and only come down at night. The recognize their cots and are particularly home-loving. As an example of their intelligence, I once had a tumbler pigeon which was caught by someone who clipped its wings. Three years later, when it got the chance and its feathers had grown again, it returned, went straight into its cot and attacked the pigeon which was occupying it at the time.

Tumbler pigeons only fly in flocks of ten or twelve. Those who preferred to fly flocks of one or two hundred kept goleys. Pigeon-flying reached such perfection in Delhi that they say when the last Mughal king, Bahadur Shah, went out in procession, a flock of two hundred pigeons would fly in the sky above his head, thus providing shade.

Pigeons are extremely fond of their homes. In Delhi pigeons were taken by cart to distant places where they were released at any given spot and then called back to the cot.

From the start, the ruling family in Lucknow practised pigeon-flying. Shuja ud Daula took a great interest in it and an inhabitant of Bareilly, Saiyyid Yar Ali, was attached to his court as an expert in pigeon-flying. Navab Asaf ud Daula and Navab Sadat Ali Khan were also interested and at the time of Ghazi ud Din Haidar and Nasir ud Din Haidar the art had reached a high level. Mir Abbas, an expert pigeon-flier, could go by invitation to anyone's house and for a payment of five rupees would release a pigeon from a cot and call it back by whistle. By no chance would the pigeon come down anywhere else.

So great was the interest taken in this sport that some rich men would fly up to nine hundred female pigeons in one flock and some nobles flew the same number of male pigeons without mixing the two flocks.

From Khost, a territory on the borders of Afghanistan, came some special-coloured pigeons called *tipait*. They were extremely costly and nobles used to spend thousands of rupees on these birds.

An old gentleman in Lucknow who was fond of novelties produced an unusual phenomenon with two young pigeons. He cut off the right wing of one and the left wing of the other and then stitched them together, thus making them into a double pigeon. Then he nurtured it with such care that when it matured it was able to fly. He produced many such composite pigeons.

Nasir ud Din Haidar, when he crossed the river from the Chattar Manzil in his boat and sat in his kothi, Dil Aram, admiring the river scenery, would make a habit of watching these extraordinary composite pigeons. They were released from the far bank by this same old gentleman. They used to fly across the river and sit near the King. This pleased him very much and the owner was rewarded.

Another old gentleman named Mir Aman Ali invented something fantastic. He could produce a pigeon of any colour he desired. Pulling out most of the original feathers and inserting different coloured feathers into the apertures, he could fix them so that they held as firmly as the original ones. Sometimes he used to paint them in a way that was so lasting that the colours did not fade throughout the year. But after the bird had moulted, the original feathers returned. These pigeons used to be sold for fifteen or twenty rupees each and wealthy men were eager to buy them. The same man could produce coloured designs and floral patterns on the wings which were extremely rare in nature and unequalled in beauty.

Navab Paley was a great expert. He used to fly tumbler pigeons like goleys. His specialty was that with a given signal at any place he could make his pigeons show off their tricks and turn somersaults in the air.

Wajid Ali Shah collected many new varieties of pigeon in Matiya Burj. It is said that he paid twenty-five thousand rupees for a silk-winged pigeon, and that he developed the breed of a certain form of green pigeon. When he died he had more than twenty-four thousand pigeons with hundreds of keepers to look after them. Their darugha was Ghulam Abbas who was unrivalled as a pigeon expert.

Interest and skill produced the most wonderful varieties of pigeon, not only for flying but for raising as well. I myself saw a shirazi pigeon which was large enough to fill the whole of a cage a yard in breadth and a guli pigeon which could pass through the bracelet of a twelve-year-old girl.

ABDUL HALIM SHARAR, *Lucknow: The Last Phase of Oriental Culture*

Edward Hamilton Aitken (1851–1909), humorist, naturalist, and a writer on the wildlife of India, well-known to Anglo-Indians by the pen-name of 'EHA', worked in the service of the Government of Bombay. He discovered a new species of anopheline mosquito, which was named after him as 'Anopheles aitkeni'. *An indefatigable worker in the museum of the Bombay Natural History Society, which he helped found, his works include* The Tribes on My Frontier, An Indian Naturalist's Foreign Policy, Behind the Bungalow, A Naturalist on the Prowl, *and* The Common Birds of Bombay.

Peter and His Relations

For the last two months the rain has been simply ridiculous. Last week the weather did seem to vacillate for a few days, but I rashly planned an excursion for Gunputtee day, and the deluge returned with renewed resolution. We have had nearly eleven feet already, but the total up to date goes on rising at the rate of several inches a day. The people say that this rain is particularly good for the crops, and so I find it. The crops of mould and mildew have grown rank beyond all precedent. If I neglect my library for a few days a reindeer might browse upon the lichens that whiten my precious books. The roots of these vegetables, penetrating the binding, disintegrate the glue underneath, so the books gradually acquire a limp and feeble-minded aspect, and presently the covers are ready to come away from the bodies; and the rain has undoubtedly some effect of the same kind on ourselves. How is it possible to keep up any firmness of mind or body in such weather? It is too dark to do anything inside the house and too wet to do anything out of it.

Peter, the Parrot, enters deeply into the general dulness. If it were fair he would be sitting in some shady bush in the garden, nipping the leaves off one by one and strewing the ground with them, but now he is confined to his cage with nothing to do. He looks at me so longingly as I pass, that my heart, already flabby from the effects of the weather, is quite softened, and I have to open the door and let him get on my shoulder. This consoles him, and he grows cheerful and soon sets out on a tour of inspection. He is wonderfully interested in my watch-chain, and more so in the buttons of my coat. He has not succeeded in getting them off yet, but I can see that he does not mean to be beaten. Nine-tenths of the pleasure of a parrot's life lies in the use and misuse of its beak, which is a wonderful instrument, quite unlike the beak of any other bird. The upper part is not firmly dovetailed into the skull, but joined by a kind of hinge, on which it moves up and down a little, so that the points of the upper and lower parts play freely against each other and can do very neat work in the way of shelling peas, or husking grains of rice. The muscular and sensitive tongue works like a finger, holding the grain in its place, or turning it round, as the operation goes on. With such artistic apparatus, feeding becomes an art, and a parrot's meals take up half the day. He will not bolt his food like a gross crow, to whom fresh meat and putrid fish, dead rats and hens' eggs all come alike, but tastes every morsel, and eats one part and throws away ten.

Peter's special luxury is bread and butter, and he eats the butter and throws away the bread. He is fond of rice too, and puddings and dry grain and nuts and buttons and the ends of pencils. I think he is the most lovable pet I have ever had. He was brought to me as an infant, neglected, dirty, and raged. Torn from his parents at that tender age, his affectionate nature clung to me, and he seemed as if he could not live without me. He would follow me from room to room and sit under my chair, and if I called to him from any distance, he would answer me. He is older now and more independent, but he still feels lonely without human society. He keeps up an affectation of a very bad temper, rushing at my fingers with barks and threats, but he is never rude towards my face. He treats my lips with touching tenderness, and I often allow him to amuse himself trying to draw my teeth.

Parrots are almost always spoken of in the feminine gender, but half of them are masculine. Peter has not determined his sex yet, so I have given him the benefit of the doubt.

There can be no question that parrots have more intellect than any other kind of birds, and it is this that makes them such favourite pets and brings upon them so many sorrows. Every cold season you will see in the Crawford Market in Bombay large basket cages, made for carrying chickens to the bazaar, but now filled with unfledged parrots, crowded as close as bottles on a shelf, all bobbing their foolish heads incessantly and joining their hundred voices in meaningless infant cries. Men will buy them for two or three annas each and carry them off to all quarters of the native town, intending, I doubt not, to treat them kindly: but 'the tender mercies of the wicked are cruel,' and confinement in a solitary cell, the discipline with which we reform hardened criminals, is misery enough to a bird with an active mind, without the superadded horrors of poor Poll's life in a tin cage, hung from a nail in the wall of a dark shop in Abdool Rahman Street. Her floor is tin and her perch a thin iron wire, so her poor feet are chilled all night, and if her prison chance to hang where the sun can reach it, then for a change they are grilled all day. Why does the Society for the Prevention of Cruelty to Animals never look into the woes of parrots?

On this side of India we have four kinds of parrots, or, more properly, parakeets, for they have all long tails. The commonest is the Rose-ringed Parakeet, neat but not gaudy in its bright green suit, with a necklace of pink and black ribbon and a beak of red coral. I suppose that three-fourths of the inmates of the jails of our Bombay fanciers belong to this race. In its native state it lives a joyous and social life, tasting every good thing in the garden and field while the day lasts, and at sunset resorting to the club, where hundreds of its kind meet to pass the night in company. Then there is screaming and shrieking indescribable. Under adversity it nurses malice and becomes implacable. You may say 'Pretty Polly' in your gentlest tone and chirrup winningly, but she will just stiffen a little, and her eyes will grow all white, and you had better not put your finger too near the bars of her cage. Yet Polly can love as well as hate if you only give her as much reason.

She is an apt pupil too, and can learn to speak so plainly that, if somebody tells you what she is saying, you can make it out quite well. Thomas Atkins often whiles away the monotony of barrack life in the tuition of parrots, and when you hear his pupils talk you almost fancy they are quoting from the works of Rudyard Kipling. A friend of mine, who wanted very much to take an educated parrot home with him, went to the Soldiers' Industrial Exhibition at Poona, where he saw a handsome bird for sale, with a sheet of foolscap hanging from its cage, in which were detailed all the pretty things it could say. Standing nearby was the owner and educator, pretending to be on duty. My friend knew something by experience of the danger of introducing Barrack parrots into polite society, so he asked the solider, 'Does this parrot say any bad words?' With a faultless salute Mr Atkins replied, 'I can't say now, sir; she has been three days in this place.' My friend decided not to buy the pupil of such a master.

But teach her what tongues you will, Polly never forgets her own, and this is my one objection to her as a pet. However happy you make her captivity, imagination will carry her at times to the green fields and the blue sky and she fancies herself somewhere near the sun, heading a long file of exultant companions in swift career through the whistling air. Then she opens her mouth and rings out a wild salute to all parrots in the far world below her. Like arrows through your cloven head those screams chase each other. If you try to stop her, she gets frantic and literally raves in screams. There is only one thing you can do that will avail at all. Dash a whole glass of water over her and she will stop instantly. But this is bad for the drawing-room carpet.

E.H. AITKEN, *A Naturalist on the Prowl*, 1917

Well-known author, historian, and novelist, William Dalrymple wrote his first book In Xanadu *when he was only twenty-two. In 1989 he moved to Delhi where he lived for six years researching his second book,* City of Djinns, *which won several awards. The piece below, about partridge fights in Delhi, is extracted from this book. His other works include* From the Holy Mountain, The Age of Kali, *and* White Mughals. *His most recent work* The Last Mughal, *on the life and times of Bahadur Shah Zafar, won wide critical acclaim.*

Partridge Fight in Delhi

If poetry, music and elephant fights were the preferred amusements of the court, the humbler folk of the age of Safdarjung had their partridges. Again and again in the letters and diaries of the period there are references to both partridge and elephant fights; they also feature prominently in the miniature illustrations of late Mughal manuscripts.

Both sports were clearly popular and well-established traditions; but when I asked my Indian friends about their survival in modern Delhi, they all shook their heads. As far as any of them knew, the last elephant fights had taken place around the turn of the century in the princely states of Rajputana; and as for partridge fights, said my friends, those sorts of Mughal traditions had all died out at Partition. I might find the odd partridge fight surviving in Lahore or somewhere in Pakistan, they thought, but not in Delhi.

It was Balvinder Singh who one day in late January suddenly announced, quite unprompted, that he would not be on duty on Sunday as he and his father Punjab Singh were going to watch what he called a 'bird challenge'. The fights apparently took place every Sunday morning in a Muslim cemetery in Old Delhi: '*Bahot acha* event *hai*,' said Balvinder. 'All very good birds fighting, very good money making, all very happy people enjoying.'

I asked whether I could come along too; and Balvinder agreed.

The following Sunday at six in the morning the three of us set off from International Backside into the thick early morning mist. As we neared the cemetery, the streets began to fill with people, all heading in the same direction. Some were carrying bulky packages covered with thick quilted cloths. Every so often one of the packages would let out a loud squawk.

The cemetery lay within a high walled enclosure at the back of the Old Delhi Idgah. Despite the early hour, the arched gateway into the cemetery was already jammed with chai wallahs and snack sellers trying to push their barrows through the narrow entrance. On the far side a crowd of two or three hundred people had already gathered: craggy old Muslims with long beards and mountainous turbans; small Hindu shopkeepers in blue striped lungis; Kashmiri pandits in long frock coats and Congress hats.

The crowd milled around chatting and exchanging tips, hawking and spitting, slurping tea and placing bets. As the partridge enthusiasts pottered about, three elderly men tried to clear a space in the centre of the cemetery. They strutted around, sombre and authoritative, clearly in charge of the proceedings. These, explained Balvinder in his (characteristically loud) stage whisper, were the *khalifa*s, the headmen of the partridge fights.

'Very big men,' said Balvinder approvingly.

'How do you become a khalifa?' I asked.

'Experience and market value,' said Punjab.

'You must be very good fighter,' added Balvinder, 'and you must have too many titar (partridges). This man here has one hundred fighting birds ...'

The khalifa whom Balvinder had pointed out came up and introduced himself. He was a very old man; he had blacked his eyelids with collyrium and his teeth and upper lip were heavily stained with paan. His name was Azar Khalifa and he lived in Sarai Khalil in the Churi Wallan Gulli of the Old City.

'We khalifas live for the bird challenge,' he said. 'We have no other occupation.'

Azar and Punjab Singh agreed that Delhi was the best place to see the partridge challenge in the whole of the subcontinent. 'I have seen the partridge fights of Lucknow, Jaipur and Peshawar,' said Azar. 'But never have I seen anything like the fights in Delhi. Khalifas come here from all over India and Pakistan to participate.'

Some of Azar Khalifa's partridges were fighting in the match that day, and the old man showed us his birds. From behind the headstone of an old Muslim grave he produced an oblong parcel trussed up in flowery chintz. Unbuckling the fastenings, he removed the wraps and uncovered a wickerwork cage. Inside, separated by a dividing trellis, were two fine plump partridges.

'This one lady. This one gent,' explained Balvinder.

The birds responded with a loud cry of 'Ti-lo! Ti-lo!'

Of the two birds, the male was the more beautiful. It had superb black markings as precise and perfect as a Bewick wood engraving running down its spine; lighter, more downy plumage covered its chest. Half-way along the back of the lower legs you could see the vicious spur with which the birds fought.

'I feed my birds on milk, almonds and sugar cane,' said Azar, sticking his finger through the cage and tickling the female under its neck. The males I train every day so that they can jump and run without feeling too much tired.'

As we talked there was a shout from one of the other khalifas; the first flight was about to begin.

Azar called me over and with a flourish sat me down on a plastic deckchair at the front of the ring of spectators. The open space in the middle had been carefully brushed and in its centre squatted two men about five feet apart; by their sides stood two cages, each containing a pair of birds. The spectators—now arranged in two ranks, those at the front squatting, those behind standing upright—hurriedly finished placing their bets. A hush fell on the graveyard.

At a signal from Azar, the two contestants unhitched the front gate of the cage; the two cock partridges strutted out. As they did so, their mates began to squawk in alarm and encouragement.

The males responded by puffing up their chests and circling slowly towards each other. Again the hens shrieked 'Ti-lo! Ti-lo!' and again the males drew closer to one another.

Then, quite suddenly, one of the two cocks lost his nerve. He turned and rushed back towards the cage; but finding the gate barred, he skittered off towards the nearest group of spectators, hotly pursued by his enemy. At the edge of the ring the cock took off, flying up amid a shower of feathers into the lower branches of a nearby tree. There he remained, shrieking 'Ti-lo', his chest heaving up and down in fright. The rival male, meanwhile, strutted around the deserted lady partridge in ill-disguised triumph. The hen averted her head.

This first short fight obviously disappointed the connoisseurs in the audience. Balvinder shook his head at me across the ring: 'This one very weakling bird,' he shouted. 'This one very weakling.'

Money was exchanged, the two contestants shook hands, and their place in the ring was taken by a second pair of fighters: a Rajput with a handlebar moustache, and a small but fierce-looking Muslim who sported a bushy black beard. To my eyes the new pair of partridges were indistinguishable from any of the other birds I had seen that day, but the rest of the audience clearly thought otherwise. A murmur of approval passed through the crowd; Balvinder got out his wallet and handed two 100-rupee notes to his neighbour.

At another signal from Azar Khalifa the two gates were pulled back and again the cocks ran out. This time there was no bluffing. Encouraged by the raucous shrieks of their mates, the two birds rushed at each other. Handlebar's bird, which was the lighter of the two, gave the Muslim's bird a vicious switch across its forehead; the darker bird responded by ripping at its rival's throat. The two then fenced at each other with their beaks, each parrying the other's thrust. A frisson went around the crowd: this was more like it. Despite the violence, the blood and the cloud of feathers, I was surprised to find it strangely thrilling to watch: it was like a miniature gladiator contest.

The two birds had now broken loose from each other and withdrawn to the vicinity of their respective cages. Then Handlebar's cock suddenly jumped into the air, flew the distance that separated it from its rival and came down on the darker bird with its neck arched and talons open. The spurs ripped at the Muslim's bird and drew blood on its back, just above the wing. The bird rolled over, but on righting itself managed to give its attacker a sharp peck on its wing tip as it tried to escape. Then, scampering up behind Handlebar's bird it grabbed its rival by the neck, gripped, and forced it down on to its side. The first bird lay pinioned there for four or five seconds before it managed to break free and fly off.

Handlebar's mate had meanwhile broken out in a kind of partridge death-wail. The cry was taken up by the loser of the previous fight who was still watching the new contest from his tree. Soon the whole graveyard was alive with the squawking of excited partridges.

The action had also electrified the crowd who were pressing in on the ring despite the best efforts of the khalifas to keep them all back. Someone knocked over the tray of a chai wallah and there were loud oaths from the squatting men over whom the tea had fallen. But the incident was soon forgotten. The Muslim's bird had gone back on to the offensive, swooping down with its spurs and ripping a great gash along its enemy's cheek. It followed the attack up with a vicious peck just above the other bird's beak. Handlebar's bird looked stunned for a second then withdrew backwards towards its mate.

The ring which had originally been twenty feet across was now little wider than seven or eight feet; the squatters were now standing and getting in everyone else's view. In the middle, handlebar was looking extremely agitated. Although the rules laid down that he could not directly intervene he hissed at the hen who dutifully shrieked out a loud distress call. This checked her mate's retreat and the bird turned around to face the Muslim's partridge with his back against her cage.

The proximity to his hen seemed to bring the cock new resolution. For a few seconds the two birds stood facing each other, chests fully extended; then Handlebar's bird flew at its rival with a new and sudden violence. He dealt the Muslim's bird a glancing blow with the hook of his beak, then rose up, wings arched, and fell heavily on the lighter bird's head. As he hopped out of reach he again cut the darker bird with his spurs.

The reprisal never came. The Muslim's bird slowly righted itself, got unsteadily to its feet, then limped off through the legs of the crowd. There was a great cheer from the spectators. Balvinder jumped up and down, punched the air, then promptly confronted the man with whom he had made the bet. The latter grudgingly handed over a stash of notes. All around the ring wallets were being slapped open and shut; fingers were being angrily pointed. Everywhere arguments were breaking out between debtors and creditors, winners and losers. The outsider had clearly won.

Suddenly there was a cry form the gateway; and the khalifas started ushering everyone to one side. All the spectators frowned.

'What's happening?' I asked Punjab, who had come up beside me.

'This khalifa is saying one dead body is coming. We must leave the graveyard for an hour.'

'Now?'

'Yes, immediately. Some people are wanting to bury some body here.'

As we were leaving, we passed the Muslim who had lost the fight. He was cradling his partridge in his hands and kissing the bird. He was close to tears, but the bird looked surprisingly perky.

'Will the bird live?' I asked Punjab.

'Yes, yes,' he replied. 'This Mahommedan will bandage the wounds with herbs and feed the titar with special food. In a few weeks the bird will be back in the fighting ring.'

Outside the cemetery we came across Azar Khalifa.

'You liked?' he asked.

'Very much,' I said.

'Everyone likes,' said Azar. 'For the people of Delhi this partridge fighting has always been a happiness.'

'It is true,' said Punjab Singh. 'People are coming here drunk, worried or tired of the chores of the world, but always they leave this place refreshed.'

WILLIAM DALRYMPLE, *City of Djinns*, 1994

FALCONS AND FALCONRY

Alongside shikar, falconry has been a recurring passion down the ages. One of the earliest British ornithologists to record the hunting qualities of Indian falcons and hawks was T.C. Jerdon (1811–72) in his book The Birds of India. *Jerdon was a physician in the British East India Company. He was instrumental in conceiving the 'Fauna of British India' series.*

Falcons of India

THE PEREGRINE FALCON

I have seen the *Bhyri* strike down various water birds, teal, duck, etc. and on one occasion I saw a pair pursue and kill a snipe. Often a large flock of duck has been forced to come within reach of my gun at some small tank by the downward swoop of a *Bhyri*, which the hapless fowl dread more than man even, and I have often had wounded teal, snipe, and other birds carried off by them.

The *Bhyri* has particular haunts that it frequents for days or weeks together, and near some of their feeding grounds there is often a particular tree to which they invariably resort to eat the birds they have caught. In their untrained state they seldom fly at larger birds than duck, to which however, they are very partial, so that their representative in America, *F. Anatum*, is there popularly called the Duck Hawk.

The *Bhyri* does not breed in this country, nor even, I believe, in the Himalayas, but migrates to the north in April, and returns about the first week of October. Mr Layard mentions the Peregrine as breeding in Ceylon in January, and Dr Adams says the he found the nest on a tree on the banks of the Indus below Ferozepore; but I imagine in both cases an old *Laggar* has been mistaken for the *Bhyri*. The Peregrine breeds in Europe and

Northern Asia, on high cliffs, often on the sea coast, or overhanging a river or lake. The eggs, three or four in number, are reddish coloured with brown spots.

The *Bhyri* is still trained in some parts of the country for the purposes of falconry, and used to be so much more extensively than now. The birds were mostly captured on the coast, and sold for a few rupees, from two or three to ten, to the falconers who came to purchase them. It is trained to strike egrets, herons, storks, cranes, the *Anastomus, Ibis papillosa, Tantalus leucocephalus,* &c. It has been known though very rarely to strike the Bustard. Native falconers do not train it to hunt in couples, as is done in Europe sometimes. I may here mention that the idea of the Heron ever transfixing the hawk with its bill is scouted by all native falconers, many of whom have had much greater experience than any Europeans. After her prey is brought to the ground indeed, the Falcon is sometimes in danger of a blow from the powerful bill of the heron, unless she lays hold of the Heron's neck with one foot, which an old bird always does. Whilst on this subject, I may state that our best artists, Landseer included, represent the Falcon, when stopping on her quarry as striking with her beak, whereas, as is well known, she strikes only with her talons, and chiefly with the powerful hind claw, backed by the impetus of her stoop, when she contracts the foot, and thus clutches her prey. When the *Kulung* (*Grus virgo*), is the quarry, the *Bhyri* keeps well on its back to avoid a blow from the sharp, curved, inner claw of the crane, which can, and sometimes does, inflict a severe wound.

THE SHAHIN FALCON

The *Shahin* breeds on steep and inaccessible cliffs. I have seen three eyries, one on the Neilgherries, another at Untoor, and a third at the large water-fall at Mhow. It lays its eggs in March and April, and the young fly in May and June, when they are caught by the Falconers.

The Royal Falcon of the East (as its Indian name implies) is very highly prized by the natives for hawking, and it is esteemed the first of all the Falcons, or black-eyed birds of prey (as they are called in native works of Falconry), the large and powerful *Bhyri* (the Peregrine) even being considered only second to it.

Although hawking is now comparatively at a low ebb in India, yet many individuals of this species are annually captured in various parts of the Peninsula, and taken for sale to Hydrabad, and other places where the noble sport of Falconry is yet carried on, and they sell for a considerable price. The *Shahin* and other falcons are usually caught by what is called the Eerwan. This is a thin strip of cane of a length about equal to the expanse of wings of the bird sought for. The ends of the stick are smeared with bird-lime for several inches, and a living bird is tied to the centre of it. On observing the hawk, the bird, which has its eyes sewn up to make it soar, is let loose, and the Falcon pounces on it, and attempts to carry if off, when the ends of its wings strike the limed twig, and it falls to the ground. The birds usually selected for this purpose are doves, either *Turtur risorius* or *T. humilis*.

The *Shahin* is always trained for what, in the language of Falconry, is called a standing gait, that is, is not slipped from the hand at the quarry, but made to hover and circle high in the air over the Falconer and party, and when the game is started, it then makes its swoop, which it does with amazing speed. It is indeed a beautiful sight to see this fine bird stoop on a partridge or florikin, which has been flushed at some considerable distance from it, as it often makes a wide circuit round the party. As soon as the Falcon observes the game which has been flushed, it makes two or three onward plunges in its direction, and then darts down obliquely with half-closed wings on the devoted quarry, with more than the velocity of an arrow. This is of course a very sure and deadly way of hunting, but though infinitely more exciting than the flight of short-winged hawks, is certainly not to be compared in interest to the flight of a *Bhyri* from the hand after the heron, or the Douk (*Tantalus leucoephalus*). The *Shahin* is usually trained to stoop at partridges and florikin (*Otis aurita*), also occasionally at the stone plover (*Oedicnemus crepitans*) and the jungle fowl. It will not hover in the air so long as the Laggar, which being of a more patient and docile disposition, will stay up above an hour.

THE GOSHAWK

The *Baz* is the most highly esteemed bird of prey in India, and a trained bird used to be sold for a large sum in former days.

They are caught when young, and sold on the skirts of the North-West Himalayas, to falconers from different parts of India, for prices varying from Rs 20 to Rs 50 for the female, and from 10 to 20 or 30 for the male. The *Baz* is trained to strike the *Houbara* bustard, Kites, *Neophrons*, Duck, and many other large water birds, as Cormorants, Herons, Ibises, etc. It is, however, chiefly trained to catch hares. For this purpose she is booted or furnished with leather leggings to prevent her legs being injured by thorns, as the hare generally drags the hawk some yards after being struck. She strikes with one leg only, and stretches the other one out behind to clutch grass, twigs, or any thing on the ground, to put the drag, as it were, on the hare. The *Jurra* is trained to strike partridges, rock pigeon, crows, teal, &c. The Goshawk flies direct at its prey, and gets its speed at once; and if does not reach the quarry within a reasonable distance, say from 100 to 200 yards, it generally gives up the chase; and either returns to the falconer's fist, or perches on some neighbouring tree, or on the ground.

THE BESRA SPARROW-HAWK

The *Besra*, or Jungle Sparrow-hawk, is comparatively rare, though well-known throughout India to all who take an interest in hawking. It is found in all the large forests of India; in the Himalayas, on the slopes of the Neelgherries, in the Malabar forests, and here and there on the Eastern Ghats, and the forests of Central India. It extends to Assam, Burmah, Malayana, and the Isles. After the breeding season is over, about July, a few birds, usually young ones, straggle to various portions of the more wooded parts of the country. Mr Elliot says he has only met with it in the Soonda Jungles (in Canara).

The Besra and other short-winged Hawks, as well as occasionally the Laggar and some of the falcons, are usually caught by what is called among Falconers the Do Guz. This is a small thin net from four to five feet long, and about three feet broad, stained of a dark colour, and fixed between two thin pieces of bamboo, by a cord on which it runs. The bamboos are fixed lightly in the ground, and a living bird is picketed about the middle of the net, and not quite a foot distant from it. The Hawk makes a dash at the bird, which it sees struggling at its tether, and in the keenness of its rush, either not observing the net from its dark colour, or not heeding it, dashes into it, the two side sticks give way, and the net folds round the bird so effectually as to keep it almost from fluttering.

T.C. JERDON, The Birds of India
(Vol. 1), 1862

R.S. Dharmakumarsinhji (1917–86), scion of the royal house of Bhavnagar in Gujarat, made significant contributions to natural history studies in India. His major ornithological works include The Birds of Saurashtra *and* Sixty Indian Birds *(co-authored with K.S. Lavkumar). Dharmakumarsinhji was also a keen sportsman and falconry expert, and the following piece provides a vivid description of falconry in the princely states of India.*

Falconry Flights

The great, kingly sport of falconry, seen at its grandest in days of the Emperor Genghis Khan and the Kublai Khan, the Great Mongols and of kings of Persia, was brought to India during the Moghul period of which Emperor Shah Jehan was the champion. Thereafter, falconry was seen during the British period when many of the Indian Princely States indulged in it, maintaining hereditary falconers, some of whom were of outstanding talent.

Here, I intend to narrate some flights of hawking worthy of mention, seen by me before Independence, with falcons trained by two falconry maestros, father and son.

'Bazdar' Makekhan Fatekhan who served as hereditary falconer to the Maharajahs of Bhavnagar was a falconer of very high calibre. One of his achievements was training the saker falcon (cherrug) on the chinkara. This is considered one of the highest standards to be reached in Indian falconry. Here is then a day of such sport.

We had motored to an area where a few chinkara had been reported. An area devoid of hills, ravines and scrub, a terrain which is of handicap to the falcons.

Makekhan had been given *carte blanche* by his master to select and buy saker falcons suitable for gazelle hawking from the famous hawk market at Amritsar (Punjab). Makekhan had bought six large female passage saker falcons in juvenile plumage, after bidding the highest price in the auction. The very large sakers of this size are known as 'harani' cherrugs (meaning that such saker falcons were trained on the Indian antelope in the early Moghul period). Although Indian artists have depicted such scenes of sport, I feel I must stretch my imagination to believe the possibility of such achievement with the fully adult blackbuck. However, I do not wish to doubt the records of such past performances.

Out of the six pairs of sakers that had been selected, Makekhan commenced training two pairs and after two months of training he had one pair to his satisfaction.

The day of gazelle hawking had arrived. It was after sunrise when we reached the open flat ground near the coast where the chinkara were undisturbed and regularly seen. There were two and we manoeuvred the car to where the chinkara stood. Without causing them suspicion, we separated the pair, and then from about 100 yards, Makekhan rose in the open car and unhooding the falcon with prayers to Allah, he released one saker. As she saw her quarry and hesitated, Makekhan, by a forward thrust of his arm, launched the falcon off his gloved fist. As soon as the first falcon was released, the second one was quickly passed to him and he did exactly the same.

The sakers are fast falcons, the size of ravens, and since chinkara had to be put on the run before releasing the falcons, the car was kept ready for the watching and follow up of the chase. The car, packed full contained two long dogs at outside for later release. As the falcon made his first stoop on the gazelle and punched its nape, the gazelle realized it was being attacked and began galloping fast. The second falcon, for some reason, had not properly seen the quarry and was flying in the wrong direction. This was a great handicap. As we followed and watched in excitement we saw the first saker swooping and missing its target and rising above in its momentum of speed and then with closed wings re-attacking it with greater force form above. Two to three strikes on the head and then we saw the fatal stroke which unbalanced the gazelle and caused it to fall. This was the signal for the dogs to be released, and there again, somehow only one dog caught sight of the gazelle, which ran to follow the chase. In the meanwhile, the gazelle had risen from its headlong fall and stood for a moment bewildered and bleeding near its eyes. In this position, the saker again came crashing down and hit the face with its strong feet with a mighty blow causing it to stumble, then turning round quickly she bound on to its nape. Immediately following this attack the dog had reached the gazelle and was pulling down the animal. Seeing the success against heavy odds, we sped to the thrilling scene for further assistance and stopped the car suddenly as we reached the scene. A young shikari jumped off from the car immediately and running to the gazelle held it down and gave the *coup de grace*.

As we got down from the car, old Makekhan knelt down at his master's feet with praise to the Almighty. It was indeed a most touching finale to a unique performance which brought us back to what we had heard and read of gazelle hawking of the past.

That success came with the action of one falcon and one dog was not expected, and this added to the credit of both. Makekhan received all praise from his master and we congratulated him with affection for his grand achievement. Later he was fully rewarded for this outstanding performance by his master Sir Krishnakumarsinhji.

Normally with two 'harani' sakers in action, repeated strikes on the head and nape have greater and quicker effect to achieve success, and two dogs can hold down a chinkara with greater ease than one. In former days, the spectators rode on horseback to follow the sport.

Makekhan specialized in training sakers and goshawks, the latter the most prized hawks during the Moghul period. In modern falconry parlance, Makekhan was Falconer-cum-Austringer as most Indian falconers were. Makekhan lived to a ripe old age of

90 years and I have portrayed his life in an obituary printed in the Illustrated Weekly of India.

Gulam Hussain was Makekhan's eldest son and doubtless the most able of his many sons. Literate and knowing a little English, he had, in fact, achieved greater success in falconry than his father, as he could train falcons to 'wait on' as well as for 'ringing flight' and also hawks of the 'fist', and no bird of prey was beyond his training. Notwithstanding, his forte lay in training peregrines, shikras, bashas (sparrowhawks) and eagles.

During one season of hawking, he undertook for his master the training of the peregrine to capture cranes, a sport ranking very high in falconry. As with most of our hawks, Gulam Hussain was given full scope to exercise his art in falconry and was ordered to purchase the best hawks from Amritsar (Punjab) and Masulipatam (Madras State) markets.

As with the selection of our hawks, the finest passage red falcons are selected and Gulam Hussain had procured about ten such peregrines with the view of training them for crane hawking. The endemic shahi falcon, a smaller peregrine can also be trained to catch cranes.

However, not all peregrines are suitable for crane hawking but at a certain stage of training, the best out of them are given a trial, out of which, usually one or two are selected. And this Gulam Hussain had done by choosing one which came from Masulipatam. The falcon had tremendous speed, sincerity of purpose and determination if not courage. It was finally a matter of building up its stamina and giving it the secret potion which eliminated inhibition.

In western India, we have with us three distinct cranes. The sarus with the red head and pink legs, the largest of all cranes, is a resident species, not hunted owing to local sentiment. The eastern common crane or kunj, which is slightly smaller than the sarus, is of a dove-grey colour with blackish legs. The head and neck is mixed with black and white, which run down along the side of the neck; the hind crown has a dark reddish bare patch not normally seen in the field. Juvenile birds are rufous-grey without the black and white patterns of adults or the red patch. A smaller crane, the demoiselle or karkara is recognized by its ash-grey colour and black legs. It has white elegant plumes on the sides of the head. The crown is grey, the rest of the head, neck to breast is black or bluish slate grey with lanccolate plumes hanging down over the upper breast. It has scarlet-red eyes.

Both the above cranes are migratory, arriving in September and departing in March or April in vast numbers. Both can be recognized by their size, patterns and calls.

The eastern common crane was selected as the quarry, and Gulam Hussain was now ready and confident with one of his peregrines. At this time it was being rumoured by some neighbouring Princes that flight at crane was just another falconer's brag, as the experiment had once failed before.

In crane hawking, especially at the early stage, it is desirable to release the falcon (the female is always used owing to its size and strength) on a solitary crane. Now this is not easy, because cranes are normally seen in flocks, some of them very large, and it is only when they feed that the kunj may be seen spread out and in pairs, often with their juvenile young. If, however, the birds are feeding a little away from each other, one may be able to get between the two but this is not easy.

The final day had come. It was dawn, flocks and flocks of crane were flying in V formation towards their feeding ground which was spread throughout the countryside. We followed their direction and often watched them settle in the open fields. The morning had been cold and the winter sun was now becoming warm. As we searched for a single crane and sighted numerous groups of kunj feeding in stubble or open cropped groups of fields, strangely, we were not able to find a pair or solitary crane. It was getting late, the falcon had commenced to bate, which tires the falcon, and the effect of the drug was wearing off. Then at midday as we were despairing, we came upon a couple of kunj which had not finished their foray. We were in our open V-8 ford tourer and slowly were able to get within 60 yards of them and as we did so we slowly edged in between them.

The day was ideal, sunny with practically no wind. As we stopped our car, the selected crane stretched its neck out and took a few steps forward before flying. As the crane flew, Gulam Hussain stood up, unhooded the peregrine, and off she went like a bullet towards her target which had taken wing. I was driving the car as usual and I opened the throttle full and sped behind as fast as the car could go over the rough fields. The peregrine had, as soon as it had caught up with the crane, which had been flying low, bound to her quarry on the back and brought it down to the ground, but the crane was too powerful.

Within minutes I had brought the car within forty yards of the scene that was taking place, the crane trying to rise, turning its head and neck to attack the falcon, the sharp claws of the legs of the crane (which are its main weapon of defence and if used may fatally wound or kill the falcon) could not be operated as the falcon had clung on the back. The danger therefore, was the sharp pointed bill which, at the time, it could not effectively use, but the huge bird was trying to rise to its feet with the help of its long and powerful wings. As the car came to a head, a young Vaghari 'Nanko' leaped out of the car, fell, rose and ran as fast as his legs could take him towards the crane. In the meanwhile, the crane, by its effort to escape was about to shake off the falcon from its grip and was about to rise in the air when 'nanko' threw himself forward like a Rugby player to catch hold of the legs of the kunj. No sooner had this been done, than the peregrine, which was a foot off the ground, quickly bound again to its prey. This dramatic end was a matter of seconds as it took place before our eyes with much anticipation, if not anxiety, from Gulam Hussain. This exciting finish was witnessed by R.K. Chandrabhanusinghji of Wankaner.

Subsequently, a very exciting and extraordinary flight at the crane occurred when the Durbar Saheb, Ala Khachar of Jasdan was present. By some misapprehension, misjudgement or by coincidence, Gulam Hussain having struck the hood of the peregrine to a nearby pair of cranes, it so happened that the peregrine's eyes fixed themselves on the crane in a flock which in the background had taken wing at the same time as the nearer pair, though slightly to one side. Immediately, to our horror, it was noticed that the falcon was speeding low towards the distant flying cranes. Seeing her pass the two flying cranes to her side and directly aim at the group of cranes gave us mixed feelings. Yet, as we watched in amazement, we saw the falcon fly rapidly with great determination. It then suddenly dawned on me, late as I was, to give the V-8 Ford a racing start. The peregrine was flying close to the ground and there was a slight breeze from the side against her. The flock of cranes never suspected that a small falcon was on its attack and the cranes, also flying low on account of the wind, flew in the usual manner. However, as the falcon followed the course of the cranes and had closed in to about forty yards from a distance of about two hundred yards, the leading crane, noticing that the flock was threatened, emitted an alarm note, and all at once, the whole flock commenced a rapid beat of wing, as if a cox of a boating crew had given the signal for a quicker stroke. With necks fully outstretched, the cranes had accelerated and yet the courageous and determined falcon was keeping up, trying to get under the belly of the last birds of the echelon. As soon as she was under the lagging crane, the crane separated from the flying formation, turning sharply to the side downwind, leaving the peregrine slightly behind.

I was pressing the car to negotiate over a ploughed field and tried to keep up with the chase as near as possible, but with clods of earth getting in the way I had to put the car into second gear and again speed on through cotton stubble, watching the exciting flight as best as I could. As the crane had taken a sharp turn, the peregrine having greater manoeuverability and taking advantage of the following wind, was again able to get below the crane and as she did so, she swooped up and struck the thighs of the crane with force. This attack bought the crane down with the falcon binding tightly and the crane on its back trumpeting with bill wide open and flapping its long wings attempting to rise. Pressing on the accelerator, and trying to keep up with the chase as best I could, we arrived in time for the runner to jump out of the car to catch hold of the neck of the crane, which was screaming and struggling hard. This he did with dexterity. We were all very happy over this successful end, because no falconer would want to release his falcon at a group of cranes at such distance. We later learnt that this belly attack culminating with both the legs of the crane being incapacitated by the feet of the falcon is an effective hold since the flying crane is over-balanced by the upward thrust of the falcon, causing it almost to somersault and come to earth on its back with its main weapons, the legs tied down and the falcon facing opposite to the crane lying on its back.

This famous peregrine we saw often bound on the back of crane, nevertheless it is possible for a crane, which is nearly six times as heavy and stronger, to sustain the weight of the falcon and keep flying for some distance. Once, to our disgust, we saw an entire flock of crane turn back and attack our falcon which had downed a crane from the group. The falcon, though injured, escaped death.

Gulam Hussain was well-rewarded at the close of the hawking season, having achieved full success in crane hawking. He had indeed earned the title of Master of Falconry given by his master.

A FALCONRY MEET

I had been invited to a falconry meet by a certain Sheikh in Kathiawar, a small chieftain who was very keen on hawking and hunting wildlife. Although, we had some of our hawks ready we had that season lacked enough peregrines and since the invitation came suddenly and towards the end of the season, I was not fully prepared. Nevertheless, the invitation was accepted and I drove out in my red sporting Lagonda car to look out for falcons. It must have been a lucky evening, for within an hour and a half, just at sunset, I saw a beautiful passage peregrine (juvenile) chasing a flock of rosy pastors. I made a note of the place and then driving a little further I saw another peregrine of equal quality flying low and alighting on a solitary tree to roost.

Both these falcons were seen within a mile of each other. I was happy, and sped back to Gulam Hussain and told him to go early next morning to where the falcons were last seen, with expert trappers to catch them. It was a chance to be taken in view of the impending falconry meet. The next afternoon I received news that one of the peregrines was snared and the other had been seen nearby. The following day, the trappers were again sent with the result that the second peregrine was also brought to hand. When I saw these fine birds, they looked almost alike as if they had been sisters. Gulam Hussain was full of joy but we had only twenty days to train them before we had to leave for the meet. Both birds were immediately put to rigorous training, night and day service, by four men.

On the fourteenth day, I watched both the birds at the crows' wing lure and they were swooping superbly. The very next day a bagged live black ibis was given to enter each of them. And then just two days before we left, we were glad to see the final training completed.

When we reached the meeting place our team of hawks consisted of two male goshawks, a lagger, a red-naped shaheen and the two peregrines.

This small team was selected out of many of our trained hawks. At the meet, all did very well with the shaheen never missing a grey partridge from a 'waiting-on' high pitch, which had to be reduced to enable her to be successful. Similarly, the lagger falcon, but unluckily the partridges, although hit fair and square, got into nearby thorn cover and thus a few points were lost.

The outstanding flights were of the two sisterly peregrines, which were released last on the black ibis. The climate at the meeting place being much cooler than at home all our hawks were feeling better. This good news given to me by Gulam Hussain was most encouraging.

All the main events were over and when the time for releasing our peregrines came it was midday, but being quite cool, the peregrines had stood to the late time well. Soon we found a group of black ibises in an open field; even they had not gone to water as it was still cool. Then as the flock of ibis was flushed, the first peregrine was released and she was able to bring one down from the flock after a good ringing flight. She was then quickly fed and hooded. Now the rest of the black ibises were circling high in the sky. The Sheikh then gave his challenge and said, 'Now if your second peregrine can bring down one from those ibises soaring above I admit defeat'.

I looked at Gulam Hussain and he in his usual unassuming style asked me, 'What is your order?' and I nodded and he understood. The hood was struck but the peregrine could not see the quarry. She was looking straight ahead and the tiny black spots of ibis circling directly below the rays of the midday sun could not be seen. Replacing the hood on the peregrine immediately, Gulam Hussain was somewhat puzzled, then he knew what to do next. The hood was taken off again and by placing the half-cupped palm of his left hand under the chin of the peregrine, made her look straight up into the sky. For a moment the peregrine hesitated and we feared she may not recognize her quarry. But with falcon's eyes she knew what they were and off she went, climbing steeply at her normal speed, rising in circles. When she was above half-way to the group of ibises, they saw the enemy and the leader gave the alarm call. At once all the birds started beating their wings rapidly and began rising in the thermal current. Stimulated by hearing the call of the ibises the peregrine commenced a rapid beat of wings and climbing at her full speed in short circles, her bells ringing to the sound of every wing beat as challenge to the ibises, she ascended with determination. Then finding that the quarry was rising more rapidly than herself, she started making wider circles and taking advantage of the leeward side of the wind where she began gaining height. Though her advantage was slight and she was having to fly further away from her quarry, she found herself gaining height slowly, and then when she knew she had got slightly above the ibis group, we saw her, a tiny speck, making straight for the flock. I had my binoculars on here all the time and I relayed this new exciting position to Gulam Hussain who was anxiously looking up, facing the blinding glare of the sun, wondering whether the falcon was still pursuing her prey.

The black ibises now knew that they were being attacked on equal level and as is their habit, one out of the flock peeled out to sacrifice itself, with a call note of defeat (which could not be heard by us) and then closing its wings now and then descended rapidly at a steep angle. The peregrine followed in a steep dive close behind with half closed wings, flapping them at times to close the gap in hot pursuit but not quite able to do so.

The ibis was doing its best to reach the river below but before doing so the peregrine had bound to it and both birds came tumbling down, wings flapping, spinning as they landed on the opposite side of the river. A Pathan horseman ready for such eventualities was at once sent charging across the river at a shallow crossing, to protect the successful falcon from eagles or pariah dogs that may have interfered before our falconer reached the spot. We had won the final match. This last performance was the one that was remembered most. The Sheikh had shown excellent flights with his hawks at kite, and with his saker falcons and his goshawks and shikras at partridges but his laggar had failed at 'waiting on'.

In celebration at the conclusion of the falconry meet, there was shooting of blackbuck and crane by stalking and watching of a pair of panthers from a 'machan' built like a huge nest on top of a huge *Euphorbia* cacti clump. Chinkara were plentiful and they disturbed golf players at the course by making their scrapes (beds) on the 'browns'. Finally there was feasting and Indian music for all of us.

The next day no one was seen until the evening, when a joyful retrospection of the previous day was underway by all who participated in the hawking events. Surprisingly, in none of the flights had an eagle intervened and it was said that the Sheikh had destroyed most of the eagles in the vicinity, a measure very destructive, but nothing in comparison with the destruction of birds by insecticides and pesticides as seen today. There is now hardly a hawk or eagle left in the countryside.

When we returned home the climate had become warm and we were considering releasing our trained hawks, allowing them sufficient time for them to regain their natural condition.

EAGLE HUNTING

In the Moghul days some eagles were trained but they were not as popular as the hawks and falcons. Most of our eagles are not as agile as the hawks or falcons but they can be trained. No one has really tried them out seriously except at the lure. I might consider myself lucky in the case of the Bonelli's or slender-legged hawk-eagle (*Hieraaetus fasciatus*). This eagle is found in most parts of India and by some ornithologists classed as hawk-eagle while others as in the *Aquila* class or true eagles.

The adult Bonelli's eagle is not a very large eagle but it excels in courage and footwork. The female is slightly larger than the male. Both have whitish underparts with brown streaks on the breast. The upper parts are amber brown with a white patch on the back. The eyes are pale yellow. The juvenile is brown above and buffy-rufescent below, the eyes brown. The long quills are blackish.

Two fine qualities of this eagle stand out: its courage and its power of gripping. The central toe is long, almost like in the falcons and its slender, long feet enable it to use them at advantage to catch its prey. The adult birds are usually seen hunting in pairs and it is a delight to see them doing so, systematically soaring a hillside, one bird high up, the other lower down, at times seemingly beating its wings to flush its prey. Once the prey is flushed, both eagles combine in its capture, diving headlong upon it in turn.

Training eagles is exasperating work owing to their recalcitrant nature, their heavy weight and capacity to fast for days at a time. But once trained the eagle shows remarkable performance. To give one example, I let loose an immature female Bonelli's eagle on a blackbuck and the eagle overturned it by a hold of her talons on the muzzle with one foot and as the antelope came crashing down and tried to kick the eagle with its powerful hind legs and sharp hoofs the second foot was swiftly and dexterously used in which both hind legs, and one foreleg at the hocks were caught and bound as in a vice in the talon. The blackbuck, to my utter amazement, lay on the ground as if its legs had been tied. This revelation of the gripping powers and dexterity of Bonelli's eagle was beyond my imagination and reminded me of a hawk I had, a female goshawk, catching a hare using her feet alternately to her advantage to overpower the much heavier prey and then lying on her back with the hare on top, thus saving her tail feathers from beings broken.

Juvenile or adult Bonelli's eagle tackling gazelle or antelope in treeless flat country is a spectacular sight. However, the juvenile and adult birds need to be trained at separate times of the year owing to the discrepancy in their moult.

The adult should be caught and trained from September to February and the juvenile from March to September or October. The juvenile birds are better but the season for them less so. A pair of intermewed Bonelli's eagles are an asset for large prey. They should only be released on a solitary animal and on a calm day.

R.S. DHARMAKUMARSINHJI, *Reminiscences of Indian Wildlife*, 1998

Naturalists on the Prowl

3

The famous sketch by R.A. Sterndale in Edward Hamilton Aitken's (EHA) *A Naturalist on the Prowl* (1917)

Of the Englishmen who lived in India it can be said that one facet of their involvement with birds and natural history was the exhibition of a certain 'attitude' when it came to experiencing nature. They did it with style and gusto: sometimes with a camera or a butterfly net in hand and at other times with a tiffin of sandwiches. Edward Hamilton Aitken's writings are reflective of that attitude as the following extracts from his well-known book The Common Birds of Bombay, *amply demonstrate.*

The Common Birds of Bombay

THE VULTURES

If the city of Bombay had a tutelary bird, there is no manner of doubt what bird that should be. I do not know why the ancient Egyptians deified the Ibis, but if Bombay bore the proud figure of a Vulture *rampant* on her shield, everybody would know why. Of all the unsalaried public servants who have identified themselves with this city and devoted their energies to its welfare, no other can take a place beside the vulture. Unfortunately the vulture has never lent itself to the spirit of heraldry. The eagle has, strangely enough, though the difference between the two is not very clearly marked in the popular mind. The translators of our Bible had no notion of it. Modern natural history has disentangled the two names and assigned them to two very different families of birds, the distinction between which in its essence is just this, that, while the eagle kills its prey, the less impatient vulture waits decently till its time comes to die. Popular sentiment persists in regarding the former as the more noble, but there can be no question which is the more useful. It is not easy indeed to realise to oneself the extent and beneficence of the work carried on throughout the length and breadth of India, from year's end to year's end, by the mighty race of vultures. Every day and all day they are patrolling the sky at a height which brings half a revenue district within their ken. The worn-out bullock falls under the yoke, never to rise again, and is dragged off the road and left; or the old cow which has ceased to be profitable and has therefore ceased to be fed, lies down in a ditch for the last time. Before the life has left the old body some distant 'pater-roller' has seen it, and, with rigid wings slightly curved, is sloping down at a rate which wipes out five miles in a few seconds. A second sees the first and, interpreting its action follows with all speed. A third pursues the second, and so on till, out of a sky in which you could not have described two birds half an hour ago, thirty or forty dark forms are converging on one spot. When they get right over it, they descend in decreasing spirals and settle at various distances and wait for the end like American reporters. When the end comes, if you are squeamish or fastidious, go away. All that will corrupt, everything in short but the bones, is to be removed from the carcase within twenty-four hours, and the vultures have taken the contract to do it. Such work cannot be made artistic and the vulture is not an aesthete. That bald head and bare neck are not ornamental, but they mean business; they are the sleeves tucked up for earnest work. It is a merciful and, I suppose, a necessary provision of nature, that every creature gets reconciled to its task and is able even to take pleasure in that which would be painful to others. The vulture enjoys the full benefit of this provision. It is in fact an enthusiast in its profession, and these funeral wakes become scenes of riotous and ghoulish glee to which I confess that even philosophic reflection fails to impart moral beauty. The gourmands jostle and bump against each other, and chase each other round the board with long, ungainly hop and open wings. One has no sooner thrust its head well into the carcase than another leaps upon its back with loud laughter. Two get hold of opposite ends of a long strip of offal and dance before each other with wings outstretched. And the cackling and grunting and roaring that go on all the while may be heard for half a mile. When darkness overtakes the revellers some of them have so shamefully over-eaten themselves that they cannot rise from the ground and are forced to spend the night where they are. They seem to be quite safe, however. The jackal is not a fastidious feeder, but it draws the line at vultures. These scenes used not very long ago to be enacted regularly on the Flats, where the carcases of horses and cattle were skinned and left.

OWLS

Owls were classed by Cuvier with eagles, hawks, and vultures, and Jerdon followed him, as all the old naturalists did. But a more careful examination of their anatomy has shown that they differ widely from all other birds of prey in many respects, and resemble parrots; so they are now placed by most in an order by themselves, midway between the hawks and the parrots. The outward and visible characteristics of this order are a short, parrot-like beak, the outer toe reversible (in parrots it is permanently reversed), very large eyes directed forwards, and uncommonly well developed ears. They make their nests in holes and lay white eggs like parrots. Their plumage is peculiarly soft, even the quills, so that they fly noiselessly. If you want more, I may tell you that there is no ambiens muscle, but basypterigoid processes are present. On the other hand, the accessory femoro-caudal and the semitendinosus and the accessory semitendinosus are wanting. Now all this is very important and not to be laughed at. These solemn words were not invented only to bamboozle the unlearned, but represent facts in the plan on which the frame of an owl is constructed. And the question on which these facts bear is more than curious. Expressed in popular language the question is this. Is the owl only a weak-eyed hawk that cannot bear the light of day, or is it a bold and bad parrot which has taken to night-walking and murder?

The Screech Owl is more common in our island than in any other part of India with which I am acquainted. This statement may surprise people who have lived for twenty years in Bombay without seeing one, but the Screech Owl does not ordinarily put itself much in the way of being seen. A dark object, like a Flying Fox, passing overhead as you drive home from dinner, and a loud, harsh, husky screech, suggesting sore throat and loss of voice, are all the indications you will commonly have of its presence. But should a pair take up their residence in any deserted building, or old ruin, in your neighbourhood, then you will know more about them. I often wonder what the Screech Owls did before man was created, for they cannot get on without him now. If he did not build churches with steeples and belfries, and forts and castles with towers, and barns with roomy lofts, where would they live? In this Presidency they are under deepest obligations to the Portuguese. Under one of the remaining walls of an ecclesiastical ruin in Bassein Fort, Mr Phipson and I once noticed the ground glittering with small white bones. We gathered a handful of them and brought them home for examination, and could scarcely believe in ourselves or each other when they proved to consist chiefly of the jaw-bones of Muskrats! In a high niche of that old wall a worthy pair of Screech Owls had, for who knows how many years, brought up an annual family of 3, 4, or 6 insatiable owlets on this nutritious food, varied only with an occasional house rat or field mouse. As is well known, owls swallow their prey whole, and after digesting all that is digestible, throw up the bones and hair rolled up into little balls. Why the bones we found were chiefly jaw-bones I cannot tell, unless the parent birds were in the habit of snipping off the heads of little animals as delicacies for their offspring and consuming the bodies themselves.

THE NIGHTJARS

How shall I describe a Goatsucker? If you are walking by day in scrubby ground on some still unreclaimed part, say, of Cumballa Hill, and a brownish bird starts from under a bush at your foot and flies, with jerky strokes of its very long wings, for a distance of twenty or thirty yards, and then drops under a bush again, it is a Goatsucker. You have disturbed it in its sleep. Or after sunset, in the dusk of the evening, you may come upon it sitting in the dust, right in the middle of the road, in some unfrequented neighbourhood. It will jump up suddenly as often as you approach it, and fly before you for a little distance, then drop into the middle of the road again and squat, looking just like a large frog, or toad, dimly seen. This is how it spends the night, or rather, I should say, the times of dusk and dawn, for I believe it sleeps at midnight. At intervals it springs up and takes a circuit, performing somersaults and other antics in the air. It is catching moths or beetles.

Sometimes it perches on a bough of a low tree, not across it, as any other bird would, but along it. Such is a Goatsukcer in the bush. In the hand it is a weird thing, with a flat head and very large, lustrous, dark eyes, like those of the heroine in a penny dreadful. Its feet are small and its bill is a mere apology, but its head is almost split in two by the width of its gape. Its soft plumage is very beautiful, but hardly describable. It consists of earthy and ashy and reddish shades, mottled, barred, or curiously pencilled with darker tints.

This bird is called a Goatsucker from its wicked habit of milking domestic goats. In modern books of natural history you will find this habit denied and the bird called a Nightjar, but they cannot get rid of its Latin name, *Caprimulgus*, with which it has been branded from the days of Pliny.

THE CUCKOOS

Cuckoo is properly the name of a particular migratory bird, which spends the spring and summer in Europe and the winter in warmer latitudes (India, for example), and is notorious for shirking its parental responsibilities and foisting its offspring upon other birds to bring up. But the name is applied to a whole group of birds which resemble the European Cuckoo in structure and have the same disreputable habit. There are many species of the family in India, and all, like the home bird, are better known to the ear than the eye. The most familiar of them all is the Koel (*Eudynamys orientalis*, or *honorata*). It is a great black fowl almost as large as a crow, with a much longer tail and a green bill. That is the male. The female is of a dark-greenish dusky hue, spotted and banded with white. But the Koel is seldom seen. It is—

No bird, but an invisible thing,
A voice a mystery.

Early in the morning, through the hottest hours of the day, late in the evening sometimes in the dead of night, its loud and mellow voice calls to us in a rising crescendo, 'Who-be-you? Who-be-you? Who-be-you?'

FLYCATCHERS

The Flycatchers are a distinct and important branch of that standing army of birds which nature keeps to make war upon the insect horders that threaten to eat her up. Their duty is well defined and they keep to it. They hunt for no caterpillars among leaves, nor tap trees for grubs, nor rummage about the ground for beetles and worms. There are others whose office it is to do all these things. The Flycatchers concern themselves only with things that fly, and they catch these on the wing. The King Crow and the Bee-eater, as we have seen, do business in that line too, but they take their stations on high places and pursue their quarry into the sky. The Flycatcher haunts sylvan shades and darts about among the branches, snapping up its tiny prey.

Indian Flycatchers may be divided into two sorts, the plain and the fancy. Of the fancy we have two species in Bombay. The first is the Paradise Flycatcher (*Tchitrea paradisi*), which wears two streamers of white satin ribbon in its tail and looks like a meteor as it flits from tree to tree. Its body and wings are white too, exquisitely white, but its head and throat and distinguished crest are glossy black, with green reflections. It is a bird that would catch the eye of a blind man, and everybody who has roamed about Matheran or Mahableshwar must be familiar with it, but I daresay some will be surprised to hear that it is a Bombay bird.

From a Mahomedan tradition we learn that the Paradise Flycatcher belongs to that unhappy class who are spoken of as having 'seen better days'. At one time it was a truly glorious bird, clad from tip to toe in dazzling white and adorned with a magnificent tail of snowy plumes. But it gave way to pride and got so puffed up at length that it presumed to compare itself with the Birds of Paradise and claimed a place among them. For this it was shorn of its tail and utterly disgraced. It repented, however, and Allah was merciful and allowed it to retain two of the feathers of its tail, but he blackened its face that it might never forget its shame.

BABBLERS

We are not accustomed to speak of autumn in India, but there is a time of year in this country, as much as in any other, when each tree puts off its old clothes and gets a new suit. The only difference is that tropical trees for the most part manage the matter more decently than those of cold countries. They do not strip themselves before the new suit is ready and stand naked till it arrives. They undress and dress at the same time, as respectable people do. In this transaction avaricious Mother Earth plays the part of Moses. She receives the 'old clo' and opens a shop, and her customers are numerous and beggarly. The earthworm sneaks up from the ground and draws a rotten leaf down into its burrow, the white ants swarm everywhere, bargaining for remnants; earwigs and vagabond cockroaches wander about, examining everything and taking nothing. In such a crowd it goes without saying that there will be no lack of sharpers, pickpockets, and cut-throats, making victims of the ignorant and unwary. These are called centipedes, scorpions, predacious beetles, wolf-spiders and so forth. In short, the carpet of dead leaves which is spread in every forest, grove, and neglected garden, affords a habitation and a livelihood to a vast and very varied multitude of creatures, which have this special interest for us to-day, that there are many kinds of birds whose sole business it is to look sharply after them. Among these are many species of long-legged ground Thrushes, and foremost among them is the Babbler. The Babbler is seldom spoken of in the singular. The natives call it *Satbhai*, the Seven Brothers; in other parts it is known as the Seven Sisters. You cannot think of it except as a member of a small party. It may be a family party, father, mother, and grown-up children; but I do not think so. I believe it is simply a social party. Among animals there is not the same diversity of individual character as among men, or the same variety: all the individuals of one species are cast pretty much in the same simple mould. But for this very reason each species exhibits more distinctly some one or other of the elements that go to make up the complex human character. Every virtue and every vice in the moral catalogue may be found typified in some beasts or bird. So I hold. And if this be true, then the phase of character which is expressed by the Babbler is jolly-good-fellowism. Not being acquainted with the method of distilling spirits, it does not pass the flowing bowl, but a large portion of its life is devoted to *al freco* eating parties, in which the excitement of finding the viands is combined with the pleasure of consuming them, and the utmost conviviality prevails.

These parties are not too large for true sociality. They consist of about half-a-dozen, whence the popular name of the bird. There is no distinction of host and guest: all are equal. They begin under some tree where the leaves have fallen thick, and proceed as humour leads. Each helps himself to what he can find, turning over the dead leaves and pouncing on any tempting morsel that tries to hurry away. If one is lucky and lights on a particularly fat lot, his neighbours come to his aid, and there is a good-humoured squabble over the partition of it. There is a regular flow of small talk, a good deal of mirth and laughter, occasionally an eager dispute, but never a quarrel; 'Fighting?' says Phil Robinson, 'Not at all; do not be misled by the tone of voice. That heptachord clamour is not the expression of any strong feelings. It is only a way they have.' They will fight for each other, but not with each other. Woe to the sparrow-hawk that thinks to make a prey of any one of that party. Only a rash young fool would attempt such a thing, and it will be taught wisdom. But, though the Babblers dine together, they do not live together. Each pair makes its nest apart, affecting great secrecy and deluding the egg-collector with mingled impudence and wiles.

THE ROBINS AND CHATS

We come to a group of birds bound together by certain well-marked family features. They are small birds, usually dressed in black and white, or brown and white, always neat, but never gaudy. They are all afflicted with some form of St Vitus' dance in the muscles of the tail; they are either twisting it, or throwing it up over their backs, or doing something else than letting it hang down decently. Lastly, they are all groundlings, collectors of crickets and beetles and other small hard-backed insects that run upon the face of the earth, but taking little interest in caterpillars, or flies of any kind, and seldom touching fruits. In all these respects they differ from the Thrushes.

I feel that the one which ought to head the list is the Indian Robin; but you must not let your thoughts run on the bird which is begging for crumbs at our windows in the old country. Mr Phil Robinson, speaking of the difficulty of getting up anything like a Christmas feeling in this land of regrets, complains that the very

Robin instead of wearing a red waistcoat, wears a red seat to its trousers. This is true if not expressed with prudery: but it is not the only difference between the two birds. The Indian Cock robin (*Thamnobia fulicata*) is a jet-black bird, with the exception of the rusty patch above-mentioned and a narrow band of pure white across the wing, which scarcely appears except when it flies. Nevertheless it is by nature a robin, making a friend of man, sitting on his house top, coming into his verandah, or even singing to him from his own window sill. You will not find it in orchards or shady gardens, for it has a prejudice against perching on a tree; but wherever there are old fields, there it is at home with its smoke-coloured mate, running a few steps on the ground. Perching on some point of rock, tossing up its tail till it almost touches the back of its head, and throwing out snatches of cherry song. No more description is needed. Everybody knows the Indian Robin. In March or April it makes its nest in a niche in a wall, or in some recess, in a pile of stones, never very far above the ground; and there it lays three dingy looking eggs, of a greenish white colour, speckled with brown. You will not find the nest very easily, for the Robin is cunning, like all birds that build near the ground, and will not come or go in sight of an enemy. And in that connection man is an enemy.

With the exception of one bird, which haunts the deep forests of the ghauts, the Magpie Robin is the finest songster that we have in Western India. In March and April, when the Thrush and Blackbird are singing to our friends as they lie in their beds, the Magpie Robin at the same hour is pouring forth a continuous torrent of far-reaching song from the top of some palm or old mango tree. And we scarcely say, 'Thank you'. Whether it is that we leave our ears at home when we come out here, or that we leave our hearts at home and the ear counts for little without the heart, I do not know; but it is a melancholy fact that there are many Englishmen in this country on whom the music of its birds appears to be wholly lost. I have been assured by a man who had spent many years in India that the birds here never sang, but only cawed, or shrieked, or jabbered. When I told him that skylarks, scarcely distinguishable from the 'embodied joy' of English fields, were singing every morning in the blue sky above the very road by which he went to his work, he scoffed at me.

He had never heard a skylark in India. There are of course more birds of song in this country than in England, because there are more birds altogether, and because the sun that cheers them is brighter and the sky that inspires them more blue. As to the quality of their songs, comparisons are odious and unprofitable, because we cannot invest Indian birds with the associations which endear those of England. The voice of the Blackbird, heard in bed in the cold silence of a spring morning, will sink into one's heart in a way which is impossible in this country, where we are not much given to lying in bed of a morning, and where the cawing of crows, the crowing of cocks, the yelping of pariah dogs, and a medley of other unmusical noises come in at the open windows with the first streak of dawn.

THE WARBLERS

In the days of Imperial Rome there were, I suppose, almost everywhere large communities of humble brickmakers, who made cheap bricks for poor folks' houses, and other sorts of obscure, but necessary, people; but Tacitus does not mention them, so far as I recollect. There are birds which fill a similar place in the feathered commonwealth. The Wren Warblers and Tree Warblers do an inestimable amount of useful work and appear to enjoy as large a measure of contentment and happiness as their betters; but there is nothing about them to catch the imagination of the historian and they will never be famous. I have been perplexed as to how I should deal with them in these papers. To attempt to describe each species is out of the question, for there are many, and they are mostly so like each other that even the title 'ornithologist' does not qualify one to distinguish them at a distance. If you can distinguish them with certainty when you have them in your hand, you will fully deserve the title.

To begin with the Wren Warblers,—they are small, dingy birds with long tails, which go about among bushes and rushes and reeds, exterminating little insects. They enjoy this life so much that they moved the envy of Charles Kingsley, and you may almost recognise them from his description:

> I would I were a tiny, browny bird from out the south,
>   Sitting among the alder holts and twittering by the stream.
> I would put my tiny tail down and put up my tiny mouth,
>   And sing my tiny life away in one melodious dream.

But you must not suppose that the said 'melodious dream' is a high class composition from a musician's point of view. These little birds are not without a humble conceit of their vocal powers, all the same, and the following inimitable passage from Richard Jefferies will refresh every one who has witnessed their performances: 'He got up into the willow from the hedge parsley somehow, without being seen to climb or fly. Suddenly he crosses to the tops of the hawthorn and immediately flings himself up into the air a yard or two, his wings and ruffled crest making a raged outline; jerk, jerk, jerk, as if it were with the utmost difficulty he could keep even at that height. He scolds and twitters and cirips, and all at once sinks like a stone into the hedge, and out of sight like a stone into a pond.'

All I have said requires abatement if applied to the Tailor Bird (*Orthotomus sutorius*), which is nevertheless a Wren Warbler by nature and feature. But it is a bird of some character and holds its tail up. It is such a prominent feature of the bird life of our gardens, that, if I cannot make it recognisable, these pages may as well cease. But before describing it let me remove a popular error by stating that the Tailor Bird is not called by that name because it makes a curious nest, nor because it comes out of an egg, nor for any other senseless reason. More than twenty years ago I was shown the cup-shaped nest of a Fly-catcher, as a great curiosity, and was informed that this was the nest of the famous Indian Tailor Bird. It did not occur to my informant to ask why the maker of that nest should be called a tailor rather than a potter or a watchmaker; and I have discovered since that his kind is common. Therefore I take this opportunity to explain that a Tailor Bird is called a Tailor Bird because it *sews*. When its nesting time approaches, which is during the monsoon, it searches for a shrub or brush, with large, soft leaves, and drawing two of them together, proceeds to stitch them to one another round their edges. At that season the silk-cotton tree is bursting its pods and scattering its white clusters, so the tiny tailor has seldom any difficulty in finding cotton, which it spins into thread with its deft little feet and beak. But if it can get ready-made thread, so much the better. Jerdon tells of one which regularly watched the *dirzie* in the verandah, and as soon as he had left his seat for the day, pounced down upon his carpet and carried off his ends of thread in triumph. The bird's needle is its sharp beak. Piercing a hole in the leaf, it passes the thread through and knots it at the other side, and so on till it has joined the two leaves by their edges all round and made a neat pocket, or purse, with its mouth at the top, or a little to one side. Then a soft padding of cotton inside makes it ready to receive its treasure of three or four pretty little eggs. They vary a good deal in colour, but are generally white, thinly spotted with light red. I have often seen a nest made of a single large leaf, and, on the other hand, where broad-leave plants are scarce, the bird will use more than two; but the fewer leaves the less tailoring, as the bird knows.

Last monsoon I was standing in the verandah of a friend's house in Bombay when I saw an eager Tailor Bird tugging desperately at a coir mat. I felt sure that it must be in straits for something to make its nest of, and knowing that my friend had a kind heart for the deserving poor, I brought the case to his notice the same evening. He promptly stuck a bunch of clean cotton wool in the trellis, and almost before I was out of bed next morning the bird had noticed it and was carrying off large beak-fulls. He practised a certain amount of guile, but was easily tracked to a low, dense bush in the garden, where, with such charitable assistance, he did not take long to make his wife a very cozy house. It may encourage others in doing good to know that in due course a fine family was reared and sent out into the world in spite of the crows.

THE WATER-WAGTAILS, PIPITS, AND TITS

When I was a boy the Wagtails had a peculiar fascination for me, and the feeling has not quite faded away yet. There is something so original and droll about their idea of life! To hold a long tail horizontally behind you and wag it vigorously and incessantly, to spend your days near cool waters, running about on the ground—not hopping like a sparrow, but running with alternate steps—and catching little somethings in the air, this is the Wagtail's notion of the way to be happy. And it is happy: the vivacity and nimble eagerness of all its motions leave no doubt about that. No other bird behaves in this fashion. I feel sure that there must be some department of insect life which other birds have missed, or despised, and which the Wagtails have appropriated. There are green caterpillars on the tender shoots and little birds to seek for them, there are grasshoppers in the grass and mynas to chevy them, there are beetles and earwigs under the fallen leaves and babblers to dislodge them, there are midges in the air and swallows to hawk them, there are grubs in the rotten bough and woodpeckers to dig them out; but besides all these it appears that there are minute winged things on moist ground in great abundance, which rise like snipe when startled, and these are the game of the Water wagtail. It runs and turns and twists and leaps into the air, and you cannot see what it is after, but you distinctly hear the snap of its little bill, like the pop of a distant snipe-shooter's gun. It follows the cattle in the pastures and runs in and out among their feet; they are its beaters, which drive the game for it. Or it hunts by itself in cool places, on the shady side of the house and wherever large trees keep out the sun.

I am thinking of the Grey wagtail, which often wanders far from water, but not from coolness and shade. It is by far the commonest species we have and a very familiar bird throughout the cold weather. In the costume which it wears at the season the upper parts are bluish-grey, but its forehead and whole face are white. On its breast there is a black patch, exactly like a child's bib, and below that again it is white. In summer it dons a different costume, in which the throat and breast and the back of the head and neck are all black, but we seldom, if ever, see this, because at that time it is in Siberia or thereabouts. There is a difficulty about the name of this bird. There are in fact two species of Grey Wagtails; quite distinct from each other, but very difficult to distinguish, so much alike are they in their winter plumage. In the early seventies Mr A.O. Hume was very much exercised about these two birds, and at that time he was very innocent of any leaning towards Buddhist principles in the matter of taking animal life. He engaged all his friends and helpers in a *jehad* against the whole race of Grey Wagtails, that he might determine to which species they belonged. I never heard the number of the slain, but some survived, and I believe that by far the greater number of those which visit us are of the species known in Europe as the White Wagtail (*Motacilla alba*). In Jerdon's book this and the other are lumped together under the name *Motacilla dukkhunensis*, the Black-faced Wagtail, a most unfortunate name for a bird whose most striking feature, when it come to us, is its clean white face.

THE MYNAS

The European Starling is common enough in the north of India, but does not roam so far south as Bombay. Its place is taken, however, by a group of birds which, though very differently dressed, cannot disguise their relationship to the starling, for the family features are too plain. In the air they have the same direct, business-like flight; on the ground the same parade-step; they have the same flexible voice and talent for mimicry; they make their nests in holes and lay blue eggs. Of course I mean the Mynas, which, among all classes of natives who keep pets at all, are favourite cage-birds for many reasons, but chiefly because they can be taught to speak. The performance is rather like a Punch-and-Judy dialogue, and you need to be told what the bird is saying before you can recognise it. But that matters little; it amuses people who can find little interest in the really amusing traits of the bird's natural character.

For the Myna has a character. I once had a Myna and a canary in cages which hung at my window. A ruffianly crow came in one day and perched on the top of the canary's cage. Of course the silly bird fluttered all around the cage, clinging to the bars, and gave the crow the chance it wanted. It caught a leg in its powerful beak and tried to pull it through the bars. But the canary's body could not pass through, so the poor bird's leg was literally torn out by the roots, and it died in a few minutes. I suppose the crow swallowed the leg, and shortly afterwards it returned, thinking to have a leg of the Myna for its next course. I was in the room, but it did not see me; so, after glancing round the room with a proprietary air, it bounced on to the top of the Myna's cage. But the Myna, sitting on its perch, knew it was quite safe and felt no agitation; so it was free to take an interest in the crow, and its interest fixed instantly on an ugly black toe which hung down through the bars over its head. It caught that toe in its sharp beak and made an example of it. I tell you, it was exhilarating to observe the suddenness with which that crow jumped to the conclusion that it had urgent business elsewhere.

THE AMADAVATS AND THE MUNIAS

From 'Amidavad,' the learned Dr Fryer tells us, come small birds, 'spotted with red and white no bigger than measles,' of which 'fifty in a cage' make an admirable chorus. That was more than two hundred years ago. I do not know whether they still come from Ahmedabad, but the name has stuck to them and they still come, more than 'fifty in a cage' sometimes, to people our aviaries. They need no description, for everybody knows them. They are the tiniest of cage-birds, and have red beaks: whence they are sometimes called Waxbills. The Munias are twice as large, though still very small, and have black, or slaty, bills. But they are all one brotherhood, and will live together in amity, though you pack them so thick that some have to find a perch on the backs of others. So you will find them packed in the cages at the Crawford Market. But they are not unhappy, like most of the birds there, for their wants are small. Give them dry seed and clean water and they will look on the bright side of things. It is to this happy disposition that they owe their popularity as pets, for they have no accomplishments and are as silly and uninteresting as birds can be. The common Amadavat has, indeed a little piping song, which is sweet, though feeble, and the Brown Munia sometimes warbles a love-sick ditty to its mate, hopping absurdly with its legs straddled out, but you must put your hand to your ear to catch the sound. And the rest confine themselves to a note of one syllable, which they repeat about thirty-five times in a minute when they are in good spirits. But it is a pleasant note, and I think a cagefull of Amadavats and Munias in the verandah always adds to the cheerfulness of the house.

THE PIGEONS AND DOVES

Every system of classification puts the pigeons and doves in an order by themselves, for they are distinguished from all other birds by not one but many family features, which cannot be mistaken. Their beaks are swollen and soft at the base, but hard at the point. Their eyes are large and lustrous, and set far back in the head, which is small. Their bodies are compact and shapely, their tails neither very long nor very short, their wings generally fitted for swift and strong flight. They rarely carry any meretricious ornament, such as crests, or trains, or fancy plumes, but they are all beautiful and some of them exquisitely lovely. Yet their loveliness is not that of golden orioles and kingfishers, but rather of clouds and distant hills and soft sunsets. Nor is their beauty in their feathers only; their eyes and their feet, and even their beaks, match their plumage and complete the effect. I think also that all the motions and attitudes of pigeons are more graceful than those of other birds. But these are outward features. There are also inward characters by which the tribe is not less markedly distinguished. They are all vegetarians, some feeding on grain and some on fruit, but refusing animal-food in every shape. It is said, indeed, that they sometimes eat snails, but, if this is true, I believe they must have swallowed them by mistake for seeds. Such mistakes will happen to all of us. I knew a person whose fate it was once to mistake lizard's eggs for small white 'sweeties'. But let us leave that subject and get back to pigeons. They drink like horses, and not by sips as other birds do. They all lay white eggs, never more than two in number, and make simple, flat nests of twigs, which they generally place in trees or bushes, but sometimes in holes. They never sing, nor chirp, nor screech. Their voice is a plaintive moan, or coo, verging sometimes on a mellow whistle. But their highest distinction lies in the strength of their social affections and the purity of their domestic life. In these respects they are far ahead of the majority of the human race. Polygamy and polyandry are alike unknown among them. They are all monogamous, and, as far as my observation goes, a pair once united remain true to each other till death do them separate. Their arts of love and courtship are strangely like our own, and after they are married they are always assuring each other of their affection by pretty tokens of tenderness. They are also devoted to their children. I had a pair of pigeons of which the hen died suddenly leaving two naked and helpless infants. I thought they must die, but the father took the whole care of them on himself and brought them up successfully.

After all this, it is painful to say, what is nevertheless true, that pigeons appear to have been designed in a special degree for the food of other creatures. Being, as I have said, strict vegetarians, their lump bodies are both wholesome and tasty. In this opinion hawks and cats are at one with man. And having no means of protection and no resource in danger, except their swiftness, they are fair game. But they hold their own and multiply, for, though they lay only two eggs at a time, they go on making nest on nest all the year through—in warm countries at least. A pair of domestic pigeons, if provided with two nest boxes, will have eggs in the second before the young are out of the first.

BUSTARD QUAILS

There is another group of small game birds known as Bustard Quails, or Button Quails, which has cost the classificators (this word is not in the dictionary, but I cannot dispense with it) no small perplexity. They mostly want the hind toe (the birds I mean, not the classificators) and have other peculiarities, on account of which they are given a whole Order to themselves in *The Fauna of British India*. They are quiet, shy birds, that live solitary lives in fields and scrub jungle, creeping about among the grass and feeding on seeds and insects. If you chance to tread on one's toes it will start out of the grass and fly swiftly for a few yards and drop again. And this is all you will ever see of it. But you may hear it. In the morning and evening, and even at dead of night, it gives vent to some feeling in one of the strangest sounds ever uttered by a bird. Jerdon describes it as 'a loud, purring call'. To me it suggests a nail drawn across the teeth of a sonorous comb of endless length. If it proceeds from the lungs of the bird, then the mystery is still unsolved how the quantity of air which must be required to keep up such a sustained effort can be compressed into so small a body. One of the eccentricities of the Bustard Quail is that the female makes all the noise. The male, as far as I know, is silent. He is smaller than she, and though I cannot say whether he is literally henpecked, there can be little doubt that he is 'sair hauden doun'. He has to stay at home and mind the babies while she goes gadding about and fighting with her female neighbours. This is not scandal, but a fact. She differs from him in having a good deal of black on the head, throat, and breast. The general colour of both is reddish brown, marked with game pattern of fine, black, cross lines, with buff edges to the feathers. I am speaking of the species which Jerdon calls the Black-breasted Bustard Quail (*Turnix taigoor*). There are two others, but this is the one that makes the curious noise described above, and the only one, I think, that is likely to be found in Bombay. I once came upon its nest in June, nor far from Bombay. It was a most artistic structure for a Quail to build, completely domed over with fine grass, with only a little hole at one side for the owner to go in and out by. I did not catch the bird, so it may have been one of the other species, but there is not much difference.

RED-WATTLED LAPWING

The Lapwing, Peewit, or Plover, which has the misfortune to lay fashionable eggs in England, is not found here, but it has a near relation which is one of our most familiar birds. It has no crest, but on its cheeks there are two bright red lappets, like the wattles of a cock, and Jerdon calls it the Red-wattled Lapwing (*Lobivanellus goensis*). It is a greenish-brown bird with a good deal of black and white upon it. The head is black, with the throat, down to the upper part of the breast. Below this the under parts, with the lining of the wings, are pure white, as you see when it flies. But why should I describe the Lapwing? It needs no description and wants no introduction. It introduces itself to you; impresses itself on you; dins itself into you. Where it sprang from I cannot tell, but there it is in the air, circling round and round, now far, now very near, now high, now low, now seeming to go, but wheeling round and coming swiftly back again; for it will not go. And all the time it is reiterating, with piercing emphasis, that mysterious taunt, 'Did you do it? Did you do it? Pity to do it.' What does the creature mean? I have done nothing. Suddenly its mate springs into visibility and joins it. I have a suspicion, a strong suspicion, that somewhere on the ground, not far from my feet, there are four stone-coloured eggs, with black blotches on them and like pegtops in shape, arranged in a cross with their points inwards. But it is no use looking for them. The

Lapwing is such an accomplished liar that it will throw you off the scent one way or another. The poet has said it,

> The lapwing lies,
> Says here when it is there.

It is altogether a wonderful character. It seems to do without food and sleep. As regards food, you never find it where there is anything to eat, and as regards sleep, the natives have a saying that it sleeps on its back with its legs turned up, for it says, 'If the sky should fall, I will catch it on my feet'; but I suspect the chief point of this saying is that it cannot be contradicted, for nobody ever caught a Lapwing asleep.

E.H. AITKEN, *The Common Birds of Bombay*, 1900

Nesting is an important aspect of the life cycle of birds, and birdwatchers have written extensively about nesting, nests, and eggs. The inimitable Edward Hamilton Aitken conjures a cheerful picture of the domestic life of garden birds.

Bird Nesting

You may live in the same garden with a little bird and meet it many times a day, and never know that it is married and has a family. For weeks the courtship went on under your windows, till she accepted him and left his rival to look for another love. Then the young couple explored every tree in the garden for suitable premises. One branch was tried and rejected on account of the ants, another was fixed on but spoiled next day by the pruning knife of the *Malee*. At length a cosy little site was found close by the path which you traverse every day, materials were collected, and for many days both the birds were busy from early morning building their house. Then one happy day they sat, with mutual congratulations and endearments, admiring the first-born egg. There had never been such an egg. It was the darlingest little egg in all the world. For a fortnight after this he led a bachelor life, coming often, however, to see how she did, and once a day taking her place while she went out for a little air and exercise. Then the little ones came and family cares began in earnest. It was *ora et labora*, four open mouths and much labour to fill them. All the day long spiders and caterpillars had to be caught and dropped into those little red funnels, all stretched out and quivering with expectancy. Now they are clothed and sitting on the edge of the nest, and their parents are in a flutter of delight and anxiety. And all this has gone on without your getting the least hint of it. The fact is that familiarity with danger has taught birds to combine circumspection with an air of unconcern which would baffle a London detective. It baffles the crow, which is sharper than a London detective. The great lizard, who lives in the tree, with his ogre eye on everything, is not so easily eluded, but he can be fought. Only if both parents are away at once will he get a chance. How every collector hates that gourmand! But I scored off him once. He had swallowed the first egg in a nest, and trusted that he would swallow the rest as they were laid; but I put in a chalk egg for his special benefit, and the marks of his teeth next day showed how he had struggled with it before he gave it up in disgust.

But I must return to my walk. I had first to visit the nest of a shrike, which I had noted a week ago. This fearless butcher practises no cunning. His nest is fixed in the thorniest bush he can find, and fenced all round with thorny twigs. He is not much in the way of crows, and if any lizard should be so silly as to show itself, it will promptly be caught and impaled on a thorn till it is tender enough to eat. No provision had been made, however, against me, and I annexed the eggs without much scruple, I confess, for sentiment does not naturally attach itself to the Butcher Bird. Somehow or other, I have quite a different feeling towards the King-crow. It seems mean to take advantage of the splendid courage with which he builds his flimsy house in the most exposed situation he can find, and forbids kites to pass that way, or crows to perch on any of the neighbouring trees. At this moment a travel-worn crow, which rested for a moment on a forbidden tree in mere ignorance, is catching it from the indomitable little tyrant and his wife. Each in turn, with torrents of contemptuous abuse, drops into him from a height of twenty feet, then wheels round and is ready again. The crow holds his ground from sheer cowardice rather than obstinacy, and twists his neck in vain efforts to present an open beak at the point of attack. At last he overbalances himself and hangs by his feet for a moment in utter despair, then flies for his life. With a derisive yell, the victor makes one last descent into his back, as if it would transfix him, then returns slowly to the tree, panting but triumphant. While the Butcher Bird and the King-crow are defying their enemies, the tiny Sunbird is outwitting them. It has selected a dirty-looking tree, literally alive with blood-thirsty red ants. At the very end of a waving branch it has hung a neat little purse, with a small hole on one side, near the top, and a porch to keep out sun and rain. The lining is of cotton from the silk cotton tree, warm and smooth. When comfort has been provided for, the work of external decoration beings. First the whole outside is draped with shreds of spiders' webs. Whether this is done to discourage the red ants I cannot say; but it serves another purpose, for now the bird goes about collecting any rubbish it can find and sticking it on the glutinous web. Old scraps of moss, spiders' egg-cases, rags of white silk from the nests of the red ants, and above all, the sawdust and refuse which woodboring caterpillars shovel out of their holes, are gathered together and stuck on anyhow, till the outside of the little house is as thoroughly disreputable as art can make it. Then it is ready for occupation. Lizards cannot get into it, crows do not suspect what it is, and if they did, would not know what to do with it. The very squirrel is baffled. And now note the way the bird behaves. He is skipping about in the highest spirits, spreading his tail and flapping his wings, singing snatches of an old glee, hovering over a flower while his long tongue searches its recesses, or peering about a dirty bunch of cobweb for little spiders. Suddenly he appears to notice that bunch hanging at the end of a branch. He flies straight to it, clings to the side with his feet, and thrusts his head into the hole at the side for a moment, then darts away again as if saying to himself, 'No spiders there'. But he gave her a kiss.

    Take another instance of fraudulent simplicity suggested by the loud 'kee-ko' of the Crested Swift as he sails about in the sky overhead, or perches on that dead tree, with his crest up and that jaunty air which fits him so well. I am certain where his nest is: it is on that same dead tree without one leaf to conceal it; but for my life I cannot find it. The difficulty is that there is nothing to find; not a straw, or a fibre, or a scrap of moss has the bird collected. He simply spat hard on the same spot from day to day. As a voluminous spitter he could give points to any Kentucky man, and the result was a sun-dried lump very like the little brown fungus which grows on old trees in the rainy season. It is not the size of a rupee, and from below is scarcely visible—certainly not distinguishable from any of the numerous warts and lumps which disfigure the old tree; but it holds one egg safely and is firm enough to support the handsome bird which sits patiently over it in the blazing sun of noon and the dew of night. Now, why should the Sunbird or the Crested Swift require to practise such arts when the dove and the bulbul get on without them? The silly dove arranges a few twigs in a cactus bush and lays her eggs on them. If you pass by she dashes off in such a flutter that you cannot help looking up to see what is the matter, and there the eggs are, shining through the structure of the flimsy nest. It is a mystery; but there is this difference between the egg of the Crested Swift and the egg of the dove—that the former is a precious thing destined to produce a Crested Swift, while the latter will come to nothing but a silly dove, with just intellect enough to find its food and grow fat for somebody to eat. And that which is precious is scarce. There is only one egg of the Crested Swift, and there will not be another for a twelvemonth; but the eggs of the dove are as plentiful as they are cheap. All the year round she is making her foolish nests, and if one comes to grief and one prospers, she will still multiply much faster than the Crested Swift. So with the simpleminded Bulbul; its function in nature appears to be the same as that of the hens in my yard—namely, to lay eggs for others to eat. Every second nest at least meets with an accident, but a merry heart doeth good like a medicine, and the Bulbul is always merry. When one nest is destroyed she just makes another and lays a few more eggs. If the first fails, the second may succeed, and if the second fails, the third may succeed; and so, by paying tribute to their enemies, the Bulbuls still contrive to multiply, and keep every garden, grove, and hillside lively with their twitter.

Thus, one in one way and one in another, one by force and one by fraud, one by resistance and one by submission, they keep their place in the great struggle, and it is curious to note how the ways and instincts of each fit in with the course which it has taken. Birds which build nests in high trees have no special fear of man, and those which make their nests in holes disregard him altogether. The Coppersmith will hammer away at a branch just over the door of your tent, caring nothing who sees her. She has one enemy, the snake, and if it finds her house, neither cunning nor courage will avail anything. But birds that make their nests on the ground fear man above all things. There is a little kind of Robin, or Chat, a dapper little bird in black and white, which makes its neat nest under the shelter of a stone, or at the root of a bush. How often I have had a contest of patience with that bird and gone away beaten! 'Oh!' it seemed to say, 'you want to find my nest, do you? I haven't one.' And, with a grasshopper in its mouth, it perched on a bush, jerking its tail pleasantly and saying *Tea* in a tone which I knew was meant as a warning to its wife and little ones. In vain I sat and tried to look innocent till I was tired, or got up and seemed to walk away, looking over my shoulder as I went. Still it sat and said *Tea*. I got behind a bush and peeped cautiously through the leaves. It saw me and said *Tea*. As soon as I was really away, it would fly straight to its little brood and comfort them with the grasshopper. Of the devices by which birds have sought to secure the safety of their little ones, I think the strangest and most ingenious is that of the Red Woodpecker. I cannot forget the feelings with which I first saw it digging its hole, not in bough or trunk, but in the great, brown nests of the vicious little tree-ant. How it gets rid of the occupants, or whether it makes terms with them, I cannot say. It keeps its secret to itself and lives, I should think, in perfect security, for no enemy is likely to come prying about those nests.

Ruminating on these things and wandering on, I noticed a great dark bird sailing over a wooded hillside, and from the blackness of its ample wings I knew it must be the Black Eagle. A rare bird it is, and noble to look at; but how debased! By lineage it is an eagle and by trade a poacher. Dr Jerdon says: 'It lives almost exclusively, I believe, by robbing birds' nests, devouring both the eggs and the young ones.' For a moment there flashed on me the gleam of a hope that its nest might be in that clump of tall, dark trees. 'What a prize its eggs would be!' I asked some country bumpkins, who were working in the fields, where they thought its nest might be; but they did not appear to see much reason to suppose that it indulged in the habit of making nests, or laying eggs. So I sat down and watched it, and very soon I saw that it was on the same errand as myself—viz., bird-nesting. Methodically it traversed the whole side of the hill, sailing low and scanning every bush and tree; then crossed the little valley in which I was standing, to beat the hill on the opposite side. Suddenly it stopped, circled quickly round a small tree and plunged into it legs foremost, as eagles always do except in pictures. It evidently missed its quarry, for it rose again, but it plunged once more into the tree and remained there. With the thermometer at 90° in the shade (and what in the sun!), I ran for that tree, thinking involuntarily of Falstaff larding the lean earth as he went along; but before I could get there, the eagle rose majestically and sailed away. Pushing the branches aside and looking in, I found a Bulbul's nest, with some eggshells and a spilt yolk, and at the foot of the tree were two feathers from a Bulbul's tail. At his first plunge he had tried to catch the sitting bird, but, like Tam o' Shanter's mare, it escaped with the loss of its tail. Then he sat down to breakfast on the eggs, with an eagle's beak for his spoon. It is no wonder that he could not eat them cleanly; and, indeed, it is one of the strangest things I know in nature that a bird so armed and equipped should feed on eggs. But it was not to feed on eggs that he was so armed and equipped. I feel sure that the Black Eagle furnishes an example of very recent degeneracy. This is an intensely interesting subject, but too large a one to enter upon today.

E.H. Aitken, A Naturalist on the Prowl, 1917

Douglas Dewar wrote on a variety of subjects, and his observations on contemporary Anglo-Indian culture are very insightful. Like EHA, Dewar's writing is free-flowing and liberally peppered with amusing anecdotes.

The Great Himalayan Barbet

Barbets may be described as woodpeckers that are trying to become toucans. The most toucan-like of them all is the great Himalayan barbet (*Megalaema marshallorum*). Barbets are heavily-built birds of medium size, armed with formidable beaks, which they do not hesitate to use for aggressive purposes. As regards the nests they excavate, the eggs they lay, the pad that grows on the hocks of young birds, and their flight, they resemble their cousins the woodpeckers. But they are fruit-eating birds, and not insectivorous; it is this that constitutes the chief difference between them and the woodpeckers. Barbets are found throughout the tropical world. A number of species occur in India. The best known of these is the coppersmith, or crimson-breasted barbet (*Xantholaema haematocephala*), the little green fiend, gaudily painted about the head, which makes the hot weather in India seem worse than it really is by filling the welkin with the eternal monotone that resembles the sound of a hammer on a brazen vessel. Nearly as widely distributed are the various species of green barbet (*Thereiceryx*), whose call is scarcely less exasperating than that of the coppersmith, and may be described as the word *kutur* shouted many times and usually preceded by a harsh laugh or cackle.

The finest of all the barbets are the *Megalaemas*. The great Himalayan barbet attains a length of 13 inches. There is no lack of colour in its plumage. The head and neck are a rich violet blue. The upper back is brownish olive with pale green longitudinal streaks. The lower back and the tail are bright green. The wings are green washed with blue, brown, and yellow. The upper breast is brown, and the remainder of the lower plumage, with the exception of the scarlet patch of feathers under the tail, is yellow with a blue band running along the middle line. This bright red patch under the tail is not uncommon in the bird world, and, curiously enough, it occurs in birds in no way related to one another and having little or nothing in common as regards habits. It is seen in many bulbuls, robins, and woodpeckers, and in the pitta. The existence of these red under tail-coverts in such diverse species can, I think, be explained only on the hypothesis that there is an inherent tendency to variation in this direction in many species.

A striking feature of the great Himalayan barbet is its massive yellow bill, which is as large as that of some species of toucan. Although the bird displays a number of brilliant colours, it is not at all easy to distinguish from its leafy surroundings. It is one of those birds which are heard more often than seen.

Barbets are never so happy as when listening to their own voices. Most birds sing and make a joyful noise only at the nesting season. Not so the barbets; they call all the year round; even unfledged nestlings raise up the voices of infantile squeakiness.

The call of the great Himalayan barbet is very distinctive and easy to recognise, but is far from easy to portray in words. Jerdon described that call as a plaintive *Pi-o, Pi-o*. Hutton speaks of it as *Hoo-Hoo-Hoo*. Scully syllabises it as *Till-low, till-low, till-low*. Perhaps the best description of the note is that it is a mournful wailing, *pee-yu, pee-yu, pee-yu*. Some like the note, and consider it both striking and pleasant. Others would leave out the second adjective. Not a few regard the cry as the reverse of pleasant, and consider the bird a nuisance. As the bird is always on the move— its call at one moment ascends from the depths of a leafy valley and at the next emanates from a tree on the summit of some hill—the note does not get on one's nerves as that of the coppersmith does. Whether men like its note or not, they all agree that it is plaintive and wailing. This, too, is the opinion of hillmen, some of whom declare that the souls of men who have suffered injuries in the law courts, and who have in consequence died of broken hearts, transmigrate into the great Himalayan barbets, and that is why these birds wail unceasingly *Un-nee-ow, Un-nee-ow*, which means 'injustice, injustice'. Obviously, the hillmen have not a high opinion of our law courts!

Himalayan barbets go about in small flocks, the members of which call out in chorus. They keep to the top of high trees, where, as has been said, they are not easily distinguished from the foliage. When perched they have a curious habit of wagging the tail from side to side, as a dog does, but with a jerky, mechanical movement. Their flight is noisy and undulating, like that of a woodpecker. They are said to subsist exclusively on fruit. This is an assertion which I feel inclined to challenge. In the first place, the species remains in the Himalayas all the year round, and fruit must be very scarce there in winter. Moreover, Mr S.M. Townsend records that a barbet kept by him in captivity on one occasion devoured with gusto a dead mouse that had been placed in its cage. Barbets nest in cavities in the trunks of trees, which they themselves excavate with their powerful breaks, after the manner of woodpeckers. The entrance to the nest cavity is a neat circular hole in a tree at heights varying from 15 to 50 feet. Most birds which rear their broods in holes enter and leave the nest cavity fearlessly, even when they know they are being watched by human beings, that their eggs or young birds are securely hidden away in the heart of the tree. Not so the *Megalœma*. It is as nervous about the site of its nest as a lapwing is. Nevertheless, on one occasion, when the nest of a pair of the great Himalayan barbets was opened out and found to contain an egg and a young bird, which latter was left unmolested, the parent birds continued to feed the young one, notwithstanding the fact that the nest had been so greatly damaged. The eggs are white, like those of all species which habitually nest in holes.

The Spotted Forktail

'Striking' is, in my opinion, the correct adjective to apply to the spotted forktail (*Henicurus maculatus*). Like the paradise flycatcher, it is a bird which cannot fail to obtrude itself upon the most unobservant person, and, once seen, it is never likely to be forgotten. I well remember the first occasion on which I saw a spotted forktail; I was walking down a Himalayan path, alongside of which a brook was flowing, when suddenly from a rock in mid-stream there arose a black-and-white apparition, that flitted away, displaying a long tail fluttering behind it. The plumage of this magnificent bird has already been described.

As was stated above, this species is often called the hill-wagtail. The name is not a particularly good one, because wagtails proper occur in the Himalayas.

The forktail, however, has many of the habits of the true wagtail. I was on the point of calling it a glorified wagtail, I refrain. Surely it is impossible to improve upon a wagtail.

In India forktails are confined to the Himalayas and the mountainous parts of Burma.

There are no fewer than eight Indian species, but I propose to confine myself to the spotted forktail. This is essentially a bird of mountain streams. It is never found far from water, but occurs at all altitudes up to the snow-line, so that, as Jerdon says, it is one of the characteristic adjuncts of Himalayan scenery. Indeed I know of few things more enjoyable than to sit, when the sun is shining, on the bank of a well-shaded burn, and, soothed by the soft melody of running water, watch the forktails moving nimbly over the boulders and stones with fairy tread, half-flight half-hop.

Forktails continually wag the tail, just as wagtails do, but not with quite the same vigour, possibly because there is so much more to wag!

Like wagtails, they do not object to their feet being wet, indeed they love to stand in running water.

Forktails often seek their quarry among the dead leaves that become collected in the various angles in the bed of the stream; when so doing they pick up each leaf, turn it over, and cast it aside just as the seven sisters do. They seem to like to work upstream when seeking for food. Jerdon states that he does not remember ever having seen a forktail perch; nevertheless the bird frequently flies on to a branch overhanging the brook, and rests there, slowly vibrating its forked tail as if in deep meditation.

Spotted forktails are often seen near the places where the *dhobis* wash clothes by banging them violently against rocks, hence the name dhobi-birds, by which they are called by many Europeans. The little forktail does not haunt the washerman's *ghat* for the sake of human companionship, for it is a bird that usually avoids man. The explanation is probably that the shallow pool in which the dhobi works and grunts is well adapted to the feeding habits of the forktail. I may here remark that in the

Himalayas the washerman usually pursues his occupation in a pool in a mountain stream overhung with oaks and rhododendron trees, amid scenery that would annually attract thousands of visitors did it happen to be within a hundred miles of London. Not that the prosaic dhobi cares two straws for the scenery—nor, I fear, does the pretty little forktail. As I have already hinted, forktails are rather shy birds. If they think they are being watched they become restless and stand about on boulders, uttering a prolonged plaintive note, which is repeated at intervals of a few seconds. When startled they fly off, emitting a loud scream. But they are pugnacious to other of their kind, especially at the breeding season. I once saw a pair attack and drive away from the vicinity of their nest a Himalayan whistling-thrush (*Myiophoneus temmincki*)—another bird that frequents hill-streams, and a near relation of the Malabar whistling-thrush or Idle Schoolboy.

The nursery of the forktail, although quite a large cup-shaped structure, is not easy to discover; it blends well with its surroundings, and the birds certainly will not betray its presence if they know they are being watched. The nest is, to use Hume's words, 'sometimes hidden in a rocky niche, sometimes on a bare ledge of rock overhung by drooping ferns and sometimes on a sloping bank, at the root of some old tree, in a very forest of club moss.' I once spent several afternoons in discovering a forktail's nest which I was positive existed and contained young, because I had repeatedly seen the parents carrying grubs in the bill. My difficulty was that the stream to which the birds had attached themselves was in a deep ravine, the sides of which were so steep that no animal save a cat could have descended it without making a noise and being seen by the birds. Eventually I decorated my *topi* with bracken fronds, after the fashion of Arry at Burnham Beeches on the August bank holiday. Thus arrayed, I descended to the stream and hid myself in the hollow stump of a tree, near the place where I knew the nest must be. By crouching down and drawing some foliage about me, I was able to command a small stretch of the stream. My arrival was of course the signal for loud outcries on the part of the parent forktails. However, after I had been squatting about ten minutes in my *cache*, to the delight of hundreds of winged insects, the suspicions of the forktails subsided, and the birds began collecting food, working their way upstream. They came nearer and nearer, until one of them passed out of sight, although it was within 10 feet of me. It was thus evident that the nest was so situated that what remained of the tree-trunk obstructed my view of it. This was annoying, but I had one resource left, namely, to sit patiently until the sound of chirping told me that a parent bird was at the nest with food.

This sound was not long in coming, and the moment I heard it, up I jumped like a Jack-in-the-box, but without the squeak, in time to see a forktail leave a spot on the bank about 6 feet above the water. I was surprised, as I had the day before examined that place without discovering the nest. However, I went straight to the spot from which the forktail had flown, and found the nest after a little searching. The bank was steep and of uneven surface. Here and there a slab of stone projected from it and pointed downwards. Into a natural hollow under one of these projecting slabs a nest consisting of a large mass of green moss and liver-worts had been wedged. From the earth above the slab grew some ferns, which partially overhung the nest. Across the nest, a few inches in front of it, ran a moss-covered root. From out of the mossy walls of the nest there emerged a growing plant. All these things served to divert attention from the nest, bulky though this was, its outer wall being over 2 inches thick. The inner wall was thin—a mere lining to the earth. The nest contained four young birds, whose eyes were barely open. The young ones were covered with tiny parasites, which seemed quite ready for a change of diet, for immediately after picking up one of the young forktails, I found some thirty or forty of these parasites crawling over my hand!

There is luck in finding birds' nests as in everything else. A few days after I had discovered the one above mentioned, I came upon another without looking for it. When I was walking along a hill-stream a forktail flew out from the bank close beside me, and a search of thirty seconds sufficed to reveal a well-concealed nest containing three eggs. These are much longer than they are broad. They are cream-coloured, mottled, and speckled with tiny red markings.

DOUGLAS DEWAR, *Birds of the Indian Hills*, 1915

The Naturalist in a Railway Train

In most parts of India a kind of 'general post' of officials takes place at the commencement of every cold weather. The authorities seem suddenly to discover that the majority of public servants are stationed at unsuitable places, and thereupon seek to remedy this state of affairs, to the great profit of the railway companies. Having been an active participator in the latest 'general post', I have been afforded an excellent opportunity of studying nature from the interior of a railway carriage. It must, in truth, be admitted that there are many worse points of view, for one sees an astonishing amount of animal life from a moving train.

The railway has now become quite an important factor in the life of many birds, chiefly owing to the fact that the iron road is accompanied by telegraph wires. When first erected, these caused the death of many an unsuspecting bird. The fowls of the air enjoy so vast a space, free from obstacles, in which to move about, that when flying they are not obliged to look very carefully where they are going. If a bird wishes to reach a certain place, it forthwith takes to its wings and makes a bee-line for its destination. Its chances of colliding with other birds are infinitesimal, it is not afraid of running up against a lamp-post, tripping up over a stone, or being run over by an omnibus or cab, so it puts down its head and lets itself go in much the same way as an athlete sprints a hundred yards race.

Thus it happened that when the telegraph was first erected many a feathered creature killed itself by coming into violent contact with the wires, which, for a time, were veritable death-traps. Calamities, such as these, are now happily things of the past.

Birds profit by experience. They have learned to avoid the treacherous wires during flight. They have further discovered that a telegraph wire forms a very comfortable perch, which that incomprehensible and eccentric being—man—has erected for their special benefit. Thus it happens that the traveller by railroad sees a succession of birds perched upon the message-bearing wires, as though they were sitting for their photographs, for the passing of the train does not perturb them in the least. A telegraph wire is, however, too attenuated to form a comfortable perch for some birds. For such there are the poles and insulators ready to hand, and of these the hawks and kites are now slow to avail themselves.

Birds which feed upon flying insects are particularly addicted to the telegraph wires, for these latter constitute an ideal point of vantage from whence the bird can look out for its quarry. Thus king-crows (*Dicrurus ater*) are to be seen distributed along the whole extent of every railway, sitting on the wires until an insect comes within range, when the drongos at once take to their wings and give chase.

It is amusing to notice how the king-crow always seeks shade when the sun is very hot. In the middle of the day fully 80 per cent of the king-crow *habitués* of the telegraph wire will be seen seated quite close to a pole, so that its shadow falls upon them.

The roller (*Coracias indica*), or blue jay, as it is more commonly called in India, is another bird which is very partial to the electric telegraph. It sits indiscriminately on either wires or poles.

Doves, too, are very fond of resting on the wires. They are not insectivorous birds, and are, consequently, not on the look out for prey, but love to sit in the sun, especially in the early winter morning when the air is still chilly, and in this attitude they ponder over the problems which agitate the feathered world. The pretty little bee-eater (*Merops viridis*) is another frequenter of the telegraph wires. Very beautiful he looks in his green dress as he sits facing the line, and still more striking is his appearance when he makes a sudden dash at some Lilliputian quarry, for, when flying in the glare of the sun, his plumage assumes a golden hue.

The birds perched on the telegraph wire, although they absorb the greater part of one's attention, form but a small fraction of the species to be seen during a railway journey. It is no exaggeration to assert that a traveller by rail from Peshawar to Madras should, aided by a good field-glass, be able to distinguish fully one-third of the commoner birds of India.

The train passes through most kinds of country. It jogs along over barren *usar* lands, across fertile fields coloured emerald-green by the young shoots of the luxuriant crops, over broad rivers, past *jhils* great and small, through bushy jungle, amid long feathery grass, through forests, among bare rocky hills and green undulating down-like country. Each of these tracts has its characteristic species. Now a flock of mynas (*Acridotheres tristis*) comes into sight, chattering with delight over some newly-discovered field rich in food. These disappear and a pair of sarus cranes (*Grus antigone*) absorb one's attention. The sarus is a strange bird, which, like an Englishman, seems to take its pleasures sadly; it invariably looks depressed, although in reality it is perfectly happy in the company of its spouse. The crane and his wife form an inseparable and devoted couple. When one is taken and the other left, the survivor is said soon to die of grief at the loss of its mate.

Scarcely have these tall creatures vanished from sight than a flight of birds of a very different feather comes into view—a screeching crowd of green parrots (*Palaeornis torquatus*) on their way to commit dacoity in an orchard of ripening fruits. The train now wends its weary way through a tract of marshy country, where, here and there, a Paddy Bird (*Ardeola grayii*) may be seen, lazily gazing into the water of some murky jhil. Nearby are some duck and coots swimming on the surface of another sheet of water. Not far removed from them is a stork, and overhead are flying a number of white egrets (*Bubulcus coromandus*) and other *kuchnes*, disturbed by the noisy train.

Once again the land becomes parched, and a hoopoe (*Upupa indica*), Solomon's brilliant messenger, is seen making its way with undulating laboured flight.

And so interminable numbers of birds appear in rapid succession.

Nor are mammals wanting. These, of course, are neither so numerous nor so conspicuous as the birds. Apart from the domesticated animals, monkeys, and blackbuck (*Antilope bezoartica*) are the mammals most frequently seen from a railway train in northern India. The latter are now, alas, far less frequent than they used to be.

Writers of fifty years ago speak of the vast herds of these elegant herbivora which abounded in those days. Such multitudes are almost unknown is most parts of Upper India in this twentieth century. The companies are now few and far between, and so sadly have they diminished in size that a tiny herd, consisting of one solitary dark-skinned buck, surrounded by his little harem of fawn-coloured does, has become no uncommon sight.

As the grey mists of dawn are lifting, or when the sinking sun has become transformed into a great fiery ball, seen through miles of dust and smoke, jackals may here and there be observed sneaking furtively back to their 'earth', or from it, on their way to help their comrades from a search-party which will presently render the night hideous by its unearthly yells.

The fauna of the railway station is not devoid of interest. There *is* such a fauna, for on this little earth of ours there is no nook or cranny in which Nature has not placed some of her children. Directly the iron horse pulls up, a crowd of kites may be seen soaring overhead, waiting for some scraps of food which a passenger will assuredly cast away. Needless to say, the crows are also on the war path, and, as they hang about, most impudent beggars, close to the carriage wheels, they get the pick of the food which is thrown out.

These bold birds, however, are not dependent on the charity of man; they help themselves, being obviously disciples of Dr Smiles, whose book, *Self-Help*, is so popular in India. A goods train loaded with sacks of grain pulls up at a station, and is at once invaded by crows, who proceed to bore with their powerful beaks holes in the sacks, through which they abstract the corn.

The enumeration of the fauna of the railway station would be incomplete without mention of the ubiquitous sparrow (*Passer domesticus*). Then there is the half-starved pariah dog, who is a regular institution at every wayside station, attending all trains. Experience seems to have taught him that charity is most rife among Europeans, for he usually takes up a position on the platform in front of a carriage occupied by them; but even their charity appears to be very uncertain, for his attitude is suppliant, he wags his tail in a half-hearted manner, he gives it the undecided motion that denotes hoping against hope. His ribs are very conspicuous objects, and the wistful look in his eyes makes one feel almost sorry that one's baggage does not include an assortment of juicy bones.

DOUGLAS DEWAR, *Bombay Ducks*, 1906

R.S.P. Bates, an officer in the British Indian Army, wrote extensively on Indian birds. His works include Bird Life in India and Breeding Birds of Kashmir (with E.H.N Lowther). Here he describes the famous heronry at Vedanthangal in Tamil Nadu.

A South Indian Heronry

Some years ago when I was stationed at St Thomas Mount just outside Madras, my attention was drawn to Vol. III of the second edition of *Hume's Nests and Eggs of Indian Birds*, wherein on pages 238 and 239 is the description by an anonymous writer of a certain heronry at a place called Vaden Thaugul not far distant from the Mount. I at once decided to search out this village to see if the place was still in existence, since heronries have a habit of persisting for a great many years and I had hopes of finding it little changed in spite of the lapse of four decades. My many adventures in search of it and after I had rediscovered it, so to speak, are contained in the following pages, but first as some of my readers may not be in possession of Hume's and Oates' admirable work, I append *in toto* the said description of this breeding ground.

About fifty miles from Madras and twelve miles from Chingleput in a south-easterly direction is a small village called Vaden Thaugul ...

The Vaden Thaugul tank is situated north-north-west of the Carangooly Fort, and is three and a half miles distant in a direct line from the Great Southern Trunk Road.

The bund, whose greatest height is twelve feet, commences from a piece of high ground near the village, runs for a distance of about six hundred yards in a south-easterly direction, then takes a sharp turn almost at a right angle, and terminates in high ground about two hundred yards further on. The waterspread is limited on the north-east by slightly rising ground overgrown with low jungle, and on the east-south-east by high gravelly and rocky ground. The area comprised in the tank is about thirty-five acres.

From the north-east to the centre of the bed of the tank there are some five or six hundred trees of the *Barringtonia racemosa*, from about ten to fifteen feet in height, with circular, regular, moderate-sized crowns, and when the tank fills, which it does during the monsoons, the tops only of the trees are just visible above the level of the water.

This place forms the breeding resort of an immense number of water-fowl—Herons, Shell-Ibises, Ibises, Water-Crows or Cormorants, Darters, and Paddy-birds, etc., make it their rendezvous on these occasions.

From about the middle of October to the middle of November small flocks of twenty or thirty of some of these birds are to be seen coming from the north to settle here during the breeding season. By the beginning of December they have all settled down; each tribe knows its appointed time, and arrive year after year with the utmost regularity within a fortnight later or earlier, depending partly on the seasons. They commence immediately by building their nests or repairing the old ones preparatory to depositing their eggs. When they have fully settled down, the scene becomes one of great interest and animation.

During the day the majority are out feeding, and towards evening the various birds begin to arrive in parties of ten, fifteen, or more, and in a short time the trees are literally covered with bird-life: every part of the crown is hidden by its noisy occupants who fight and struggle with each other for perches. Each tree appears like a moving mass of black, white, and grey, the snowy white plumage of the Egrets and Ibises contrasting with, and relieved by, the glossy black of the Water-Crows and Darters and by the grey and black plumage of the Shell-Ibises.

The nests lie side by side touching each other, those of the different species arranged in groups of five or six, or even as many as ten or twenty, on each tree.

The nests are shallow, and vary in inside diameter from 6 to 8 inches according to the size of the bird.

The Ibises do not build separate nests, but raise a large mound of twigs and sticks shelved into terraces as it were, and each terrace forms a separate nest; thus eight or ten run into each other. The Shell-Ibises sometimes adopt a similar plan.

The whole of the nests are built of sticks and twigs, interwoven to the height of 8 or 10 inches, with an outside diameter of 18 to 24 inches; the inside is slightly hollowed out, in some more and in others less, and lined with grass; reeds and quantities of leaves are laid on the nests. During this time the parent birds are constantly moving on the wing, backwards and forwards, in search for food, now returning to the young loaded with the spoil, and again, as soon as they have satisfied their cravings, going off in search of a further supply. About the end of January or early in February the young are able to leave their nests and scramble into those of others. They begin to perch about the trees, and by the end of February or the beginning of March those that were hatched first are able to take wing and accompany their parents on foraging expeditions; and a week or two later, in consequence of the drying-up of the tanks in the vicinity, they begin to emigrate towards the north with their parents and friends, except perhaps a few whose young are not as yet fledged, and who stay behind some time longer. Thus in succession the different birds leave the place, so that it is completely deserted by the middle of April, by which time the tank also becomes dry; and the village cattle graze in its bed or shelter themselves under the trees from the scorching heat of the mid-day sun, while the cow-boys find amusement in pulling down the deserted nests.

The directions given here seem sufficiently comprehensive, so to find it should not have proved particularly difficult, but unfortunately I was suddenly transferred to Trichinopoly, which delayed matters until March of the following year. On March 5th, 1926, it so happened that I was called to Madras to give evidence in a court case against a bright youth who had forged my initials on a stray railway warrant in so many places that it resembled a Chinese puzzle. Having finished at the court by 11 a.m. I decided then and there to make an attempt to find the heronry, so rushed to the railway station only to be told that no available train went further than Chingleput. At Chingleput not even a tonga was procurable, let alone a car, so I had perforce to squash myself into a springless bullock-bandi drawn by an animal hardly as large as a donkey, and in this I optimistically set out about 3 p.m. to cover the supposed 12 miles to Vaden Thaugul. And now a further problem presented itself. Carangooly Fort I was able to fix, as there is now a railway station, 13 miles south of Chingleput, bearing the not very dissimilar name of Karunguzhi, close to which I ascertained that the remains of a fort did verily exist; but of Vaden Thaugul no information whatsoever was obtainable. Alas, I was without a map, and could only recall the fact that Vaden Thaugul was said to be 12 miles south-east of Chingleput which placed it in my mind somewhere between Karunguzhi and the coast and to the east of the Grand Trunk Road. Had I remembered that it was also said to be north-north-west of the Fort, I might have realized that a slip of the pen had crept in, as it could not possibly be south-east of Chingleput and yet westwards of Karunguzhi. As it was, I found myself actually on the outskirts of that village with the sun already setting, rapidly coming to the conclusion that Hume's correspondent must have had very good reasons indeed for remaining anonymous. At this point, however, I fell in with a wizened veteran carrying a muzzle-loading weapon every bit as old as himself who professed to be a shikari, and I elicited from him that some 7 miles away to the west of the road was a village named Vedan Thangal where waterfowl did actually collect. Mark the difference; *Vedan Thangal* and not Vaden Thaugul. Many times afterwards, when I was within a mile or so of the heronry, I purposely asked the way to Vaden Thaugul and always got a blank look in return. Times change and so do pronunciation and spelling, though I admit I am not a Tamil scholar, so perhaps the fault lies with myself.

It was of course impossible to follow up the clue then as night was already falling, and it was imperative for me to catch the Ceylon Mail right back at Chingleput as, alas, it did not stop at Karunguzhi. Twenty-four miles of a springless bullock-cart with a yelling Jehu keeping his reluctant half-starved animal at a shambling trot saw me back at the station almost wishing heronries had never been thought of. And so ended my first effort.

Again I was transferred—to Lahore, a far cry from Madras, but fortunately I had once more to go to Trichinopoly for the space of two months to a Territorial Battalion. November 30th, 1926 therefore saw me a second time at Chingleput with a whole day at my disposal before the training started. Early that morning I took a train to Karunguzhi and there found a boy who professed to know English and who said he could take me to Vedan Thangal. His English turned out to be scant in the extreme, and no matter what question I put to him I invariably got an emphatic *yes*. In return so the information I gleaned from him about the heronry was unreliable to say the least of it. As the rains had failed and from the train I had noted that a great many tanks were quite dry, I feared that once more I might draw a blank, but no matter whether I gently suggested that we would find no birds or yet many birds at the spot, my self-appointed guide immediately agreed with either proposal. We therefore plodded on. After crossing the main road at the very spot where I had met the doddering shikari in the previous March, we came upon Madurantakam Tank, a sheet of water some miles square with an arm about half a mile wide running out in a north-easterly direction. The water not being so deep as usual, we now proceeded to wade laboriously across this arm, sometimes nearly up to our waists, along a submerged bund or causeway. We then skirted the main tank for another mile or so and traversed a plantation of young casuarinas. Emerging therefrom, the object of my search at last appeared. There was the bund as described with the village at one end, the slightly higher ground at the other, and in the eastern half of the tank-bed the five or six hundred leafy trees of the *Barringtonia racemosa*. Water, alas, there was none, or rather a few shallow pools with one flock of three dejected Teal floating upon the largest one. Of other birds there was no sign. To say I was bitterly disappointed would be untrue, as I had at least the satisfaction of knowing that I had undoubtedly hit upon the right place, but an investigation of the trees made it hard to believe that birds ever bred there, as not the slightest vestige of an old nest could I find. However, 'when rain coming, birds coming', as translated by my guide, seemed to be the tale of all the ryots working in the fields around, so my spirits rose, especially as it was undoubtedly an ideal place for a heronry. An examination of the tree-trunks showed a well-defined watermark barely three feet from ground level in the case of the trees furthest from the bund, but seven or eight feet up those near the centre of the tank. The shape of each tree was most remarkable. They looked like so many giant umbrellas, the branches and foliage being cut clean off as it were, at water level. In height they varied from, I should say, 15 to 25 feet and were so close together that only in a few places were gaps of any size to be found in the undulating mass of foliage. Close to the edge nearest the tank's centre there was a well-defined lane running through them at right angles to the bund. There were also a few isolated trees separate from the main mass. No further useful information being forthcoming, I retraced my steps, leaving with the boy a couple of postcards to be posted to me at intervals of a month. These he actually posted on the due dates, much to my astonishment, but each bore the information that no rain had fallen so no birds had appeared.

Once more I left Madras behind, and this time, as I thought, with little prospect of seeing it again, but after enjoying a furlough at home and spending a year in the Khyber, to my great satisfaction my Battalion was actually posted to St Thomas Mount, so that in September 1928, just before the north-east monsoon broke in a furious storm, I found myself in my old bungalow there. Four days of incessant rain had practically filled all the tanks in the neighbourhood, so I felt certain that if 'rains coming, birds coming too' was really a fact, I would at last find the heronry occupied. It was not however until November 26th that I was able to go there. From the map I now possessed, I had noted a village called Pudupattu, barely two miles from Vedan Thangal, up to which I could get a car, so I arrived in the vicinity of the tank with considerably less trouble than previously. On the way from Pudupattu I saw a couple of Egrets and later a single Openbill, and when at length the tops of the trees in the tank came into view over the rising ground to the east, I at first saw no birds other than a flight of Cormorants which swooped over the bund and disappeared from view behind the crests of the further trees. However, when I came in sight of the water a very different scene appeared. Some hundreds of snowy-white Egrets were standing around the margins of the tank; many were following the ploughs in the rice fields, while a hummock projecting from the water was white with birds, and at the edge of a narrow channel a couple of Spoonbills and an Openbill were leisurely paddling about. Even now the existence of nests was still open to doubt, but when I advanced to the bund and was able to look thence across the strip of water on to those trees which had at first been hidden from me, my excitement knew no bounds, for from water level to their summits, the trees without exaggerating were completely covered with birds. Here were a couple jet-black with Cormorants; there another snow-white with Spoonbills. A colony of Open-bills occupying the crest of one of the tallest trees was outlined against the sky and the blue-grey of many Herons also impressed itself upon me; but on the whole in the words of Hume's informant 'each tree appeared like a moving mass of black, white, and grey,' but the white was not that of the Egrets who were strangely aloof standing about in groups quite separate from the other species on the isolated belt of trees. The birds to be seen besides these Egrets were Openbills, Spoonbills, Cormorants, Grey Herons and Night Herons, and even at this distance it was evident that they were hopelessly mixed together, although a definite area held a preponderance of Spoonbills and another of Openbills. The scene was not exactly one of feverish activity, though occasional bands of Cormorants swung in with a loud swish of fast-moving wings low over the toddy palms lining the bund. A Snakebird, a few Cormorants, and also Openbills, were noted coming in with sticks for their nests, but the majority of birds appeared to be standing about doing nothing whatsoever. It therefore seemed to me that laying must be the main occupation of the moment. However, the conclusions I came to from thus watching the heronry at a distance nearly all proved wrong; so I will not risk befogging my readers as to the true state of affairs by discoursing upon them. Of noise there was comparatively little, merely a subdued murmur as of a busy throng, a crowd whose number I found it an impossibility to gauge, though after further visits I estimated it at being not less than five thousand birds. One of my friends insisted on thirty thousand as his estimate, so I certainly think that my computation can be taken as in no way an exaggeration.

The occupied trees were those in the deeper water; so, a boat being essential, I went to the village to see what chances there were of obtaining one. It transpired that the chances were exactly nil. There was no boat in the village and I was given to understand that none was procurable in the neighbourhood, not even from the Madurantakam tank, added to which I was politely told that they did not at all like the idea of my disturbing the birds. On my inquiring the reason, the village Karnam produced a stained sheet of paper which proved of the utmost interest. I have often cogitated over the age of some of these heronries, and here was a document which, if followed up, might lead to my finding out the age of this one. Unfortunately the writing upon it was in Tamil, but it bore three illegible though obviously English signatures and was dated 1858. I wonder if this very copy was shown to Hume's informant? There was a man in the village who purported to be the chowkidar of the tank. He knew a very limited amount of Urdu and through him I learnt that this document gave to the village the right not only to stop shooting on or near the tank but to prevent interference with the birds in any form. It seemed that it had been issued by a Collector of the district. On my return to the Mount I therefore approached the then Collector of Chingleput, but it appeared that no trace of any such document ever having been issued existed in his office. However, shortly afterwards, when on tour at Chingleput, he very kindly offered to go with me to the place, and on March 1st, 1929 we went there together. The paper turned out to be a copy of a yet older document issued by one Mr Place who was Collector of Chingleput from 1796 to 1798, so the age of the heronry had now risen to a minimum of 130 years. To trace its existence further back will, I fear, prove impossible, as this part of the country came finally into our possession about 1780 only after Haidar Ali and Tippoo had been once for all expelled from the district. It is of course obvious that the order would never have been issued had not the village been already in a position to consider itself the special guardian of the heronry. In other words its existence stretches back into a dim and distant obscurity. The Collector also enlightened me as to the reason for this guardianship. I fear me a love of bird-life does not enter into the picture: the reason is purely a mercenary one. Owing to the water of the tank being considered to possess high fertilizing properties, the value of the land irrigated by it is assessed for revenue purposes at a higher rate than other land in the vicinity. Personally I could believe anything of such water. By March its colour and consistency are that of pea soup and its odour compares favourably—save the mark—with the most evil-smelling of Harrogate's medicinal springs. One thing I cannot quite understand: I know the Madrasi is no naturalist, but it appears to me most strange that the existence of a heronry of such antiquity and superlative interest should be utterly unknown to the European population of the district, many of whom do take an interest in bird-life and to most of whom Madurantakam tank, from which Vedan Thangal is not a mile as the crow flies, is well known.

There is one other point of interest in connection with its history. The village has a perpetual complaint to make, this being that, except in the very good rains, water does not, or is not allowed to, flow into the tank—the reason for this I do not know but doubtless the Irrigation Department know all about it. The point however is this: it is by no means uncommon for the tank to remain dry, as it did in 1926, and at times even a couple of cold seasons may pass without giving the birds a chance of breeding there. Has this always been the case? If so, how is it that the heronry has been able to persist? Lastly, where do the birds go when the tank is dry? These are questions I cannot answer, but with regard to the last one, all my inquires and expeditions have failed to bring forth the existence of any other heronry in the district, though lately I have had suspicions that there may be one roughly 40 miles out of Madras near the Bangalore road.

R.S.P. BATES, *Bird Life in India*, 1931

B.B. Osmaston (1868–1961) *was in the Indian Forest Service and spent much of his time in the Andamans. A fine observer and writer as is evident from his reports, he wrote the book* Wildlife and Adventures in Indian Forests.

Edible Birds' Nests

The Swiftlets *Collocalia unicolor* responsible for the edible nests are small birds, no bigger than a sand martin. They frequent the coast-line in the Andamans and live in communities, their nests being stuck on the roofs of caves in the rock. The nests are very peculiar, being composed of a semi-transparent colourless gelatine which is secreted from special salivary glands by the female bird. The secretion is first applied in a tacky fluid condition but it rapidly hardens on exposure to the air. The nest when completed is a half-saucer, fixed against the sloping roof of the cave in a horizontal position. The diameter of each nest is only about 2 inches. The Chinese consider them a great delicacy and buy them by paying their weight in silver. An average nest weighs about 1/2 oz. so the Government Royalty came to about Rs 1 per nest.

I told my cook to make me some soup from a few nests and I found it quite tasteless and uninteresting. The contractor who had purchased the right to collect these nests in the Andamans removed one crop in the beginning of the breeding season. A fortnight later another crop was allowed to be removed but after that no more.

The third lot usually contained traces of blood mixed with the inspissated saliva showing that there was an undue strain in the production of three nests in quick succession. The contractor's men knew most of the caves patronised by these birds. They thought they knew them all, but I discovered a small well-concealed cave on the Cinque Islands which had escaped their vigilance. It contained 50 beautiful little nests each with two fresh eggs, which is the full complement. Another cave, with much larger dimensions, had its entrance only just above the water level. The opening was alternatively open and closed at the fall and rise of the swell, and it was fascinating to watch the birds waiting their time to dart in and out as opportunity offered. To explore this cave was beyond me as there was deep water below and it was also nearly dark. I think it must have been a sanctuary as far as the contractor was concerned.

Another small kind of Swiftlet used extraneous material such as fine sea-weed, leaves, etc. glued together with mucilage to form its nest, but this nest was not edible. Large numbers of these birds built their nests under the wooden roof of the Chatham sawmills, and they darted in and out without apparently being the least disturbed by the continual roar of the machinery and the screeching of circular and frame saws. The material largely used, in addition to the mucilage was human hair. When the convicts on the island had just had their annual haircut, the birds found it a very suitable material. In one nest examined by me the bird had inadvertently formed a noose of one long hair and I found her with her head in the noose, suspended lifeless below the nest.

B.B. OSMASTON, *Newsletter for Birdwatchers*

To the Englishmen who ventured into the dark and mysterious Indian Subcontinent, the East must have seemed topsy-turvy at the first encounter. The familiar little Robin was red under the vent rather than on the breast! No wonder, the Hawk-Cuckoo's incessant calls, which heralded the season of love for Indians, annoyed the Europeans so much that they gave it the name 'Brain-fever Bird'. Thomas Bainbrigge Fletcher and Charles McFarlane Inglis's article on the Brain-fever bird illustrate the point.

The Common Hawk-Cuckoo or Brain-fever Bird

The common name 'Hawk-Cuckoo' conveys a good description of this bird, as it is really a Cuckoo which looks very like a hawk. It is about the size of a Myna, but with a longer tail, greyish-brown in colour, whitish beneath, the breast tinged with pink, each feather with darker cross-bars, eyes and legs brilliant yellow. When on the wing, it looks very much like a small hawk but, when it alights, it at once assumes a slouching, cuckoo-like attitude, with the wings dropped forward so as to touch the perch and the tail slightly raised and expanded, thus presenting an aspect very different from the compact and alert look of a hawk. Seen thus, at rest, this bird can hardly be mistaken for a true hawk, as it has the furtive, peering ways of common Cuckoos, constantly jerking itself from side to side and puffing out its throat.

The appearance of the Hawk-Cuckoo is probably less familiar to most people than is its note, which has aptly earned for it the notorious title of the 'Brain-fever Bird'. Our Indian gardens and groves contain many sweet-voiced singers amongst their avian denizens and a few whose voices are less grateful to the ear, but there is not one whose notes consist of such ear-splitting and nerve-racking cries as do those of the Brain-fever Bird. With the most annoying persistence and reiteration this bird repeats its cry, which bears a remarkable resemblance to the word 'brain-fever' repeated in a piercing shriek running up the scale. The cry may also be written as 'Pipiha' and in some districts the vernacular name of the bird is given as *Pupiya*. Another rendering of the call, which includes the overture preceding the triple note, is, 'O lor'! 'O lor'! how very hot it's getting—we feel it, *we feel it*, WE FEEL IT'.

The call is extremely loud and shrill and can be heard—indeed, it cannot but be heard—within a radius of several hundred yards, but one of the most annoying things about it is its intermittent character. The human ear soon becomes accustomed to any continuous and uniform kind of noise. One becomes so accustomed to the buzz of a dynamo that one awakens at once if it stops. The Coppersmith *tonk-tonking* in the garden all day is hardly heard consciously unless one listens for it. But the shrieks of the Brain-fever Bird burst their way without ceremony into one's inner consciousness, whether awake or asleep, and one cannot help but hear them. 'We feel it, *we feel it*, WE FEEL IT' go the cries, up and up the scale, and then suddenly stop, and one hopes fervently that this fiend in bird's plumage has burst its throat or at least flown away out of earshot. But no; after a short interval it begins again and may continue for hours at a stretch. Very often the performance commences just at dusk, when it has got too dark to make out the culprit, and lasts all night without intermission. When this sort of thing takes place on a really hot night, the victim, who is attempting to woo sleep after a hard day's work, may well be excused if the first dim dawn sees him sallying forth on vengeance bent. But vengeance is not always easy to attain. The bird usually perches high up in a tall tree and keeps so still and is so inconspicuously coloured that, even when its shrieks locate the very branch whereon it is sitting, it is not always easy to make out. Furthermore, it is wary and often flies off as soon as it sees that it has been detected. There are, however, usually only a few individuals in each locality and a comparatively small reduction in numbers works wonders in abating the nuisance. The call being very penetrating, it often happens that these birds call to one another across a distance of perhaps half a mile and, by shooting one bird forming a link in the chain between others on either side of it, the chain is broken, and a blessed peace reigns once more, at least until another bird invades the immediate neighbourhood. In Bihar the call of this bird coincides with the approach and duration of the hot, dry weather before the monsoon; occasionally it may be heard as early as in December but more usually commences about February and is continued, becoming more frequent and continuous, until the rains break, when there is a welcome cessation for a few months. In other districts this may not be so; thus, as regards Calcutta, Cunningham states that 'there is hardly any season at which their characteristic notes may not occasionally be heard; but, as a rule, it is during the rainy months that they are most frequent, so that the designation "hot-weather bird", that is often applied to the species in other parts of the country, is hardly applicable to it in Calcutta'.

T.B. F<small>LETCHER</small> and C.M. I<small>NGLIS</small>, Birds of an Indian Garden, 1936

C.H. Donald worked in the fisheries department in British India. His article reprinted below enlightens us about the importance of observing living birds, especially when it comes to identifying raptors.

The Flight of Eagles

When I received an invitation from our editors to write a note on Eagles for the *Journal* for this, its fiftieth birthday, I accepted with the greatest pleasure for well did I know what joy was in store for me. Would I not be going over some of the happiest days in my life in which eagles, falconry and the Bombay Natural History Society were all inextricably woven into a glorious background of the vast virgin forests of Bhadarwa and Kashmir, where I seemed to be the little tin god in command of a world of forest coolies, with plenty of leisure on my hands?

I had already embarked on falconry in the plains of India and had been most fortunate in securing the services of two old *bazdars* (falconers) who had served my father in Hissar, during the troublous times of 1857. It was impossible to live long in the company of such enthusiasts without being bitten to the bone with their craze. They were brothers and rejoiced in the names of Jhanda and Balunda, respectively. White-bearded old Jhanda, who said he was not yet quite seventy, usually stayed behind and looked after my team of falcons, and incidentally did most of the training, while *little* Balunda—a mere boy of some 50 odd summers—accompanied me everywhere and was my constant companion and *ustad* or tutor. With eyes like one of the falcons on his wrist, that man missed nothing which flew or ran, and from him I learnt lessons which have stood me in very good stead for over half a century, of how to recognise the different birds of prey by their flight, almost as far as you could see them. With a few tips from Balunda I soon discovered the process as not only interesting, but amazingly simple. It just came, and gradually you found yourself recognizing at a glance, confidently, bird after bird as it flew past or soared high up in the sky.

One day the supreme test came; we were up at about 11,000 ft. and above tree level, when Balunda came to a stop and said in awed tones, 'Sahib, what is that?'

I followed his gaze and there, a thousand feet or so above us soared a huge bird on motionless pinions. '*Burra Jumbiz*!' I exclaimed, unable to think of anything else for a very dark and large bird. 'No, no, Sahib, that is no *Jumbiz* but a mighty hunter which I have never seen before.'

I marvelled. The old man admitted he had never seen the bird before yet recognized it as a mighty hunter, a thousand feet above him. I looked and looked again. I have seen that bird before many times, in different localities but now for the first time saw what Balunda meant. The flight was entirely different to that of the *Jumbiz* or Imperial Eagle. Forceful and resolute, yet light and buoyant. 'Balunda, call him down and I'll shoot him and find out what it is.' 'That is easy, Sahib. You get in under that bush and I'll have him down in a couple of minutes.' Out came Balunda's ubiquitous bag and from it he extracted a dead pigeon, a lure composed of crows' wings attached to some 10 ft. of string. Then taking the falcon on his wrist he removed the hood and placed the bird on a conspicuous boulder and rushed back to hide under a bush to my right. Next he threw out the lure, giving the customary call for the falcon to come and bind to it, which she did immediately, and Balunda proceeded to draw her in, still holding the lure, which made her flutter not a little. None of this drama was lost to those all-seeing eyes up in the sky. 'He is coming, Sahib', whispered Balunda, a fact I had noted for myself a few seconds previously. The falcon saw her danger and picking up the lure flew under Balunda's bush just as I fired at the black ball descending at umpteen miles per hour. No. 1 shot did the trick, and the great bird fell with a dull thud, dead, where the falcon had been a couple of seconds before.

Balunda rushed to it, turned it over, and pointed to the enormous foot and claws. 'Did I not say he was a great hunter, Sahib? That bird could kill a sheep or even a man.' 'This must be the bird the shepherds call a *Muriari* of which I have heard a lot in the last few months,' said Balunda, and I too had heard a good deal of its depredations among the shepherds' flocks. But as time went on and I persisted in my search for correct information, the assertions of its killing sheep and lambs became more and more vague; and in some 50 years of wandering all over the Himalayas I do not think I met with more than half a dozen men who had *actually seen* this eagle attack a sheep, though I had myself seen one kill a tahr.

We wrapped him up in Balunda's sheet and made for camp where the eagle was skinned and filled with moss and lichen, and on the following morning the skin was on its way to Bombay. A long week of suspense and, at long last, a reply from the Hony. Secretary, acknowledging receipt of the 'lovely skin' and informing me that the bird was an Imperial Eagle. How could I break this to Balunda, the more especially that after a few talks and explanations from him as the bird, to say nothing of those claws, I was now very much of his way of thinking. By return post I replied and thanked the Hony. Secretary for his letter and asked for another examination, as I was sure the bird was *not* an Imperial Eagle whatever else it might be. Back came a reply that a committee of the leading ornithologists, then in India, had gone carefully over the bird and come to the unanimous conclusion that the bird *was* an Imperial Eagle. This was getting serious, so what should we do next? I again replied very politely and asked if it would be possible to send the bird to the Natural History Museum, London. It went, and three months later came the reply: 'The bird is a young Golden Eagle in transition stage of plumage.' Good old Balunda! He had the unfailing key to the identification of accipitrine birds—Flight. A falconer, born and bred from many generations of men who had watched every phase of flight, and had not confined themselves to their hawks and falcons.

Now it must not be supposed that I have written the above introduction merely to praise Balunda, but when I quote from a well-known book, which many members must have read, a paragraph which completely misled me, and must have similarly put off many a young tyro like myself, thirsting for knowledge, it will be conceded there is some method in my madness. The paragraph reads: 'As far as I am aware this bird is of such excessive rarity in the Himalayas, south of the snows, as scarcely to deserve a place in our lists. Every so-called Golden Eagle which has yet been sent to me, has proved to be *A. imperialis* in the dark 3rd stage of plumage.' The author had, at Kotgarh (Simla Hills), a regular establishment for shooting and preserving birds, from which he received over a thousand specimens and who had special injunctions to shoot all large eagles. From them he apparently received several Imperial Eagles but not one single Golden.

Later, he modifies the above in his 'Nests and Eggs', Vol. III, pp. 130–1, by saying the Golden Eagle occurs and breeds sparingly in the Himalayas from Sikkim to Afghanistan. In the eastern and central portion of this tract it is confined to the immediate neighbourhood of the snowy ranges, but in the extreme, N.W. it comes nearer down towards the plains.

Another well-known ornithologist once wrote to inform me that in 20 years collecting his collectors had never found a Golden Eagle in Kashmir. Some months later I happened to be in Srinagar, and paid a visit to the museum, and the very first thing, on entering the door, I was confronted by was a magnificent specimen of a female Golden Eagle, in its first plumage, labelled '*Aquila heliaca: The Imperial Eagle. This bird sometimes catches Chukor.*'

Further comment seems superfluous, except the emphasis on the fact that if an illiterate old man is able to identify a bird which he was never even seen in his life, at about 1,000 ft above him, as a mighty hunter and *not* an Imperial Eagle, it is obvious there must be something in his system of identification which is entirely lacking in the make-up of most good ornithologists; and that something is the key in the study of the birds of prey, viz. their *very distinctive flight* which varies considerably from the one to the other of the various species.

All Indian falconers are extremely good at recognizing birds on the wing, but Balunda had made of this hobby a fine art, and in the five years or so he was with me I never lost an opportunity of asking him what any particular species that might be passing at the time was, and, as a rule, his reply came pat without the least hesitation, but very occasionally he seemed to look very carefully before replying and in such cases it was generally *Astur badius* or *Accipiter nisus* that caused the slight momentary doubt in his mind, and that only when the light was against him, and no colouring or markings could be seen.

I would not like to say that this method is infallible, but it is certainly 95 per cent correct, and where it goes wrong is probably due to the specimen in hand rather than the system, as aberrant specimens are by no means unknown among the Raptors, and a very obvious Tawny Eagle in the air might turn out to be a Steppe in the hand, or vice versa; extremely rare, I should say, but just possible.

C.H. DONALD, *Journal of the Bombay Natural History Society*, 1952

Natural History and Science

4

MEMOIRS OF THE
DEPARTMENT OF AGRICULTURE IN INDIA

THE FOOD OF BIRDS IN INDIA

BY

C. W. MASON, M.S.E.A.C.,

Lately Supernumerary Entomologist, Imperial Department of Agriculture for India

EDITED BY

H. MAXWELL-LEFROY, M.A., F.E.S., F.Z.S.,

Imperial Entomologist.

AGRICULTURAL RESEARCH INSTITUTE, PUSA

PUBLISHED BY
THE IMPERIAL DEPARTMENT OF AGRICULTURE IN INDIA

THACKER, SPINK & CO., CALCUTTA
W. THACKER & CO., 2, CREED LANE, LONDON

Facsimile of the title page of *The Food of Birds in India* (1912) by C.W. Mason and H. Maxwell-Lefroy. Like other important studies on birds of this period, this book was also published by Thacker, Spink & Co. which had offices in Calcutta

The observations of Zahir-ud-din Mohammad Babur in his memoirs, The Baburnama, *are some of the earliest recorded ornithological descriptions of the Indian subcontinent.*

Birds of Hindustan

BIRDS[1]

Peacock. The peacock is a colourful and ornamental animal, although its body, like that of a crane but not so tall, is not equal to its colour and beauty. On both the male's and female's head are twenty to thirty feathers two to three fingers long. The female has no other colourful plumage. The male's head has an iridescent collar and its neck is a beautiful blue. Below the neck its back is painted yellow, blue, and violet. The eye on its back are very, very small. From its back down to the tip of the tail are much larger eyes in these same colours. The tails of some peacocks are as along as a human being. Under the eye feathers are shortish feathers like those of other birds. Its true tail and wing feathers are red. The peacock occurs in Bajaur and Swat and lower; farther up, in Kunar and Laghman, it occurs nowhere. It is less capable of flight than even the pheasant and cannot do more than one or two short flutters. Because it is all but flightless it sticks to mountains and forests. It is strange that in the forest where peacocks are, there are also many jackals. With a tail a fathom long, how can it run from forest to forest and not fall prey to the jackals?[*]

[1]Precise identification of all the birds that follow is impossible. The identifications given in the annotations are the best educated guesses.
[*]Editor's note: The question, why does the peacock have such a long tail, which is plainly a handicap has intrigued modern day biologists. For more, see Madhav Gadgil's piece 'Ornithology in Bandipur' on pp. 305–15.

Hindustanis call the peacock *mor*. In the sect of Imam Abu-Hanifa it is licit to eat it. Its meat is not without flavor, rather like the partridge, but one eats it, like the camel, only with reluctance.

Starlings. There are many of them in Laghman. Farther down in Hindustan they are quite numerous and occur in many varieties. One kind is the one found so often in Laghman. Its head is black, its wings are spotted, its body is a bit larger and rounder than a lark.[2] It can be taught to talk. Another kind brought from Bengal is called *baindawali*. It is solid black, and its body is much bigger than the former starling. Its beak and feet are yellow, and on each ear is a yellow skin that hangs down and looks ugly. It is called a myna, and it too can be taught to speak both well and eloquently.[3] Another kind of starling is more slender and has red around it eyes. This kind cannot be taught to speak. It is called a 'wood starling.'[4] When I made a bridge across the Ganges and crossed to rout my enemies, in the vicinity of Lucknow and Oudh a kind of starling was seen that had a white breast, spotted head, and black back. It had never been seen before. This kind probably cannot learn to speak.[5]

Lucha.[6] This bird is also called *buqalamun*.[7] From head to tail it has five or six different colours that shimmer like a pigeon's throat. It is about the size of a snow cock. It most likely is the Indian snow cock, for it runs about mountaintops like a snow cock. It occurs in the Nijrao mountains of Kabul Province and the mountains farther down, but not higher than that. The people there tell an amazing story about it: in winter it descends to the mountain foothills, and if it is made to fly over a vineyard, it cannot fly any more and can be caught.[8] It is edible and has quite delicious flesh.

[2]The Himalayan starling (*Sturnus humii*).
[3]The Indian grackle, or hill myna (*Eulabes intermedia*).
[4]The glossy starling, or tree stare (*Calornis chalybeius*).
[5]The pied myna (*Sturnopastor contra contra*). [probably Asian Pied Starling *Sturnus contra*—Ed]
[6]The reading of this word, as well as the precise identification, is still a mystery.
[7]*Buqalamun* is the modern Persian word for turkey, but its sense here is multicoloured, iridescent.
[8]Presumably because it has eaten too many grapes in the vineyard.

Partridge. The partridge is not peculiar to Hindustan but is found in warm climates. However, because certain kinds are not found anywhere except in Hindustan, I mention it here. The black partridge's body is the size of a snow cock. The male's back is the colour of a female pheasant. Its throat and breast are black, and it has bright white sports. Red lines come down either side of its eyes. It has a fantastic cry. *Sher daram shakarak*[9] can be heard from its cry. It says *sher* like *qit*, but it pronounces *daram shakark* quite correctly. The partridges in Astarabad say *qat meni tuttilar*,[10] and those in Arabia and thereabouts say *bi'sh-shukri tadumu 'n-ni'am*.[11] The female's colouration is like that of a young pheasant. They occur below Nijrao.

Another kind of partridge is the gray partridge, which has a body the size of a black partridge. Its cry greatly resembles that of the partridge, but its sound is much shriller. The male and female have only a slight difference in colouration. It occurs in the Peshawar and Hashnaghar regions and farther down, but not higher up.

Phul-paykar.[12] It is as long as a Himalayan snow cock. Its body is the size of the domestic chicken. Its colour is that of a hen. From its gullet down to its breast it is a beautiful red. The phul-paykar is found in the mountains of Hindustan.

Wild fowl. The difference between this fowl and a domestic one is that the wild fowl flies like a pheasant and, unlike the domestic fowl, does not occur in all colours. It is found in the Bajaur mountains and lower but not farther up than that.[13]

Chalsi.[14] Its body is like that of the phul-paykar, which has, however, a more beautiful colour. It occurs in the Bajaur mountains.

[9]'I have milk and a little sugar' (Persian).
[10]'Quick, they have seized me' (Turkish).
[11]'With gratitude good things endure' (Arabic).
[12]The horned monal (*Trapogon melanocephala*), a type of pheasant. [Western Tragopan *Tragopan melanocephalus*—Ed]
[13]May be the red jungle fowl (*Gallus ferrugineus*).
[14]May be the western bamboo partridge (*Bambusicola fytchii*) or a Himalayan pheasant (*Gallophasis albocristalus*).

Sham.[15] It is as long as a domestic fowl but has unique colours. It too occurs in the Bajaur mountains.

Quail. Although the quail is not limited to Hindustan, four or five kinds are peculiar to Hindustan. One is the quail that goes to our country.[16] Another quail, somewhat smaller than the one that goes to our country,[17] has wings and tail of a reddish colour. This kind of quail flies in flocks like the sparrow. Another kind, smaller than the one that goes to our country, has much black on its throat and breast.[18] Another, which very seldom goes to Kabul, is a tiny quail, a bit longer than a swallow.[19] In Kabul it is called a *quratu*.

Kharchal. It is as long as a bustard and probably is the bustard of Hindustan. Its flesh is delicious. In some birds the thigh is good, and in others the breast is good. All the Kharchal's meat is delicious and good.

Charz.[20] It is a little smaller than the bustard in body. The male's back is like the bustard's, but its breast is black. The female is one colour. The meat of the charz is also quite delicious. As the kharchal resembles the great bustard, the charz resembles the lesser bustard.

Indian sand grouse. It is smaller and slenderer than other sand grouses. The black of its tail is less, and its cry is less shrill.

BIRDS THAT LIVE IN AND BESIDE WATER

Among the birds that ¹ ve in water and on the banks of rivers is the adjutant. It has a huge body, and each of its wings spans a fathom. There are no feathers on its head or neck. From its throat hangs something like a pouch. Its back is black, and its breast white. Occasionally it goes to Kabul. One year an adjutant was caught in Kabul and brought to me. It became nicely tame and

[15] May be another common Himalayan pheasant (*Pucrasia macrolopha*).
[16] The gray quail (*Coturnix communis*).
[17] The rock bush quail (*Perdicula argunda*).
[18] The black-breasted or rain quail (*Coturnix coromandelica*).
[19] This may be the lesser button quail (*Turnix dussumierii*).
[20] The florican (*Sypheotis bengalensis* and *S. aurita*).

ate meat that was tossed to it, never missing as it caught the meat in its bill. Once it swallowed a shoe, another time a whole chicken—wings and feathers and all.

Saras. The Turks in Hindustan call it a 'camel crane.' It is smaller than an adjutant, but its neck is longer. Its head is brilliant red. It is kept in houses and becomes quite tame.

White-necked stork. Its neck is almost as long as that of the saras, but its body is smaller. It is much larger than a white stork, which it resembles, and its black bill is correspondingly longer. It head is iridescent, its neck is white, and its wings are variegated: the tips and edges are white and the midsection black.

Stork. Its neck is white, but its head and all its limbs are black. It is smaller than the storks that go to our country. Hindustanis call it *bag dhek*. Another kind of stork has a colour and body exactly like those of the ones that go to our country except that its bill is rather dark and it is much smaller.

Another bird resembles the gray heron and the stork. Its bill is bigger and longer than the heron's, and its body is smaller than the stork's.

The great black ibis is as large as a buzzard. The back of its wings is black, and it has a loud cry. Another is the white ibis, whose head and bill are black. It is much larger than the ibises that go to our country but smaller than the Hindustan ibis.

Ducks. One kind of duck is called the spotted-billed. It is larger than common ducks. Both male and female are of one colour. There are always some in Hashnaghar. Occasionally they go to Laghman too. Its meat is quite delicious.

Another kind of duck is called the comb duck. It is slightly smaller than a goose. On top of its bill is a hump. Its breast is white, its back black. Its flesh is good to eat.

Another is the black hawk-eagle. It is as long as an eagle, and back.

Another is the *sar*. Its back and tail are red.[21]

---

[21]The Turkish *sar* means buzzard, but buzzards do not have red backs; in Persian, *sar* means starling, but the starlings have already been dealt

Pied crow. The pied crow of Hindustan is somewhat smaller and slender than the crows of our country. It has a bit of white on its throat.

There is another bird that resembles the crow and the magpie. In Laghman they call it 'forest bird.' Its head and breast are black, its wings and tail reddish, and its eye deep red. Since it has difficulty flying, it does not emerge from the forest, for which reason they call it a 'forest bird.'[22]

Another is the great bat called *chamgiddar*. It is as large as an owl. Its head resembles a puppy's. When it is going to hang on to a tree, it grabs a branch and hangs upside down. It is strange.

Indian magpie. The Hindustan magpie is called *mata* and is slightly smaller than a common magpie. The magpie is mottled black and white; the mata is pale mottled.

There is another little bird the size of a nightingale. It is a nice red colour and has small black markings on its wings.

Swift. It is like a swallow but much larger. It is solid jet black.

Cuckoo. It is as long as but much slenderer than a crow. It sings beautifully. It is the nightingale of Hindustan. The people of Hindustan have great respect for the nightingale. It inhabits gardens with many trees.

Another bird is like the green magpie. It clings to trees. It is as large as a green magpie and is green in colour like a parrot.[23]

The Baburnama: Memoirs of Babur, Prince and Emperor

---

with. It has been suggested that the rose-coloured starling (*Pastor roseus*) is meant here, although that identification is far from certain.

[22] The crow pheasant, or Malabar pheasant (*Centropus sinensis*).

[23] Perhaps some sort of green woodpecker.

Mughal Emperor Nuruddin Salim Jahangir had a scientific temperament, and when it came to natural history he did not rely on mere hearsay but made close, personal observations, at times even trying an experimental approach to understand natural phenomena. The following extracts from The Jahangirnama *amply reveal the monarch's gift for scientific inquiry. It has been said about Jahangir that he would have been a happier man if he was the curator of a natural history museum.*

Koel

On the eve of Monday the twenty-seventh [April 5] we camped in the village of Badarwala in the pargana of Sehra. Here the sound of the *koil* [black cuckoo] was heard. The cuckoo is a bird something like a raven, but it is smaller. A raven's eye are black, but the cuckoo's are red. The female has white spots, and the male is totally black. The male has a very beautiful voice, completely beyond comparison with the female's. The cuckoo is really the nightingale of India, but whereas the nightingale is agitated in the spring, the cuckoo gets agitated at the beginning of the monsoon, which is the spring of Hindustan. Its cry is extremely pleasant. Its period of agitation coincides with the maturing of mangoes, and mostly the cuckoo sits on mango trees. It must enjoy the colour and scent of the mangoes. One of the strange things about the cuckoo is that it doesn't hatch its own eggs. When it is ready to lay an egg it finds an unprotected raven's nest, breaks the raven's eggs with its beak and throws them out, and then lays its own eggs and flies away. The raven thinks they are its own eggs, hatches them, and rears them. I have seen this strange thing myself in Allahabad.

A Quail with a Spur

One day around this time on the hunting ground Imamverdi the head scout spied a quail with a spur on one of its legs but not on the other. Since the mark that distinguishes male from female is this spur, as an experiment he asked me whether it was male or female. I said immediately, 'It is female.' Then they split open its belly, and several embryonic eggs came out. Those who in attendance asked in disbelief, 'How did you know?' 'The female's head and beak are smaller than the male's,' I said, 'and with much observation and perseverance one gets the knack.'

Windpipes of Bustard and Crane

It is an amazing thing that in all birds the windpipe, which the Turks call *chanaq*,[24] goes straight from the top of the neck to the crop, while in the bustard, unlike any other bird, there is a single windpipe from the top of the throat for a distance of four fingers, then it splits in two and goes to the crop. At the point at which it forks there is a blockage, like a knot, that can be felt with the hand. In the crane it is even stranger, for its windpipe twists like a snake through the bones of the chest and passes to the root of the tail, and then it turns around and comes back to the throat.

Close Observations on Bustards

There were thought to be two kinds of bustard, one black and spotted and the other dun coloured. Recently it was learned that they are not two types: the spotted black one is male and the dun-coloured one is female. The proof was that testicles were found in the spotted one and eggs in the dun-coloured one. The experiment was made repeatedly.

The Jahangirnama: Memoirs of Jahangir, Emperor of India

Nuruddin Salim Jahangir seems to have been particularly fascinated by the Sarus Crane. The Jahangirnama *records several instances of his observations on cranes.*

The Nesting Cranes

On Monday the twenty-first [August 2] the saras crane whose mating I have recorded on a previous page gathered sticks and twigs in the small garden and laid an egg. The third day it laid

---

[24]Emending the text's *hanaq* (Arabic for throat) with *chanaq* ('originally something like "a hollow conical object,"' Clauson, *An Etymological Dictionary*, 425).

another egg. This pair of sarases were captured when they were one month old, and they have been in my establishment for five years. After five months and a half a year [?] they mated and continued to mate for a month.[25]

On the twenty-first of Amurdad, which the people of India call Savan, they produced an egg. The female sits on the eggs by herself all night, and the male stands next to the female, keeping watch. He is so watchful that he doesn't allow any animal to come near. Once a large weasel° appeared, and he ran at it with great vehemence and didn't stop until the weasel had put itself back in its hole. When the sun illuminates the world the male goes to the female and scratches her back with his beak. Then the female gets up, and the male sits down. Later, in the same fashion, the female comes, has him get up, and sits down herself. When they are getting up and sitting down they take great precaution lest damage be done to the eggs.

THE NESTING CRANES, CONTINUED

Friday was the first of Shahrivar [August 13]. It rained on Wednesday the sixth [August 18] and Wednesday night. A strange thing is that on other days the pair of saras cranes took five or six turns sitting on their eggs, but during this twenty-four period while it was raining and cold, the male sat on the eggs to keep them warm continuously from dawn until midday. From midday until the morning of the next day the female sat continuously—lest the eggs be damaged or spoiled by the cold while they were getting up and sitting down.

In short, what a human being comprehends by the guidance of his reason animals do by a instinct made innate in them by eternal wisdom. Even stranger is the fact that at the beginning they kept the eggs next to each other under their breasts, but after fourteen or fifteen days had passed they made enough space between the eggs so there wouldn't be too much heat and the eggs wouldn't be spoiled.

[25]The wording is peculiar here. The Persian has umabiguously 'after five months and half a year' (*ba'd az panj mab u nim sal*). 'Five months or half a year' would make sense, but he doesn't say after what.

°Editor's Note: This could be a mongoose according to Mr Toby Sinclair (personal communication).

THE CRANES HATCH CHICKS

On the eve of Thursday the twenty-first [September 2] the saras cranes hatched a chick. On the eve of Monday the twenty-fifth [September 6] they hatched a second chick, that is one after thirty-four days and the second after thirty-six days. In size they were a third larger than a gosling or about the size of a one-month-old peafowl chick. Their down is blue. The first day they didn't eat anything. From the second day on, the mother brought small locusts in her beak, and sometimes she fed them as doves do and other times, like chickens, she threw it down in front of them to peck at themselves. If the locust was small she left it whole, but if it was larger she would divide it into two or three pieces so that the chicks could eat it easily.

Since I was most anxious to see them, I ordered them brought very carefully so they wouldn't be hurt or injured. After looking at them I ordered them taken back to the same small garden in the palace and guarded very carefully. Whenever it was possible they were to be brought to me.

MORE ON THE CRANES

At first the male crane used to hold its chicks upside down in his beak. Fearing that this might be out of enmity and that he might kill them, I ordered the male kept apart and not allowed near the chicks. Around this time, by way of experimentation, I ordered the servants to let it go near the chicks to see whether it was affectionate or not. It showed great love and affection, no less than the female, so it was obvious that its action had been affectionate.

THE SAGA OF THE CRANES, CONTINUED

On this day a strange sight was seen. The pair of cranes that had produced young had been brought from Ahmadabad on Thursday and left to wander around with their chicks in the courtyard of the palace beside the tank. As it happened both the male and female made cries and a pair of wild saras cranes heard them and cried out from the other side of the tank. Then they flew over and the male began to fight with the male and the female with the female. Although several people were standing around, they paid not the slightest attention to them. The eunuchs who were in charge of the cranes ran forward to catch the wild ones, and one of the eunuchs seized the males and another the female. The one who had seized the male managed to keep a struggling hold on it, while the one who had caught the female couldn't hold her, and she escaped his grasp. With my own hand I put rings in their noses and on their feet and turned them loose. They both returned to their place and quieted down. However, every time the tame cranes cried out, the others cried out in answer.

Something similar was once seen in a wild antelope. I had gone hunting in the pargana of Karnal, and there were around thirty persons with me, huntsmen and servants. A black buck antelope came into view along with several does, and we set out a decoy antelope to fight with it. They butted horns two or three times, and the decoy turned around and came back toward us. We wanted to fasten a *phana* on the decoy's horns and put it back out again so that the other one would get tangled in it.[26] Just then the wild antelope, in a territorial rage, disregarded the men and came charging, butted horns two or three times with the decoy antelope, and disappeared.

The Jahangirnama: Memoirs of Jahangir, Emperor of India

Allan Octavian Hume (1829–1912), a civil servant in British India, was a political reformer, best remembered for his role in founding the Indian National Congress. However, Hume is also regarded as the father of Indian ornithology because of his

---

[26]The *phana* is a weighted cord attached to a decoy antelope's antlers that gets tangled in a wild antelope's antlers and locks the two together. This method of catching antelopes is described by Babur, *Baburnama*.

several important contributions, which included My Scrapbook: Or Rough Notes on Indian Oology and Ornithology *which had accounts of 81 species of Indian birds. His second book* The Game Birds of India, Burmah, and Ceylon, *in three volumes was a comprehensive piece of work compiled with contributions and notes from a network of more than 200 correspondents from all across India. This monumental work was co-authored by Charles Henry Tilson Marshall (1841–1927), Hume's close associate. Hume also started an ornithological journal,* Stray Feathers, *which till today is reckoned as a veritable goldmine of information about Indian birds. The accounts of the three birds from Hume and Marshall's* The Game Birds of India, Burmah, and Ceylon *reproduced here, have been carefully chosen. The Siberian Crane (known as the 'Snow Wreath' then) is an endangered Indian bird. Till some years ago small numbers of this crane used to spend their winter in the marshes of Keoladeo Ghana National Park in Bharatpur, Rajasthan—the only surviving wintering ground for the western population of this species. The Pink-headed Duck is now believed to be extinct. The Mute Swan, commonly depicted in contemporary calendar art as a vehicle of Hindu gods and goddesses, is a rare winter visitor at some lakes in the upper reaches of the Himalayas.*

The Snow-wreath or Siberian Crane

As a rule, the Snow-Wreath does not, I think, put in an appearance even in the sub-Himalayan tracts before the middle of October, and they are at least a week later further south, as at Etawah, but in 1879 one was shot somewhere near Kurnal on the 3rd, and Mr W. Forsyth notes having seen a large flock at Dehree-on-Soane on the 6th of October.

The distribution of this species in India must be, to a great extent, governed by its peculiar habits. It affects only good sized sheets of water, large portions of which are shallow, and which contain a considerable growth of the rushes and aquatic plants on which it seems to feed exclusively. Necessarily, therefore, the localities in which it can occur in India, and especially in northern and north-western India, (and we have no reason to suppose that it ever goes far south,) are comparatively limited in number; and though I can name a good many tracts of country which are yearly visited by small flocks or parties, still, taking Upper India as a whole, they are excessively rare birds, and I should greatly doubt as many even as five thousand birds of this species yearly visiting India proper. We gather that during the summer they are more or less abundant over incredibly vast tracts of northern Asia, and it is pretty certain that they do not spend their winter in Turkestan, Kashgar, or Tibet, but only a very small portion of those that migrate from the north can be accounted for in India, and it seems to me probable that they will prove to go further east, and that when we know more of the fauna of these tracts, we shall find that they occur in Assam, Yunan, the Shan States, Independent Burma, etc., where broads and lakes suited to their peculiar habits exist.

No plate, that has ever been given of this species, does any justice to its extreme elegance of form, or to the dense, snowy, Swan-like character of its plumage. To judge by the pictures, Gould's, Dresser's, our own, the bird appears a gaunt, gawky, ill-proportioned creature, whereas, in reality, it is the lily of birds, and stand in what position it may, the entire outline of its form presents a series of the most graceful and harmonious curves.

No one else appears ever to have watched these birds carefully, or to have recorded anything about their habits, haunts, or food, and I myself have seen but little of them for the last ten years, so that I am constrained to reproduce, with a few verbal alterations, the account I published of this species in the *Ibis* for 1868.

Many years have now elapsed since I first shot one in Ladakh. This was late in September, and the birds were doubtless on their way to the plains of India. They arrived at a small fresh-water lake, near the Tso-khar, in the Ley district, besides which I was encamped, towards nightfall; and though after I had fired at them and secured a specimen they again (contrary to their usual custom) settled at some little distance, I did not molest them further. They remained there all night, and I saw them again up to nine o'clock, but they had left the place when I went down to the lake again about noon.

After this, through constantly shooting both in the Himalayas and many parts of the North-Western Provinces, I did not again meet with this species until 1859, when I succeeded in shooting one out of a flock of some five and twenty, which I found in a large *jhil* in the north of the Etawah district.

Later again, during the winters of 1865–6 and 1866–7, I procured numerous specimens, and had opportunities of watching the habits of the species rather closely. The locality in which, during these two winters, I saw and procured, comparatively, so many of these beautiful birds, is somewhat peculiar. A broad straggling belt of Dhak (*Butea frondosa*) jungle, some ten miles in width, at one time doubtless continuous, but now much encroached upon and intersected in many places by cultivation, runs down through nearly the whole of the 'Doab', marking, possibly, an ancient river course. Just where the northern and southern boundaries of the Etawah and Mynpoorie districts lie within this belt, the latter encloses a number of large shallow ponds or lakes (jhils as we call them), which, covering from two hundred acres to many square miles of country each, at the close of the rainy season, are many of them still somewhat imposing sheets of water early in January, and some few of them of considerable extent, even as late as the commencement of March. Mohree-Sonthenan, Mamun, Sirsai-Nawur, Kurree, Beenan, Soj, Hurrera, Suman, Kishnee, Phurenjhee, are some of the largest of these rain-water lakes, many of which abound with rushes and sedges, and as the waters gradually dry up or are drawn off for irrigating purposes, become successively the favourite haunts of the White Crane.

There will always be, at any particular time, two or three 'jhils,' that for the moment they particularly affect, and these are, as a rule, just those that then happen to average about eighteen inches to two feet in depth, and that have a good deal of rush (*Scirpus carinatus* amongst others) somewhere in the shallower parts.

To this tract of country they make their way as early as the 25th of October (and possible sooner, though this is the earliest date on which I have observed them), and there they remain at least as late as the end of March, or perhaps a week or two longer. During the whole of our cold season they stay in this neighbourhood; and, though growing more and more wary (if possible) each time they are fired at, and disappearing for a day or two from any 'jhil,' where an attempt has been made to kill or capture them, they never seem to forsake the locality until the change of temperature warns them to retreat to their cool northern homes. Week after week I have noticed and repeatedly fired at, sometime even slightly wounded, particular birds, which have nevertheless remained about the place their full time—nay, I have twice now killed the young bird early in the season, and the parents, one by one, at intervals of nearly a couple of months.

The Buhelias, a native caste of fowlers, (and, I fear I must add, thieves) of whom there are many in the neighbourhood, and who are keen observers of all wild animals, assured me that, as far back as any of them could remember (namely, for at least the previous fifty years), parties of the White Crane, or as they call them *karekhurs*[27] have been in the habit of yearly spending their winters in the same locality.

Though occasionally seen in larger flocks, it is usual to find either a pair of old ones accompanied by a single young one, or

---

[27]Professor Max Muller justly ridicules the excessive length to which what he denominates the 'bow-wow theory' of the origin of words, has been pushed by some comparative etymologists; but, in the case of the Cranes, the Hindu names in use, in this portion of northern India, clearly owe their origin to the cries of the several birds. Thus *Grus communis* is called *Kooroonch*, or *Koorch*, *Anthropoides virgo*, *Kurrkurra*, and *G. leucogeranus*, *Karekhur*, each of these names, when pronounced by a native, conveying to my idea an appreciable imitation of the cry of the particular species it serves to designate. Not so, however, thinks Mr Brooks. He says: 'With regard to the notes of *Grus leucogeranus* how the native can imagine that their name Karekhur, or, as I should call it, 'Carecur', expresses any one of them, I cannot conceive. The notes are all simply whistles, from a mellow one to a peculiar feeble shrill shivering whistle, if I may so express it. No written word will express the note of this species, nor give the faintest idea of it. I watched a flock of these fine birds for a long time, yesterday, as they fed in a marsh, in company with about a dozen of *G. antigone* and three of *G. cinerea*. I found it impossible to get within shot of the White Cranes, nor could I get them driven over me as I sat in ambush; for, as soon as they take wing, they immediately begin to soar, and circle round and round till they attain a height far above the reach of any shot; they then fly straight away, uttering their peculiar whistle, which, though weak, compared with the call of other Cranes, can still be heard a mile off, or even more. It is a magnificent bird, and I think, the most graceful of the group in its attitudes. The species is abundant, being found in large flocks; and the eggs might be obtained from Russian sources. The plumage is so very compact and Swan-like that it must go very far north to breed, where perhaps its snowy plumage harmonizes with the still unmelted snow as it sits upon its nest.'

small parties of five or six, which then, as far as I can judge, consist exclusively of birds of the second year.

The fully adult birds are even, when they first arrive, of snowy whiteness, and each pair is, almost without exception, accompanied by a young one, which, when first seen, is of a sandy or buffy tint throughout, and very noticeably smaller than its parents. The males are considerably larger and heavier than the females, the adults of the former weighing up to 19 lbs., but of the latter only, as far as my experience goes, to about 16 lbs. Of the young birds, however, when they first arrive, the males do not exceed about 10 lbs. in weight, and the females 9 lbs., though generally very fat and well cared for by the parents.

When we first see them, they cannot, I estimate, be more than six months' old. The testes and ovaria of adults, examined on the 20th of March, were still, if I may use the term, quite dormant; and allowing for the 'passage home', the pairing season, and incubation, they can scarcely hatch off before the middle of May.

They never appear to have more than one young one with them; but it does not at all follow that they do not lay more than one egg. The Sarus, which usually lays two, and sometimes, though rarely, three eggs, and which has no longer or arduous journey to perform, constantly fails to rear more than one young one.

The watchful care and tender solicitude evinced by the old birds for their only child is most noticeable. They never suffer the young one to stray from their side; and, while they themselves are rarely more than thirty yards apart, and generally much closer, the young, I think, is invariably somewhere between them. If either bird find a particularly promising rush tuft, it will call the little one to its side, by a faint creaking cry, and watch it eating, every now and then affectionately running its long bill through the young one's feathers. If, as sometimes happens, the young only be shot, the old birds, though rising in the air with may cries, will not leave the place, but for hours after keep circling round and round high out of gun or even rifle shot, and for many days afterwards will return apparently disconsolately seeking their lost treasure.

Like the Sarus, these birds pair, I think, for life; at any rate a pair, whose young one was shot last year, and both of whom were subsequently wounded about the legs, so as to make them very recognizable, appeared again this year, accompanied by a young one, and were at once noticed as being our wary friends of the past year, by both the native fowlers and myself. I was glad to see they were none the worse for their swollen, crooked, bandy legs, and this year at least they have got safe home, I hope, with their precious charge.

Throughout their sojourn here, the young remain as closely attached to their parents as when they first arrived, but doubtless by the time the party return to their northern homes, the young are dismissed, with a blessing, to shift for themselves.

Long before they leave, the rich buff or sandy colour has begun to give place to the white of the adult plumage, and the faces and foreheads, which (as in the Common Crane) are feathered in the young, have begun to grow bare. This, I notice seems to result from the barbs composing the vanes of the tiny feathers falling off, and leaving only the naked hair-like shafts. Even when they leave us, however, there is still a good deal of buff about the head, upper back, lesser and median wing-coverts, longer scapulars, and tertials of the young, while the dingy patch along the front of the tarsus is still well marked.

Each year several small parties of birds are noticeable, unaccompanied by any young ones, and never separating into pairs. These, when they first come, still show a few buff feathers, and have a dingy patch on the tarsus; and, though before they leave us, they become almost as purely white, and have almost as well-coloured faces and legs as the old ones that are in pairs, they never seem to attain to the full weight of these latter. From these facts I am disposed to infer that these parties, which include individuals of both sexes, consist of birds of the second year; that our birds do not either breed or assume their perfect plumage till just at the close of their second year; and that, like Pigeons and many others, they do not attain their full weight until they have bred once at least.

Unlike the four other species of Crane with which I am acquainted, the Snow-Wreath never seems to resort during any part of the day or night to dry plains or fields in which to feed, and unlike them, too, as far as my experience goes, it is exclusively a vegetable eater. I have never found the slightest traces of insects or reptiles (so common in those of the other species) in any of the twenty odd stomachs of these White Cranes that I have myself examined.

Day and night they are to be seen, if undisturbed, standing in the shallow water. Asleep, they rest on one leg, with the head and neck somehow nestled into the back, or they will stand like marble statues, contemplating the water with curved necks, not a little resembling some White Egret on a gigantic scale; or, again, we see them marching to and fro, slowly and gracefully feeding amongst the low rushes.

Other Cranes, and notably the common one and the Demoiselle, daily pay visits in large numbers to our fields, where they commit great havoc, devouring grain of all descriptions, flower, shoots, and even some kinds of vegetables. The White Crane, however, seeks no such dainties, but finds its frugal food, rush-seeds, bulbs, corms, and even leaves of various aquatic plants, in the cool waters where it spends its whole time.

Without preparations by me for comparison I hardly like to be too positive on this score; but I am impressed with the idea that the stomach in this species is much less muscular than in any of the others with which I am acquainted. The enormous number of small pebbles that their stomachs contain is remarkable. Out of an old male I took very nearly sufficient to fill an ordinary-sized wine-glass, and that, too, after they had been thoroughly cleansed and freed from the macerated vegetable matter which clung to them. These pebbles were mostly quartz, (amorphous and crystalline,) greenstone, and some kind of porphyretic rock; the largest scarcely exceeding in size an ordinary pea, while the majority were not bigger than large pins' heads.

I have found similar pebbles in the stomachs of the Common and Demoiselle Cranes, but never in anything like such numbers as in those of the present species.

When not alarmed, the White Cranes' note is what, for so large a bird, may be called a mere chirrup; and even when most alarmed, and circling and soaring wildly round and round, looking down upon the capture of wounded offspring of partner, their cry (a mere repetition of the syllables *karekhur*) is very feeble as compared with that of any other of the Cranes (including even *Balearica pavonina*) whose notes I have myself never heard.

An examination of the trachea of a fine male that I dissected on the 22nd of February 1867, at once explained this feebleness. Instead of a convolution entering and running far back into the sternum, there is merely a somewhat dilated bend just where the windpipe enters the cavity of the body; and it is only after the pipe has divided, which it does symmetrically into two very nearly equal tubes, about three inches before entering the lungs, that the rings are at all strongly marked, or that the tube impresses one as at all powerful.

I have already noticed that it is not easy to get at these birds (possibly due in part to a keen sense of hearing, accompanying their large ear-orifices); and, as far as my experience goes, there is only one way of shooting them with a shot gun. With a rifle it is not difficult to get within two-hundred and fifty to three-hundred yards of them, at which distance, with a heavy .442 match rifle, one ought to knock them over every time. The melancholy fact, however, is, that habitually one only succeeds in missing them, and thoroughly scaring them with a rifle; so nothing remains but to have recourse to a long single eight-bore with B.B. wire cartridge. This will easily knock them down up to seventy, or, if a shot tells well in the neck, up to eighty yards; but getting within eighty or even a hundred yards of them can only be managed, as a general rule, in one way. You obtain from one of the native fowlers the loan of a trained buffalo, and enter the water a good quarter of a mile, away from the birds, under cover of the quadruped. It has, as usual, a string run through the nostrils, and tied tightly together behind the horns. You hold this string where it lies across the cheek with the left hand; your extended left arm is hidden behind the neck; your whole body is bent, so that your head and neck are covered by the buffalo's shoulders, your body and the greater part of your legs, by its body. Only your legs to a little above the knees show close to the hind legs; and, as far as possible, you always keep the beast up to his belly in water. Thus covered you slowly sidle up towards the Cranes, making the buffalo, now put his head up, nose in air, now stop and lower his head to the water, and generally dawdle and meander about with apparently no fixed idea in his head, according to the natural manners and customs of a free and independent buffalo. With a little practice it is easy thus to get within shot. You softly let the cheek string go, and at once fire below the buffalo's neck. Before your gun is well off, your sporting companion, who has a marked distrust of Europeans and white faces, and has been incessantly endeavouring to kick you throughout your whole promenade, knocks you head over heels, and rushes off towards his dusky owner, bellowing as if he, and not you, were the injured party. This is first-rate sport; but, after trying it once or twice, nearly catching my death of cold, losing a power flask, and realizing a stock-in-trade of bruises enough to last the rest of my natural life, I have preferred sitting quietly on the bank and allowing my native coadjutors to shoot the birds I wanted.

When shot they were worth nothing as food, which considering their purely vegetable diet, is surprising.

I ought not to omit to notice that, out of more than twenty specimens of the White Crane that I have procured (between October and the middle of March), none had the tertials at all conspicuously elongated; and in no instance did these, when the wings were closed, exceed the tail feathers or longest primaries (which usually reach just to the end of the tail) by more than 3 inches. It is possible that at the breeding season the tertials may be *much* more developed; but such is not the case with the Sarus, nor, I fancy (to judge from the magnificent trains of plumes with which we here shoot them in the spring,) with the Common Crane.

The feathers of the hind head and nape are somewhat lengthened, so as to form a full and broad, though short, subcrest, very noticeable when a wounded bird is defending itself against dogs or other assailants. It is a brave bird, and fights to the last, striking out powerfully, at times with bill, legs, and wings, but most generally defending itself chiefly with its bill, with which it inflicts, occasionally, almost serious wounds.

The Pink-headed Duck

Despite strenuous efforts I have been quite unable to clear up conclusively the question of the distribution of the Pink-headed Duck.

NATURAL HISTORY AND SCIENCE 217

I have no record of its occurrence in Ceylon or the Madras districts south of Mysore, or in this latter; or in the Konkan, Gujarat, Cutch, Kathiawar, Sindh, Rajputana, the Central India Agency, or Central Provinces.

In the Punjab I only certainly[28] know of its having been once procured, and that near Delhi, in the easternmost portion of the Province. In the Doab and Rohilkhand, of the North-Western Provinces, it is so excessively rare that during nearly twenty years' fowling I never once saw or heard of it; but Anderson shot one female near Fatehpur, and a writer in the *Asian* professes to have obtained them in the Dun and at Bareilly.[29] In the western portions of Oudh outside the sal forests it is certainly rare; Major Maurice Tweedie, an ardent sportsman, who was for five years

---

[28]Major Alexander Kinloch writes: 'I shot two Pink-headed Ducks on the banks of the canal leading to the Najjafgarh *jhil* near Delhi, during the winter of 1862–3, and a brother officer shot another.'

Mr R.W. Rumsby *thinks* that he once shot it on an exposed jhil southwest of Umballa, and also in the Gurdaspur district; but on further investigation he is clearly not certain of the species. No one else that I can hear of *thinks* even that they have procured it in the Punjab. Adams never met with it there; neither have I myself, nor any one of the very numerous friends who have collected there for me (some of them for years and most exhaustively), so that I can only (at present), consider it as a rare and accidental straggler during the cold season into the easternmost portions of the Punjab, Cis-Satlej.

[29]It is impossible to attach much weight to anonymous communications by writers who admit knowing very little of birds. Still I quote what was said, *quantum valeat*:

'Some time before Christmas, I was out shooting in the Dun, and accidentally came across the very bird, I think, he means. There were only five, and I shot two of them—a male and a female. Had I known that it would have been of any use I would have preserved them, but now alas! they have been eaten, for as Jerdon says, they are 'excellent eating', and I knew that. These birds I found in a large pool, formed in the river Asun, made for irrigation purposes. It was a very cold morning, for the night before the water froze in my tent. These Ducks I have come across but seldom in the North-West Provinces, principally about Delhi and Bareilly. At the former place I have often bagged 50 head of Ducks, but it was rarely I found one to two of the Pink-headed among the bag. I do not think I can be mistaken in the bird. Although I am not a naturalist, I follow Jerdon's description.'

stationed in the Kheri district (now Lakhimpur) never so much as heard of it. But Mr Battie shows[30] that even in the west it is not very uncommon in the forests, and in the central portions of the Province, though rare, it also occurs. From time to time specimens are netted by fowlers in the neighbourhood of Lucknow; and Mr Geo. Reid has himself observed it near Mohunlalganj on the Rai Bareilly road during the cold season. In the eastern portions of Oudh it is still rare, but appears to be a *regular*, though scarce, cold weather visitant to the jhils of the Fyzabad (where Anderson shot it) and Gondah districts. Again it is reported from Gorakhpur and Basti and from the Nepal Terai, but in all these it is scarce; and as far as I can learn, in all but the latter, a cold season visitant only.[31] It probably occurs in Azimgarh and Ghazipur also, as it certainly does in Arrah, where Mr Doyle informs me that he shot one on the 22nd of November 1879, at the Bhojpur jhil, near the Dumraon Railway Station.

Further east in Behar, Purneah, and Maldah[32] it would seem to be a permanent resident, and in special localities in Tirhoot and Purneah to be comparatively common. Throughout the rest of Lower and eastern Bengal (except Tipperah and Chittagong, from whence it has not been reported), it occurs, but is everywhere said to be rare. So too both from Sylhet and the entire valley of Assam up to Sadiya, (and in Munipur, where Damant saw it), it is reported by one correspondent or another, but always as a bird very rarely met with.

[30]I shot a Pink-headed Duck this year, in May or June, up in the sal forests in the north of the Kheri district. Another was shot some time afterwards in the same jhil and you often see it in pairs in *nullahs* in the forests.

'I am told by the natives that this bird breeds in the sal forests, but I have never found its nest. I know for a fact however that it stays down in the forests all the year round.'—*J. Battie*.

[31]But too much stress must not be laid upon this as the question has not been properly worked out, and it may, though rarely, breed in all these as also in Oudh, where Irby says that he saw it three times (apparently the only occasions on which he met with it) towards the close of the rainy season.

[32]Mr H. Millett kindly informed me, under date the 2nd of May, that Mr Herbert Reily had then recently 'shot four or five specimens of this Duck in the Maldah district,' and that his brother had also previously shot one there.

South of the Ganges, already mentioned, it is occasionally found at Arrah, and as Ball tells us, in the Rajmehal hills, near Hazaribagh, near Sahibgunge[33] on the Ganges, and in Manbhum of Chota Nagpur.

To the Deccan it is an extremely rare and accidental visitor. Neither Davidson nor Wenden ever met with it there, but Fairbank saws it once near Ahmednagar, Colonel McMaster shot it once about twenty miles from Secunderabad after the rains had set in, and Jerdon heard of it at Jalna.

But along the east coast it is less rare. It certainly occurs in the Pulicat lake, as I have a specimen shot there, and Jerdon, years previously, had obtained a specimen in the Madras market caught there, and another from Nellore. Again north of Nellore it appears to occur in suitable situations in Vizagapatam[34] and Ganjam, north of which again in Cuttack, as in the rest of Lower Bengal, it is an occasional straggler.

So far as I yet know, this species does not occur in either Pegu or Tenasserim, but Blyth gives it from Arakan.

Outside our limits I have only heard of its occurrence north of Bhamo, in Independent Burma. It is never found anywhere in the Himalayas, and its therefore not likely to cross them, but

---

[33] It is nearly opposite Sahibgunge, in the neighbourhood of Caragola (at the south of the Purneah district) that the Pink-headed Duck, to judge from what Mr F.A. Shillingford, Captain W.T. Heaviside, R.E., and others tell me, is specially abundant.

[34] Lieut.-Colonel W.J. Wilson kindly favours me with the following note: The Pink-headed Duck used to frequent a piece of water near Condakirla, about 27 miles south of Vizagapatam, and in all probability is still to be found there, as well as at similar places in the northern Circars, although I do not now remember having actually seen it except at Condakirla.

The lakes in question are extensive and thickly covered by aquatic plants, so that the birds have plenty of cover, and the only way of shooting them is from a long narrow canoe punted through the weeds.

These lakes seem to be peculiar to the Circars, and are called 'AWA' in Vizagapatam, and '*Tumpera*' in Ganjam. They are resorted to by wild fowl of most kinds.

To the best of my recollection the Pink-headed Duck I shot were killed in November and December. I think I saw about 15 or 20 on each occasion of my visit.

it may extend via Assam and Upper Burma into south-western China (Yunan), though as yet this fact has not been ascertained.

Summing up the meagre information at my command, I am disposed to consider Behar and the rest of Bengal north of the Ganges and west of the Brahmaputra as its head-quarters; I include the Nepal and Oudh Terai, the central and eastern portions of Oudh, the Benares division of the North-Western Provinces, the whole of the rest of Bengal, Assam, and Munipur, and the east coast littoral as far south as Madras, within its normal range, throughout which, however, it is, except in certain special isolated localities, very rare. Its occurrence elsewhere in any part of the empire I look upon as quite exceptional and abnormal.

***

Never having myself met with this Duck alive in a feral state, what little I have to say of its habits will be based solely on the reports of others.

Essentially a lake and swamp species, this bird is never seen on any of our large rivers, or indeed, so far as I can learn, on running water of any kind.

Tanks and pools, thickly set with reeds and aquatic plants, swamps dense with beds of bulrushes and the like, and nullahs and ponds hemmed in by forest, appear to be its favourite, if not its only, haunts. During the cold season it keeps commonly in small parties of from four to eight or ten, but is sometimes seen in flocks of from twenty to thirty. During the breeding season they are found in pairs. Mr F.A. Shillingford, who has rendered me more assistance than any one else where this species is concerned, writes, that it 'may be freely found throughout the year in the southern and western portions of the Purneah district. From November to April they are to be met with in flocks, numbering as many as twenty, along the swamps adjoining the rivers Great Coosee and Ganges; and during the rainy season (June to September) I have observed that they are usually seen in pairs, and are to be met with generally in the higher parts only of the district. Though not to be met with in such numbers as the commoner species, they are not considered at all rare in this district, but they are difficult to get at, remaining, as they do during the cold season, in large swamps fringed with dense jungle.'

Mr J.C. Parker writes: 'Years ago I have fired at them when passing with other Ducks, when out shooting in the *bhils* of Kishnaghur and Jessore. They were easy to distinguish by their beautiful pink heads and salmon-coloured wing-linings. The flight of this Duck is very powerful and rapid.'

'Its call,' says Mr Shillingford, 'resembles that of the common drake, with a slight musical ring about it.'

Hodgson notes: 'Lives and breeds below always. Avoids flowing waters; shy; resides in remote jhils and feeds at night.'

Jerdon says: 'It shows a decided preference for tanks and jhils well sheltered by overhanging bushes, or abounding in dense reeds; and in such places it may be found in the cold season in flocks of twenty or so occasionally, but generally in small parties of from four to eight. During the heat of the day, it generally remains near the middle of the tank or jhil, and is somewhat shy and wary.'

Mr Shillingford says that the gizzard of one specimen that he examined contained 'half-digested water weeds and various kinds of small shells.'

Beyond this there is absolutely nothing on record.

***

Mr F.A. Shillingford and his brother had found the eggs of this species in former years; but the egg he sent me was so very peculiar that I hesitated to accept it as genuine, and at my request he, and several of his friends, set to work to discover a nest and he was soon able to send me the following note:

On the 3rd of July Mr T. Hill, of Jouneah Factory, succeeded in finding a nest of the Pink-headed Duck near the Dabeepoor Factory.

The nest contained nine such incubated eggs, of which I send you four. These, as you will observe are of precisely the same type as the one I formerly sent you.

The nest was well hidden in tall grass (*Andropogon muricatum*), and both male and female were started from the vicinity of the nest, which was about 400 yards from a *nullah* containing water. The nest was well formed, made of dry grass, interspersed with a few feathers, the interior portion being circular, and about 9 inches in diameter and 4 to 5 inches deep.

To the *Asian* he sent the following further interesting note:

During the cold weather, November to March, the Pink-headers remain in flocks, varying from 6 to 30, or even 40 birds, in the lagoons adjoining the larger rivers, and have been observed by myself in considerable numbers in the southern and western portions of this district, that portion of eastern Bhaugulpore which lies immediately to the north of the river Ganges, and the south-western parts of Maldah. They come up to the central or higher parts of the Purneah district in pairs, during the months of April, begin to build in May, and their eggs may be found in June and July. The nests are well formed, (made of dry grass interspersed with a few feathers,) perfectly circular in shape, about 9 inches in diameter, and 4 or 5 inches deep, with 3 to 4 inch walls, and have no special lining. The nests are placed in the centre of tufts of tall grass, well hidden, and difficult to find, generally not more than 500 yards from water. They lay from 5 to 10 eggs in a nest. Both the male and female have been started simultaneously from the vicinity of the nest; but whether the former assists in incubation is uncertain, though, judging from the loss of weight during the breeding season, the male must be in constant attendance at the nest. The weights of five males, shot between 13th February and 28th June 1880, in consecutive order being—(1), 2 lbs. 3 ozs. (13th February); (2) 1 lb. 14 ozs.; (3) 2 lbs.; (4), 1 lb. 13 ozs. and (5), 1 lb. 12 ozs., (28th June). When the young are fledged in September and October, the Pink-headers retire with the receding waters to their usual haunts—the jungly lagoons.

The following account, as indicating their strong attachment to their young, may prove of interest. On the 17th July 1880, whilst searching for Pink-header's nests with T.H. at the northern extremity of Patraha Katal, where nests were reported, we flushed a female Pink-header in the grass jungle on the banks of the Patraha jhil. T.H. fired with his Miniature Express at a distance of about 300 yards at the bird which had settled at the other end of the jhil. The ball was seen by both of us to strike the water some distance above, and a little to the left of the bird which did not rise. Upon going up to the spot, to our surprise she fluttered about and dragged herself along with loud quackings. Being closely pursued, she flew along at an elevation of about 6 feet from the ground in a manner that led us to believe that she was badly wounded, and one of her wings damaged, and she fell rather than settled in a patch of grass on dry land. Upon approaching this a similar manoeuvre was gone through, and she deposited herself some 100 yards further on. Having decoyed us thus far she flew up into the air with such facility that our old *shikaree* mahout could not help exclaiming *phair jee gya* (it's come to life again), and directed her flight in a direction away from the piece of water. After describing a considerable circuit, she came back to the jhil on the banks of which we were still standing. Two more bullets were fired at her from the same gun, which only made her rise after each shot, and settle down again some 10 yards further on. Seeing that her tactics had failed in withdrawing us from the vicinity of her young, she again took to the grass jungle, and all endeavours to flush her again proved futile, though she was observed in the same piece of water subsequently.

The Mute Swan

This species may be considered a pretty regular, though somewhat rare, cold-weather visitant to the Peshawer and Hazara districts, and an occasional straggler to the Kohat and Rawal Pindi districts and to the Trans-Indus portions of Sind. It has also, *perhaps*, occurred on the Runn of Cutch.

Outside our limits, this species has been seen in the Kabul river near Jellalabad, and is known to visit northern Afghanistan pretty regularly. It is abundant on the Caspian.[35] It occurs and breeds in western Turkestan and central Siberia, and is found also in Kashgar, where it is said to be plentiful at Aksu, and further east at the Lob series of lakes. But specimens from this latter locality have yet to be compared, and it is not impossible that these eastern birds, as well as Radde's and Prjevalski's supposed *olor* from south-east Siberia and south-east Mongolia, may really prove to have belonged to the more eastern species with *feathered* lores and orange-red bill and feet named by Swinhoe, *C. davidi*.

---

[35]In 1877 Captain Butler learnt from some of the telegraph officers in the Persian Gulf that Swans had been occasionally seen about the head of that gulf and the mouths of the Euphrates. It is impossible to say to what species these birds may have belonged.

The present species is also found pretty well throughout Europe, but becomes very rare towards the north, and in Great Britain never seems to occur in a truly wild state. It extends in winter to northern Africa, Egypt, and Asia Minor.

***

This is the tame Swan of Europe, so well known to all that it is needless to quote, from European writers, accounts of its habits which, here in India, I have never had any opportunity of observing in a wild state.

This species has been, however, so seldom recorded as killed in India that it may be well to enumerate every instance of this which has come to my knowledge.

The first occurrence of this species, of which I have a record, was near Peshawer, in 1857, when a small flock were seen, and one shot and placed in the Peshawer Museum, whence it was sent to me by Sir F. Pollock in, I think, 1867.

This Swan was shot by W. Mahamed Oomer Khan, who wrote to me about it as follows:

'In the month of January 1857, I shot this Swan in the Peshawer district on the Shah Alum River, about a mile and a half on this side of the Kabul River. Neither before nor after have I seen other Swans, but a few years after I killed it, I heard from the shikaris of Hashtnagar (also in the Peshawer district) that they had recently seen five of these birds in the Agra (?) village lake, in this same district, but had failed to shoot any.'

The specimen had been so entirely ruined by exposure and insects that I could not, at the time, decide positively to which species it belonged, but from what remains of the bill and head I have since satisfied myself that it was *C. olor*.

In 1871 Captain Unwin, of the 5th Goorkhas, sent me the skins of a pair of young Swans of this species with the following extract from his diary, under date 17th January 1871:

Today, while Duck-shooting on the Jubbee stream, on the border of the Hazara and Rawal Pindi districts, during a short halt for breakfast on the banks of the nullah, I was attracted by seeing two large white birds flying over the stream some 250 yards lower down.

The Jubbee has here a wide stony bed, with a small stream in the centre, forming occasional pools, in one of which the birds seemed inclined to alight. Changing their intention, however, they came flying up, and passed me at a distance of about 60 yards; to my surprise and delight I recognised in them most undoubted wild Swans. Firing with loose shot at that distance was useless, so I watched in the hope that they would settle in some of the pools higher up the stream, and thereby afford a stalk, but they continued their slow, heavy, flight until I lost sight of them in the distance.

Concluding that they would not stop until they reached the Indus, some 20 miles off, I was returning to my breakfast, a sadder and a wiser man, when, in taking a last look in their direction, I saw them returning. I hastily got into the centre of the nullah, in their line of flight, and as they rose slightly, to avoid me, fired both barrels, No. 3 shot, at the leader. She (for it proved to be the female) staggered, but went on, slowly sinking, till she settled in a large pool, about 400 yards off, accompanied by her mate, which alighted close beside her.

The pool, being commanded by a high bank, offered an easy stalk, and getting round into a favourable position, I found the Swans within 20 yards of me. A crowd of Gadwall (*C. streperus*), which was close by, took flight on seeing me, but the male Swan stuck nobly by his mate, and paid dearly for his fidelity, and shortly I had the satisfaction of landing them both.

The villagers who collected to see the birds gave the local name as 'Penr' (pronounced with a nasal *n*), and told me that the birds came there occasionally once in every three or fours years.

I may here notice that in other parts of Upper India this name 'Penr' is usually applied to Pelicans.

In the cold weather of 1871–2, Dr Stoliczka, when in Cutch, thought he saw Swans there. He says, J.A.S., B., 1872, 229: 'While crossing the Runn from Kachh to Pachain early in November (1871), I noticed several Swans, but at too great a distance for it to be possible to form an idea as to the species the birds belonged to.'

Until recently I had always considered (S.F., IV., 33) that Stoliczka, being very short-sighted, has mistaken pelicans (the white *P. crispus* abounds there) for Swans; but the recent occurrence of Swans in Sind renders it not improbable that

Stoliczka was right after all, and if so they would almost certainly have belonged to the present species.

Between 1872 and 1876 I received notices of Swans being killed on three occasions, on the Swat and Kabul rivers, in the Peshawer district, and in Kohat near one of our salt mines, in November, January, and February. In one case a pair, in another three, and in the last case five, were seen, one being shot in each case, but none preserved. All would seem, from what was noted of the tails and colours of the bill, to have been *olor*.

During the cold season of 1877–8 Swans were numerous in the far North-West. One was killed near Attock on the 17th of January by Lt. Hill, of the Rifle Brigade, and I heard of two others being killed in the Peshawer district in February, and of many others being seen.

On the 12th of February, Mr H.E. Watson killed three Swans in the Sehwan district in Sind.

He first saw birds of this species in January, at the Manchhar Lake, and later saw five, and actually procured three, in a small broad in the same district. He writes:—'I shot three Swans this morning. As far as I can judge they are identical with the English species' (that is the tame Swan); 'there were five on a small "dhand" or tank, about half a mile or less in length by a quarter of a mile or less in breadth. I went to shoot Ducks, but seeing these large white birds, I went after them and recognized them to be the same as those I had seen on the Manchhar. They let a boat get pretty close and I shot one. The other four flew round the tank a few times and then settled on it again. I went up in the boat and fired again, but without effect. They flew round and then settled again. The third time I shot another; the three remaining again flew round and settled, and the fourth time I fired I did not kill. Exactly the same thing happened the fifth time, the birds flew round and settled close to me, and I shot a third. The remaining two flew a little distance, and settled, but I thought it would be a pity to kill them. I considered that there would be more than I could skin myself (for I have no one that can do it for me) so I began to shoot Ducks, and then the two remaining Swans flew by me, one on the right and one on the left, so that I could easily have knocked them over with small shots. However I spared them and came home with three.'

These specimens proved, as surmised by Mr Watson, to belong to the present species and to be adults—a noteworthy fact—it being almost exclusively birds of the year that visit India.

But the most remarkable instances have yet to be noticed.

On the 3rd of June 1878 Major Waterfield telegraphed to me from Peshawer that a swan had just been shot.

Later he wrote: 'The Swan was killed on the Ojca jhil on the 3rd of June; there were a pair, but the other flew away. The bird that I have had preserved for you measured exactly 5 feet in length and 7 feet 5 inches in expanse. The feet and legs were black; the upper mandible is reddish white; its edge, lores, and lower mandible black.'

A few days later Mr D.B. Sinclair wrote to say that he had killed another swan, a male, on the 1st of June at the Gulabad jhil, 12 miles north-east of Peshawer, and on the 7th July he wrote to say that there was still at least one Swan left on this same jhil.

The specimen sent by Major Waterfield proved to be a nearly mature *C. olor*, but Mr Sinclair's bird, unfortunately imperfectly preserved, decayed so rapidly in the hot weather that then prevailed, (the temperature was over 100° Far. in the shade at 10 a.m., in Peshawer at the time,) that it shortly grew a mass 'to make men tremble who never weep;' and though, from what was said, I believe it also to have been *olor*, I cannot be certain.

What *could* possibly keep a number of Swans down in the middle of June in one of the hottest places in India, I cannot pretend to say.

Looking to the uncertainty that exists at present as to the number of species that visit us, and to the difficulty apparently experienced by many (a difficulty in which until I had studied the group I fully shared) in discriminating the young birds, it is very desirable that sportsmen should preserve every specimen they shoot, and submit them for examination to some competent ornithologist.

***

Naturally this rare and *normally* only winter visitor does not breed with us. Many of us have taken the eggs of tame birds at home, and know well the huge nest that they build of rushes, reeds, and coarse aquatic herbage, on the bank of some island or shore of a lake, or in thick reed beds. The nidification of the wild birds in Turkey, south Russia, in Sweden and Denmark, about the Caspian, in western Turkestan and central Siberia, what little has been recorded of it, seems to differ in no way from that of their domesticated brethren, except that the wild birds are said by some to breed gregariously, many nests being placed in close proximity to each other. They lay from five to eight eggs, (and the domesticated birds at times as many as eleven), with but little gloss of a rather coarse texture; in shape rather elongated, very regular, obtuse ended ovals; in colour a dull pale greenish grey or white, and they average nearly four and a half inches in length by nearly three in breadth.

ALLAN OCTAVIAN HUME and CHARLES HENRY TILSON MARSHALL, The Game Birds of India, Burmah, and Ceylon, 1879–81

The introductory remarks in Mason and Maxwell-Lefroy's landmark book The Food of the Birds of India *make for interesting reading. These early pioneers of the scientific method discuss the importance of studying the food and feeding habits of birds in an attempt to lay the foundations of economic ornithology in India. They talk of the biological control of insects and pests and the useful role of birds in this regard. C.W. Mason was a British ornithologist, and Harold Maxwell-Lefroy (1877–1925), who was a professor of entomology at Imperial College, London, was appointed as Entomologist to the Government of India. He was involved in the creation of the Imperial Agricultural Research Institute in Pusa, Bihar.*

Economic Ornithology

Ornithology is a subject which naturally appeals to most men whose work or leisure takes them into various districts and especially the wilder and little known ones. The observations of these men have naturally tended in a few directions, namely, a definite knowledge of what species of birds occur within Indian limits, a knowledge which is all-important from an economic side of the question, definite localities in which the various species occur, their life-history and general habits. Very little is on record with regard to the actual food of birds, and no definite work has been done in this direction. It is now a generally recognized fact that birds play a very important part in checking ravages of insects on the farm and elsewhere. But owing to ignorance, lack of observation, and often to faulty observation, a very small percentage of the good done by birds in checking undue proportion of insect life is attributed to them, and for similar reasons some birds at present considered beneficial are injurious and vice versa, or else fall under a neutral heading, whilst others again are both beneficial and injurious depending on locality and food-supply.

Improvements in agriculture such as are now going on, naturally tend (and will continue to do so) to the introduction of new varieties of crops into districts suitable for them, and in which as yet they have not been grown. Now, a crop newly introduced into a district is grown experimentally at first, on small areas; should these small areas be attacked by insects—and newly introduced crops often are so—the people of that district will conclude such crops are not worth taking up on a commercial scale, if most of the mare to be grown to feed insects. Insecticides and practically-applied, scientific measures can play an important part in checking insect-attack. Such measure are all important on experimental areas and during sporadic insect-attacks; but it is as well to bear in mind that natural checks are quite as important, if not more so. Natural checks are always there, always keeping the balance of life more or less even, and it is these we have to thank for limiting injury to crops and orchards to a very large extent; they act as a continual check on injurious insects and insects which are generally regarded as harmless, but which may at any time change their habits somewhat to the injury of crops. These checks consist of parasitic and predaceous insects, animals, frogs, reptiles and, above all, birds. As man upsets the balance of nature by extending cultivated areas and by a more or less artificial production of crops, he lays himself open to attack from all sides, and must make as much use as he can of the help given him by nature against these attacks.

From the most casual field observations, much can be learnt in a general way about the food of certain birds during some parts of the year. We can see Mynahs catching moths, crickets, etc., and eating maize, the Hoopoe probing the ground for caterpillars, the Rose-ringed Paraqueet pulling wheat and mustard to pieces and taking more than his share of *lichi*s; and many other similar notes can be made about these and other species of birds. It is therefore quite an easy matter to state that the food of such and such a bird consists of, say berries, beetles, and grubs, and it is interesting to know such is the case. Such sweeping statements are, however, valueless to any one in a practical way except as showing vaguely what class of food a bird may be expected to take at certain seasons, and merely show how little is known about the bird's food. Scientifically such information is practically valueless, and for all practical purposes can only be used as a doubtful basis for future work on the subject. But it must always be remembered that field observations, if first hand and made by capable men, are invaluable as supplement to laboratory examination and determination of stomach contents of birds, and should be recorded whenever possible, however vague and useless they may appear.

In India, at present practically nothing is known about what birds do eat. From the economic point of view, the scientific identification of birds' food is of the utmost importance, and especially with regard to the insect portion. Economic ornithology is, therefore, a sister science to economic entomology, just as much or perhaps even more so than botany.

To aid agricultural interests, nature is called in practically and artificially, and every effort should be made to use such helps from every possible source. Wild birds are the source in question here. We should therefore know the value of every separate species of bird to man, i.e., know what insects, what seeds, what fruits and other vegetable and animal materials are taken as food by birds at all times of the year under all conditions, climatic and physical. We can then, by encouragement of useful species and destruction of harmful ones, check the attacks of insects on crops, and enable the country to increase crop outturns, and in every way benefit agricultural and therefore the country's interests.

C.W. MASON and HAROLD MAXWELL-LEFROY,
The Food of the Birds of India, 1912

Hugh Whistler (1889–1943), well-known for his book Popular Handbook of Indian Birds, *served with the Indian police in the Punjab province. His collection of bird skins is now at the Natural History Museum in England. The following piece was a letter of advise to Sálim Ali on how to run a bird survey.*

Suggestions on How to Run a Bird Survey

'We are all rather agreed that certain recent surveys, particularly those of the Americans and Germans, have been examples of how *not* to go to work. They send a collector to an area and their aim is simply accumulation of a vast number of specimens, largely in order to get new forms and to get duplicates for exchange. They make no endeavour to furnish information of general or biological interest, they teach us practically nothing and their reports are merely critical remarks on skins. The result has been chiefly discredit. I want you to work on much more general and useful lines and I know that that is your own particular bent. You are a biologist and not merely a dry-as-dust closet worker, so I feel sure that we see eye to eye in this matter.

First of all we want actual specimens (1) for identification of the forms which occur in the state (2) in corroboration of your field notes (3) for record and comparison with specimens in other areas (4) for studies of plumages and moult. With Henricks as a skinner you are relived from the manual drudgery and expense of time that the preparation of skins implies. On my own trips I always have to waste my own time in skinning which therefore lessens my time for more important work. You will be free.

Half an hour round the camp in the morning will therefore suffice to provide Henricks with work on the series of common birds. Anyone who can fire a gun can produce half a dozen birds for him to get on with, pending the arrival of more important things. You will then be free to work the surrounding terrain properly to make sure that no species are overlooked. In Ladakh I used to get out early and get home about noon; and generally have a short evening turn as well. If the common stuff is dealt with by an underling at the camp you yourself can confine your own attention to bringing in more important things. If you are doing a five-mile round it is waste of opportunity for you to be getting the babblers and bulbuls which are common by the camp.

On arrival at a camp you want to study a large-scale map very carefully and see what types of terrain are in the vicinity. Most birds are distributed according to terrain—especially the more interesting ones—so if you only go the same old round again and again you will miss half the interest of the neighbourhood. The map will show you perhaps that in one direction there are low rocky hills, in another an open wide river bed, two miles off is a large *jheel*; a reserved forest and open cultivation fill in the other areas. Each of these terrains will hold certain special species in addition to the generally adapted forms which are capable of flourishing in all the types of terrain. It is important therefore to establish for each camp (1) which species are able to flourish throughout the area (2) which species only inhabit certain parts of the area. You are then in a position to start to establish the biological factors which are responsible for the differences of distribution. Everyone knows for instance why the Snipe will only be found along the margin of the *jheel* but there are innumerable similar factors which regulate the distribution of other forms—food, cover, special adaptations, breeding requirements, etc. Each type of terrain requires to be worked until you are satisfied that you know all about its inhabitants. Watching and thought are of as much importance as killing. I propose in my next letter to give you suggestions as to the points to consider, species by species. Here I am only generalizing, so I will not say more about this study of terrain than to give the larks as an example. Round your camp in a five mile radius you will perhaps find 4 forms of Lark irregularly distributed. Now there must be an explanation behind the irregularity of their distribution. Its ultimate basis is probably food, but food will express itself in external form— one lark may be confined to black cotton soil: another may need open ground under *sal* trees: a third may only be patchy because from some ecological reason it is numerically scarce and so there are not enough individuals, and its numbers never increase, to populate the region. We do not therefore want merely on the American model a report:

Mirafra cantillans.	Abundant.	75 specimens	Camps A B C D
Mirafra assamica.	Very rare.	1 specimen	Camp B
Alauda arvensis.		21 specimens	Camps A & B

We want some hint of the factors which included their comparative abundance and their difference in distribution. That is what should lie behind ornithology—the specimens and the correct name should not be means to an end.

I realize of course that in the time at your disposal you will not be able to settle all these points, but we want *your* observations both as a contribution to the problems and as a stimulus to workers in other areas who can then proceed to corroborate or disprove your suggestions.

All the time ask yourself the question WHY. Why is the distribution patchy? Why is the bird in the sandy dry river bed and not in the cultivated plain alongside? Why is it in the roadside avenue and not in the forest? And why has it special modification? Any bird with special modification, the scimitar-bill of *Pomatorhinus*, the racket-tail of *Dissemurus*, the heavy beak of *Pyrrhulauda*, must inspire you with a desire to see why it has the modifications which separate it from others of its family. All such points will occur to you naturally as the survey progresses.

Let your notebook be just as important as your gun. I should recommend you to keep several notebooks. First of all you should have a large general diary to be kept day by day after the lines of those kept by Hume in his Sind trip (Vol. I *Stray Feathers*) and Scully in his trip to Eastern Turkestan (Vol. IV *Stray Feathers*). This will describe the localities, terrain, chief forms met, with special points of interest. At the end of the Survey it can then easily be polished up as an introduction to the Survey report.

Then I personally keep another notebook under species heading. This I run through daily noting each date and place where each species is met with and all points of interest. Each new species is given a space and heading as it occurs. Running through the pages daily serves to ensure that nothing is forgotten. It also gives one the distribution clearly. It was annoying often in the E. Ghats survey to have nothing to show whether common birds did or did not occur at a camp—the absence of skins often probably really meant that the bird was not collected as sufficient had been obtained at other camps—there was nothing to show whether it did or did not occur. The amount given under these species headings varies of course. The Jungle Babbler for instance gets off with 'May 2–31 Camp Hylakandy common and general in all types of terrain'; whereas with a migrant or irregularly distributed species there are daily records with full details. I have daily records extending over years for the migratory species in the Punjab which show the waxing and waning of their passage periods.

Then I should keep a small notebook for soft parts of specimens. Do not write the colours of soft parts on the labels. With each fresh species obtained start a separate page for it in a notebook: write down very carefully the soft parts of the first specimen with the serial number of the skin. Each fresh specimen would then be compared with that entry, the similarity or the differences being noted under its serial number. This will ensure uniformity of description and then when the skins are worked out we can see if differences in the soft parts are correlated with sex, age, and seasonal differences. The usual hackneyed formula writing label by label takes far more time and gives far less value— the specimens are divided up, the results are never correlated, and if the colours given on two labels differ one does not feel sure that the difference is not merely two different days' versions of the same colours. This of course implies that your first act in bringing in the day's specimens is to list them up in your serial register of skins and fit each bird with its serial number before it is skinned. The soft parts should of course be noted as soon as possible. This little register may well go out in your knapsack to the field. Be sure to include the colours of the inside of the mouth, which are usually quite neglected but often tell one a great deal.

Don't trouble to measure birds in the flesh. It is however of interest to weigh the larger forms.

Your labels should give information on the following points (1) The state of the organs (2) State of skull (3) Fat (4) Moult.

Regard correct sexing as the most important part of the preparation of the specimen. Henricks will not be able to sex every specimen—shot marks, heat, immaturity, off season, will all make it impossible to sex certain specimens. But it is essential for me to know that when you mark a bird as male you do so because you have had absolute proof *by dissection* that it is a male. Do not guess from the plumage—I can do that, and also the plumages are far less safe a guide than you may realize. Because people have guessed or sexed wrongly for 100 years many facts about the sequence of plumages have been unknown to us. Describe the organs as you find on the label—give a drawing of the size of the testes or of the ovaries where possible. If you are doubtful say so—'organs obscure but apparently ?'. The more importance you attach to this point the more value I shall be able to extract from the skin in due course. Mark the presence of incubation patches. Say if shot actually *off* a nest—we don't know which sexes incubate the eggs.

With regard to the skull an experienced skinner can say whether a bird is juvenile or adult (within certain limits) from the degree of ossification of the skull. In the juvenile the skull is very soft, hardly more than cartilage—it takes 3 to 4 months to ossify fully. Ossification starts at the base of the skull by the insertion of the vertebral column and also behind the eye—the two areas advancing to meet each other over the brain pan. After the post juvenal moult there is still a patch of unossified skull showing as a little window in the centre, gradually decreasing till the window fades out. If notes are made in the skulls about ossification— and an experienced skinner soon knows it well—between the breeding season and November (after that it is too late) it gives tremendous help in plumage studies, as incomplete ossification at once betrays the immature bird, whatever the plumage.

Presence of fat in excessive quantities shows that the bird cannot be breeding and that it is probably on migration. Presence or absence of moult if noted helps in plumage studies. Details are not necessary.

I should like you to pay a good deal of attention to food, not of course the hackneyed remark 'insects' or 'seeds', but to any special foods which are obviously being favoured by particular species and which may help to explain their distribution. It is advisable to take a good supply of small test tubes and then stomach and crop contents can be preserved in weak spirit for later identification.

Such small test tubes are also useful for preserving small chicks. You should preserve for down studies 1 or 2 chicks of every species of which nests are found—regard chicks as more value than eggs. Downy nidifugous chicks and larger nidicolous species (e.g. birds of prey & eggs) are better skinned. Be careful to establish the identity of your chicks.

You will say to yourself in reading all this long farrago that there is nothing new in it and that all the directions are obvious. I agree, but my experience is that 9 out of 10 collecting trips and collections lose a huge proportion of their value from a neglect of these obvious details. Carry them out and we shall be able to write a first class report on the birds of Hyderabad State which will be of far more than local interest.'

HUGH WHISTLER in Sálim Ali's *The Fall of a Sparrow*, 1985

British geneticist and evolutionary biologist, John Burdon Sanderson Haldane (1892–1964) was among the founders of population genetics. On the invitation of P.C. Mahalanobis,

Haldane moved to India to join the Indian Statistical Institute, Calcutta, as a Research Professor in 1957, and subsequently acquired Indian citizenship. He later moved to Bhubaneswar to start his own Genetics and Biometry Laboratory. Haldane wrote extensively on science popularization. The title of the piece reproduced here reveals his reverence for Mahatma Gandhi since it emphasizes non-violence and stresses that birds can be studied by observation and counting, and killing them for the purpose of study is therefore unnecessary.

The Non-violent Scientific Study of Birds

I am very ignorant about birds, largely because I am unmusical and most British birds are small and inconspicuous, so that their songs and call-notes are more distinctive than their colours or shapes.

In the nineteenth century it was hard to study birds without killing them. The first job of an ornithologist is to identify species, and in order to be sure that we have, for example, three and only three species of kingfisher in the suburbs of Calcutta it is necessary to kill a number, and find that all can be assigned to one of these species. This phase is now fortunately nearly over. One can learn to assign a bird to its correct species without killing it.

What is the next step? It is, I think, to find the distribution of species and subspecies in India at different times of the year, and also their local habitat, names, and so on. Here Ogniev's great Zoology of the U.S.S.R. could be a model. Ultimately we should look forward to a time when there will be an ornithologist for every hundred or so square miles of India capable of enumerating the local species, and a central organization such as the Bombay Natural History Society to make maps showing the distribution of each species in India. As, however, this would require ten thousand or so ornithologists it is not immediately possible. But a start can be made.

The next question to be asked is, perhaps, how many birds of one or more species there are in a given area. At first sight this is a very difficult question, as birds are so mobile. But as eggs they are extremely immobile. I hope that, if we develop statistical biology at the Indian Statistical Institute, we may make the attempt to enumerate all the nests of some conspicuous species, such as vultures, night herons, and cattle egrets, in an area of ten square miles or so. When this has been done for thirty or so representative areas in India we shall be in a position to estimate, no doubt very roughly, the total population of these species in India.

The total numbers of breeding adults of a few local species are roughly known (see Fisher and Lockley 1954). Thus for the gannet, *Sula bassana*, the number of nests in the East Atlantic area (Britain, etc.) was about 70,000 in 1939 and had risen to 82,000 in 1949. In the West Atlantic (Newfoundland, etc.) it was about 13,000 in 1939. Thus at present there are about two lakhs of mated birds, and perhaps as many juveniles. They live on a small number of precipitous rocks, mostly on small islands. There are fifteen 'cities' of 17,000 to 1200 nests (*sic*), and fourteen 'villages' of 500 nests or fewer. These numbers are fairly accurately known. James Fisher had counted thousands of nests on cliffs from small boats. He was able to induce the British Naval Air Force (Fleet Air Arm) to photograph many of these sites as part of their training. The exact numbers of nests could be counted at leisure from the photographs, and the results compared with those obtained by cheaper methods. Few of the latter were incorrect by ten per cent.

This 'urbanization' is characteristic of sea birds, and is carried to great lengths in more numerous species. The extreme example is furnished by *Uria lomvia*, Brunnich's Guillemot, of which there appear to be four or five million on the coasts of Greenland, about half of which breed on a single rock Agpar-s-suit. This is one extreme of bird behaviour. Most small song birds keep a 'territory' around their nests private by singing and quarrelling with intruders, even if they are more sociable when not breeding, while others, such as the Indian weaver bird, live in 'villages' of a few tens or hundreds of nests.

Is there any possibility of counting all the breeding numbers of an Indian bird species? I suggest that the most hopeful targets are the large flamingo *Phoenicopterus antiquorum*, and the smaller species *Phoeniconaias minor*. The former breeds in the Great Rann of Kutch, the latter possibly in the Little Rann. The Lesser Flamingo, which lives on unicellular algae, is not apparently found in many other localities except Sambhar Lake in Rajasthan.

The Rann of Kutch is unsuitable for walking but, owing to the absence of trees, it should be possible to photograph nesting birds from the air. This can of course only be done by the Indian Air Force. In peace time the armed forces have to carry out exercises of various kinds. Their efficiency can be better gauged from their performance against natural forces, for example the rapid replacement of bridges destroyed by floods, or the landing on a difficult coast, than by their prowess against 'enemies' who they know will not hurt them. Hence such co-operation would, I believe, increase the efficiency of our Air Force.

So much for mere populations or densities per square mile. But how do these increase or decrease? Observations on a few hundred or even a few dozen nests of any species will tell us the average number of eggs laid per year. More careful, but not very arduous, watching will tell us how many young birds per nest survive to start flight. On the whole tropical birds produce fewer eggs in a clutch than birds of the same species or a closely related species in a temperate climate. This is at least partly due to the shorter tropical days, which do not give the parents time to feed a large brood. Most of the comparisons have been made by Moreau with African birds, but Lack (1950) points out that in India *Parus major* (the Great Tit) has an average clutch of 3 compared with 10 in England. This difference must be compensated in one of two ways. Either the average number of clutches in India must be greater or the mortality less. There must be a balance because if, for example, the numbers in an area increased by only 10 per cent per year for a century, the density would increase 13,781 times. This can of course happen when a new species occupies a country, but not with established species. In only one case has this balance been directly demonstrated by comparison of statistics. In Switzerland the Starling (*Sturnus vulgaris*, a bird very similar to the myna) lays more eggs than in England, but dies younger. It will be easy to get data on numbers of broods in India, not so easy to get data on mortality. Before I speak about mortality, let me say a few words on the feeding of young.

What do they get to eat? One can of course kill parents and examine their crop contents. Apart from ethical consideration this means that one can only get one piece of information from a bird. Several other methods are available. Lack found that if he caught parent swifts (*Apus apus*) they might desert their young. So he waited until a parent bird fed a baby and departed, and then pressed the baby's throat, getting a pellet containing about 600 insects entangled in the parents' sticky saliva. They were largely flying aphids, so swifts eat insects which compete with men for food plants, and what is more, eat them while they are moving to new food plants and invulnerable to sprays and other insecticides. Thus swifts seem to be wholly favourable to agriculture, whereas some other bird species live largely on seeds and lower agricultural output, while other insect eating birds eat some insects, such as bees, which assist in the pollination of plants and thus help human horticulture and even agriculture. We should certainly encourage the birds which are helpful to man, even if we do not massacre the others. A second non-violent method has been used in the Soviet Union. The nestlings are replaced by models which, when a watcher pulls a string, open mouths and may emit a suitable noise. The food falls into a bag, and I hope is given to its legitimate owners after the insects, molluscs, seeds, and so on, have been assigned to their correct species.

Do young birds get enough to eat? Lack (1954) found that when the brood size was less than the average, the number of young starlings surviving for a few months was roughly proportional to the brood size. However this was not so when the brood size exceeded the average. Even if the excess young survived to fly, they did not survive much longer. Presumably their parents could give them enough food to fledge, but not enough to get an adequate start in life. The technique consists of ringing nestlings. But of 15,000 starlings ringed in this research, only 346 or 2.3 per cent were recovered, that is to say found dead and the rings returned.

This ringing technique was invented by Mortensen in Denmark to study migration. As you know, several ducks ringed in India have been picked up in Siberia and vice versa, and one German-ringed stork in India. Ringing birds does not harm them. One ringed robin (*Erithacus rubecula*) in Eire lived for eleven years, though nearly two-thirds of all robins die each year, so only about one robin per lakh is expected to live so long. It is a fortunate and peculiar fact that birds' legs are fully grown before they start flying. A metal or plastic ring can therefore be put on a nestling and remain on its leg for life. The rings usually carry a request to send them to a certain address if found. There may be a small reward. In Western Europe population density and literacy are both so high that as many as 15 per cent of the rings on large birds returned. We cannot yet hope for such good results in India. But we may reach them when our children are educated.

JOHN BURDON SANDERSON HALDANE, Journal of the Bombay Natural History Society, 1959

Birdwatching and Beyond

5

Popular and highly stylized forms of representation of birds on a matchbox (TOP) and a goods-ferrying truck

In this long forgotten classic, Sálim Ali writes about the detective work and the skills required for tracking birds' nests. The piece, though written in 1930, was later reprinted in several newspapers and journals, including the English daily Indian Express *and* Newsletter of Birdwatchers.

Stopping by the Woods on a Sunday Morning

The island of Salsette, the potential Greater Bombay, is a veritable Dr Jekyll and Mr Hyde. For the greater part of the year it sleeps under the drab mantle of desiccated grass and dustladen foliage, which is only lifted here and there in patches as the hot weather advances, revealing gorgeous tints of scarlet and orange as the various flowering trees, the silk cotton, the coral, and the butea (palas), blossom forth in masses of living flame.

But what a transformation the first few showers of the monsoon bring about. It is as though some magician had, by a pass of his wand, instilled fresh life into every object in the countryside. The grass springs up everywhere, and with it a host of innumerable monsoon weeds, till soon, the whole landscape becomes one great fantasy in green.

Pools and puddles begin to form. The bull frog awakens to the song of spring after his protracted underground slumbers, and his croaking fills the air as he joyfully serenades his lady love. Soon she will lay her eggs in some sequestered pond, where a few days later innumerable multitudes of tadpoles will emerge to carry on the race and save music from extinction.

We shall select some Sunday morning late in August for a jaunt into the exquisite country surrounding the city. The heaviest blast of the monsoon is blown over, and we may now look forward without undue optimism to fine weather. The air is delightfully cool, the sky thinly overcast; banks of threatening nimbus drift across the heavens resulting only in occasional drizzles which help to subdue the uncomfortable steamy vapour that begins to rise immediately after the sun peeps out of his cloudy veil.

We leave our car by the side of the road, and loading ourselves with haversacks containing some sandwiches, a water bottle, specimen tubes, and a camera, we commence the trudge into the interior. Creepers of that magnificent lily, *Gloriosa superba*, aptly named, are growing in every hedge. A few of the flowers are out, though the majority will bloom after a couple of weeks. Every monsoon, this creeper springs into life from the bulbs lying latent underground from the previous season.

The flower itself is a picture of loveliness, yellow, red and green. The delicately shaped tapering leaves terminate in tactile tendrils which readily entwine themselves round any object that comes within their reach. Would that colour photography were easier of attainment! No ordinary photograph can ever hope to do justice to this exquisite flower. The gloriosa lily is, as it were, part and parcel of the suburban countryside of Bombay in the rains.

Large numbers of a wild gentian are also in the swampy grass fields now, while clusters of the dazzling redixora flowers are present on all sides. Everything is calling out to the naturalist; would that this state of loveliness could survive the months to come, of dust and heat and desiccation!

We follow the path leading into the 'hinterland', the main object of our ramble this morning being to locate birds' nests. We turn our footsteps towards the hills, on the other side of which lies the beautiful Tulsi lake hemmed in by verdure which might rival in magnificence that of any tropical rain forest.

The monsoon is the breeding season par excellence of insectivorous birds, and also of the numerous others who, though when adult, subsist principally on grain, yet require soft food in the nature of juicy grubs and caterpillars to nourish their young in the nest. Owing to the sprouting of fresh grass and vegetation, the caterpillars, which also appear at this time in devastating hordes, find easy sustenance. Thus, it is in birds that Providence has devised the most efficient automatic control agencies. Were it not for the check exercised by man's feathered friends at this crucial period, a time would soon come when not only crops but all vegetation would cease to be. Such is the astounding rate at which insects multiply that no power of man's invention alone would ever be capable of stemming the overwhelming tide of their numbers.

There is warbling and song on every side; courting and nest-building are in progress everywhere, and a few early birds are already catching the worm, that is to say, those who have already undertaken parental cares are now busy feeding their chicks.

Finding a nest in thick cover is by no means a simple matter. There are people who will cover miles of a morning in the most promising-looking country and complain to you later that they did not come upon a single nest; that, as a matter of fact, there were no nests to come upon. Happily, there is a knack in locating birds' nests, and to hope to come upon them accidentally is futile. If this were not so, it would result in a very serious menace to the birds and be a grave impediment to their success in rearing families.

Nests are protected from their enemies either by being built in such secluded spots that without a clue of some sort, no one would think of searching for them there; or they are built of such material and design and with so much cunning and camouflage that to the untrained eye they either become totally invisible or entirely unsuspicious looking objects.

The nest of the Purple-rumped Sunbird, common almost throughout India, affords a case in point. It is a pendulous pouch attached to the tip of an outhanging twig of the *ber* or *babool* tree, seldom more than eight or ten feet from the ground. The material employed in its construction is fine rootlets and fibres, and the whole thing is so untidily plastered over on the outside with all manner of rubbish—spiders' egg cases, pieces and shreds of pith, bark and paper, strings of caterpillar borings and droppings, and so on—as to resemble to perfection a mass of rubbish, and least likely to attract the attention of the casual passerby.

The simulation is further heightened by the fact that the entrance to the nest—a round hole near the top of the pearshaped structure, surmounted by a tiny little porch—is always on the inside, i.e., facing the tree, and therefore concealed from the intruder. Such camouflaged nests usually get detected only on account of the movements of their owners, their comings and goings with building material or with food for their young or, otherwise, by their inordinate fussiness.

The trick of locating nests, therefore lies not so much in traversing miles of likely country as in keeping an ever-watchful eye as you slowly saunter along, and patiently waiting for the birds to give away their secrets of their own accord.

An insignificant little brown and white bird, somewhat smaller than a bulbul, silently slips off a *karonda* bush at our approach, and flies into a neighbouring tree whence the field glasses disclose his apparent anxiety. This behaviour is distinctly suggestive. We walk upto the bush and peer inside. A pleasant surprise is in store.

There, concealed from view by the large green leaves and almost in the centre, is a deep cup made of rootlets and grass, slung hammockwise between the stems of two monsoon plants. It is plastered on the outside with a supply of cobwebs. The cup is so deep that we have to bend right over for a view of the contents. It holds three beautiful roundish eggs, yellow-white in colour with fine ruddy specks. Having photographed the nest, we withdraw behind a neighbouring bush and await the return of the unidentified proprietor.

Finding the coast clear, our friend approaches, he is too cautious to fly straight up to the nest. Alighting on the further side of the bush, he hops from twig to twig, peering through the tangle to assure himself that the danger is past. Soon, he comes into full view and in the twinkling of an eye, slips in and is settled on the eggs. Binoculars now disclose his identity.

He, or it may be she, for both sexes are alike and take part in incubation, is the Yellow-eyed Babbler, a chestnut brown bird with white underparts and a conspicuous white streak over the eye. Close relatives of the well-known 'seven sisters', Yellow-eyed Babblers go about in small parties, searching for insect prey among bushes and under fallen leaves. During this, their bridal season, the males constantly clamber up to the exposed tips of bushes and tussocks of grass, and burst forth into a pleasant little song of several loud and melodious notes.

Leaving the Yellow-eye to its parental cares, we proceed on our way. A great commotion set up by a pair of fussy little Tailorbirds draws us towards the thick tangle of a large-leafed creeper. The anxious couple hops around us from bush to bush, expressing the deepest concern in a series of alarmed 'pit-pit pit-pits'. Their antics lead to a search which is soon rewarded by the revelation of that beautiful little sartorial masterpiece, the Tailor's home.

A large pendant leaf is folded round in the shape of a funnel, and neatly stitched with thread of vegetable down along the edges. Within the cone so formed is a regular cup of fibres lined with cotton and down. The nest is fresh but empty, but we soon discover a trio of fluffy chicks, stumpy-tailed little mites, who have obviously just made their debut into the world. They sit huddled together on an adjacent twig, too innocent yet to have learnt anything of the wiles and treacheries of the world, and so, are perfectly fearless and confiding. Unfortunately, the light is far from satisfactory, and we have perforce to resume our tramp without having used the camera.

The Tailorbird is one of our three commonest warblers, and certainly the most accomplished nest-builder of them all. The other two are also tiny birds of about the same size with longish, loosely-set tails, and are known as the Ashy Wren-warbler and the Indian Wren-warbler respectively. Both these are also busy with family cares at the present time, and, as a matter of fact, we have not far to go before we alight on a nest of the former.

It is on the farther side of a *nullah* that lies across our path. We catch a glimpse of the occupant as he takes off from a chunk of the large leafed monsoon weeds now so abundant everywhere. Marking down the spot, we wade across and bend low to have a good look under the leaves. There it is, a structure not unlike the abode of the Tailorbird, which, to our delight, contains three tiny, polished, brick red eggs.

The remarkable thing about this nest is that it hangs directly over a used cattle path, on which the hoof marks still show— muddy puddles indicating that cattle have just gone over. Each time an animal passes this way, the nest must be brushed aside and shaken violently. The bird is undoubtedly an optimist; but it has at least the courage of its convictions, and is now well on the way to bringing up a family, mischievous herd boys permitting.

That dainty little fairy waltzer, the Fantail Flycatcher, whose cheery song and lively movements delight every resident of Bombay fortunate enough to posses a garden, has also turned his fancy to thoughts of love. In a lime tree growing in a semi-deserted garden, barely at a height of four feet from the ground, a marvellous little cone-shaped cup, two inches across, is marked down by following a bird carrying off a caterpillar. It is well plastered on the exterior with that approved cement of bird architects, cobwebs.

Three little baby birds occupy this nest. They are nearly full-fledged and will sally forth into the world in a day or two. Everybody acquainted with the Fantail knows what a fury it can become when its nest is in danger. The parents promptly launch a violent attack, pecking at our hats and uttering feverish chucks of irritation, which, no doubt, are far from complimentary language.

The harmony within the Fantail household is an object lesson in domestic give-and-take. We are astonished at the way in which three strapping, grown-up, hungry chicks can accommodate themselves amicably in this diminutive domicile. They are packed so tightly together that two of them have their wings hanging over the edges of the nest.

The Common Iora's nest is a very similar structure to the flycatcher's, with this consistent difference that while in the fantail's nest, strips of grass and rubbish are left dangling below, the Iora's is well rounded off at the bottom. Ioras also nest at this time of the year. They usually select a crotch formed by horizontal or vertical twigs, building skillfully around them so as to incorporate the supports into the wall of the nest.

The cup is composed of grass, fine roots, and fibres, and here again, cobwebs play an important role in the lacing. In addition to binding the material firmly together, cobwebs serve as an efficient waterproof covering to prevent the water from seeping in through the sides of the nest. When the bird is sitting on its eggs, the plumage of the back is frowzled out and raised to form a dome. The fluffy feathers of the lower back, moreover, overhang the sides of the nest, the tout ensemble forming a most effective protection against the heaviest monsoon shower.

Birds are loath to allow their eggs to get cooled. While we are getting the camera ready to photograph the nest, the sitting bird is alarmed and leaves. Presently, a drizzle intervenes, the hen iora takes up her position on the eggs regardless of our proximity, nevertheless keeping one eye intently on our movements.

Although male bayas (weaver birds) have begun to don their nuptial garb about the time of the rains breaking and quite a number may be seen playing at nest construction, they hardly give serious thought to parental responsibilities till the monsoon is well advanced. Around the end of August, operations begin in earnest and work is in full swing everywhere. Unlike the sunbird, the lion's share of the work appears to devolve on the cocks. The hens make their appearance on the scene only at a later stage. Their arrival invariably causes a great flurry amongst the lovelorn swains, whose strutting, impetuous advances must be quite embarrassing to the fair ones.

It is getting late in the afternoon, and we are a good way off from the car. Our way back lies through patches of open grassland, where the cattle from the neighbouring village are turned out to graze. Amongst these we find numbers of large white birds with long slender necks and pointed dagger-like bills. They run freely in and out of the animals' feet, darting forward every now and again with lighting rapidly at the insects and grasshoppers disturbed in their progress. These are the cattle egrets, found in attendance on village cattle all over India. They are now in their breeding livery; golden on the neck and the back.

A monsoon ramble through the woods will delight anyone who has the eyes to see and the soul to wonder at the romance and charm of this other world within our world. The electrification of the suburban railways has now thrown the delightful country in the environs of Bombay within comfortable and speedy reach of everybody. To the lover of the out-of-doors, the opportunities are such as might rightly be the envy of the less fortunate dwellers of almost every one of the other large cities in the country. Yet, how few are there who will sacrifice their Sunday morning sleep.

SÁLIM ALI, 1930

The Baya played a significant role in shaping Sálim Ali's career. His first major ornithological discovery was the mating behaviour of the Baya. The piece which Sálim Ali originally wrote on his discovery received an award as an exceptional of piece of writing in English by an Indian. Regrettably, the original piece could not be traced, and instead a variation of the same is reproduced here from Sálim Ali's autobiography, The Fall of a Sparrow.

The Nesting of Baya

The male baya, who in his breeding livery is a handsome little sparrow-like bird, largely brilliant golden yellow, is an artful polygamist. He may acquire any number of wives, from two to four, sometimes even five—not all at once in the harem style but one by one progressively, depending upon his capacity to provide them each with a home. The male alone is responsible for building the nest; the female has no hand in it. Males select a *babool* or palm tree to hang their compactly woven retort-shaped nests, and several males build together in a colony which may sometimes contain a hundred nests or more. At a particular stage in the construction, when the nest is about half finished, there is suddenly, one fine morning, an invasion by a party of females prospecting for desirable homes. They arrive at the colony in a body, amidst great noise and excitement from the welcoming males, and deliberately visit nest after nest to inspect its workmanship, as it were. Some nests are approved, others are rejected. While the examination is in progress the builder clings on the outside, excitedly flapping his wings in invitation and awaiting her verdict. If the female is satisfied with the structure she just takes possession of it and accepts his impetuous advances. A hurried copulation takes place on the 'chinstrap' of the helmet-like half-built nest, and the pair bond is sealed. Thereafter the male resumes his building activity and soon completes the nest with its long entrance tube. The female lays her eggs within, incubates them and brings up the family. This is entirely her responsibility, and it is rarely—and only after his building impulse has finally subsided—that the male takes a hand in foraging for the chicks. Having completed this nest the male proceeds almost immediately to start a second one a few feet away. At the appropriate half-built stage, another house-hunting female may in like manner take possession of this second nest, and the whole process is then repeated. Thus the male baya may find himself the happy husband of several wives and proud father of several families at practically one and the same time. It sometimes happens that for some feminine foible, female after female fails to accept a certain nest. Undeterred, the male abandons the half-built structure and promptly tries again. In every baya colony there are usually to be seen a number of such half-built abandoned nests. This is the prosaic explanation for them and unfortunately not the more popular lyrical one that they are for the use of the male to swing himself and sing love songs to his incubating spouse nearby!

Sálim Ali, The Fall of a Sparrow, 1985

Shivrajkumar was a scion of the royal family of Jasdan, Gujarat, and a well-known birdwatcher and conservationist. Ramesh M. Naik (1931–91), professor of zoology at University of Rajkot, was among the foremost ornithologists of India. A close associate of Professor J.C. George of the MS University of Baroda, he made important contributions to muscle physiology and histology of birds. He edited the journal Pavo *and created a school of ornithology at Rajkot. Lavkumar Khacher (also known as K.S. Lavkumar in his other writings) taught at Rajkumar College, Rajkot, and later worked with the World Wildlife Fund and other environmental organizations. He was closely associated with the Centre for Environmental Education in Ahmedabad where he served as the Director of its Sundarvan Snake Park for many years. These three birdwatchers surveyed the Rann of Kutch, and their report provides for a very interesting travelogue.*

A Visit to the Flamingos in the Great Rann of Kutch

On *16th April 1960* the three of us and P.W. Soman of the Bombay Natural History Society assembled in the Modern Hotel at Bhuj. 'A sight for the Gods' Dr Sálim Ali had written to us describing Flamingo City which he had visited on 21 March 1960 from the bird-ringing camp at Kuar Bet and, excited by what he said, we were about to set off hoping to share in the pleasure and to seize the privileged opportunity of witnessing the spectacle. However there was deep down amidst all the enthusiasm, a mild doubt which was ever so lightly perceptible in all our minds. Had not McCann said that the Rann dries up pretty fast when it does start doing so? He had found that the water receded many miles each day, forcing the breeding flamingos to evacuate their nests, some even containing hard-set eggs and newly hatched young, while countless chicks in the running stage fall out by the wayside unable to keep up with the rapidly retreating waters and lie as mute testimony to the disasters which strike breeding populations in nature. Suppose, we asked each other, that such a catastrophe had happened? Would we merely be witnesses to one of Nature's cruel acts of profligacy? Another alternative, the voice of doubt softly whispered: 'What if the birds have hatched out their eggs and taken their broods into deeper water?'

However, these unpleasant thoughts were not allowed to choke our enthusiasm, for the bus seats for Khavda had to be reserved for the next morning, essential stores had to be purchased, and as a couple of us discovered that we lacked sunglasses to protect our eyes against the glare of the Rann we had to set about acquiring these. We therefore spent a busy afternoon in the picturesque and tortuous bazaars of Bhuj behind the thick medieval walls.

17 April 1960. The following morning we left for Khavda. The bus, a new one, was not over-crowded as most State Transport buses are. We had front seats, and the morning was pleasantly cool with a fresh breeze coming from the north. We travelled fast and soon the sandstone hills of Kutch were left behind, and the level country of the Banni's edge began. At first there were dense shrubberies of *Salvadora* and tamarisk which later thinned out and gave way to that flat featureless expanse of the true Banni. The Grey Partridge was plentiful and scuttled into cover. Common Babblers chattered among the shrubberies, and chattering flocks of Rosy Pastors migrating northward flew overhead. Redvented Bulbuls were common and White-eared Bulbuls were seen in increasing numbers. The latter seem to be very partial to *Salvadora*, and it is worthy of note that in nearby Saurashtra this pretty little bulbul is totally absent from large tracts of the interior, and is found chiefly around the coastal flats where the *Salvadora* also appears. Can it be merely a coincidence that the requirements of the bird and the plant are similar, or does the plant have an ecological influence on the birds?

The Banni itself is a flat featureless expanse, a limitless plain of blue-green vegetation. The plants are short bushes with highly xerophytic characters; their fleshy leaves store large amounts of water, and are avidly cropped by camels. The Banni is a remarkable phenomenon; the plain lies but a few feet above sea-level, and is so lacking in gradient that the rain-water lies in a shallow sheet over it slowly draining into the sea, or just evaporating into the dry air. The disappearance of the water is followed by a parching desiccation during the hot season, and the soil is encrusted by salt; tall dust devils race across the expanses and mirages shimmer, tantalizing and cool, in every direction. This is stock-rearing country, famous for its fine herds of large-horned cattle and camels.

For the first time the large Franklin's Crested Lark appeared, several female Pale Harriers and a Marsh Harrier went gliding north, hunting as they went. We were all on the lookout for the splendid Desert Larks but saw none; instead, we saw a few Redwinged Bush Larks in patches of grass and *Prosopis*, and isolated pairs of Ashycrowned Finch Larks feeding beside the road. Some of the trenches on both sides of the road still contained watery mud choked with frogs and provided ample sustenance to flocks of Lesser Egrets, Small Egrets, Pond Herons, and solitary Large Egrets.

In the centre of the Banni is an extensive oasis of large shady Acacias casting deep pools of shade, cool and restful after the sun-drenched Banni. The grass was refreshingly green. Large herds of cattle and buffaloes stood around under the trees, attended by Cattle Egrets. The inhabitants are Jhads, a cattle-herding tribe. They are well-built people and, though they are Muslims, their women go around unveiled and appear to enjoy a status equal to the men.

Leaving Brindiala, as this oasis is called, and proceeding again into the dazzling brightness of the Banni, we passed a couple of small herds of Chinkara, which do not seem to be at all as plentiful as claimed; nor did we come across any Blackbuck or other game, big or small. Undoubtedly the Banni no longer enjoys its former status of a small game paradise, and is certainly not behind the times in this respect compared with the rest of India; man's predation is greater and more effective than is believed outside Kutch.

The commonest animal of the Banni was the Spinytailed Lizard, *Uromastix hardwicki*, sunning itself on the baked sand. Its performance of suddenly vanishing after a short spurt was intriguing until one realised that the disappearing trick was done down its burrow. These lizards are undoubtedly the chief provender of the large numbers of passing birds of prey during the autumn and spring migrations. Certain nomadic tribes of the area consider that this lizard's flesh has strong aphrodisiac properties and regard it as a valuable delicacy. ... And so in time for an early lunch in Khavda.

At Khavda, arrangements had been made for our stay at the police chowky. It was in a way like a home-coming, as we had been here in the summer of 1956, unsuccessfully on a similar errand, and were then as now hospitably accommodated by the local officials. Things had changed little in the intervening years and, except for a block of new orderly quarters in the large enclosure, everything was as it had been then—time might well have stood still for the period. As some high dignitaries were on a routine inspection, we saw little of the officials who had been making arrangements for guides and baggage animals to take us to Nir. The chief guide Jamal Nathu was nowhere in town, and there was no news of him. 'He might arrive this evening', they all said. Camels were also not available for hire readily. It was very depressing and it seemed that after all we were stranded. The day all of a sudden lost its charm, and we dejectedly prepared for lunch and then to await developments in the afternoon. Just then someone said Jamal had arrived and sure enough there he was standing in the doorway, a thin dark man dressed in the manner of his people. Spirits revived and lunch was forsaken to get all information out of him. He was optimistic, as there was still plenty of water in the Rann, and large numbers of flamingos fed along its edges. They had no young with them, and so it was obvious that the birds were still at the colony site and had not moved off; the camels and ponies had also arrived and were grazing on the turf of the village tank, so we could leave for Nir the first thing next morning. The well-cooked lunch, a tribute to the long catering arm of the Modern Hotel, and a short siesta were followed by an afternoon of baggage sorting. In the cool of the evening we went for a stroll to see our pack animals, and to record what birds were around. In addition to the usual birds one sees around any village tank in Kutch, we saw a Wryneck, a little Green Heron, and a pair of Laggar Falcons.

18 April 1960. The sturdiest camel was loaded with all the baggage, three others and two ponies were to be used for riding. At 7.50 a.m. the cavalcade started across Pachham towards the northwest end of the Island where the spurs of Kala Dongar subside into soft undulations and finally give way to the flat expanse of the Rann. The entire way is well covered by scrub, and there is little cultivation. Cattle-rearing is the chief occupation, and there are signs of considerable overgrazing. The little cultivation done is of a perfunctory nature and depends entirely on the vagaries of the monsoon rains. Much of the land is deeply eroded. The thickets however were well populated by Grey Partridge, Rain Quail, Purple Sunbirds, Redvented Bulbuls, White-eared Bulbuls, Rufousfronted Wren-Warblers, Franklin's Wren-Warblers, Common Babblers, Tailor Birds, Common Mynas, Brahminy Mynas, Redwinged Bush Larks, Ring- and Little Brown Doves, and Roseringed Parakeets. Common Sandgrouse were noted flighting in pairs and small parties. The heavens overhead were quartered by King Vultures, Whitebacked Vultures, Longbilled Vultures, and White Scavenger Vultures accompanied by a few Griffons. Tawny Eagles were also seen soaring in loose pairs, while in the shady Banyan over the well at Wadvala, known as Wad-vali Wav, a pair of Redheaded Merlin had three young in the nest. A pair of Laggar Falcons and a Peregrine were also recorded. At Kakrao where we spent the heat of the day under a densely shaded small tree, we heard and later saw a Raven. It was there on our return.

From Kakrao we rode north. After descending the last low ridge we came to the Rann, and then turned east along its edge. On our right the great escarpment of Kala Dongar rose steeply—great beetling crags covered by thick tangled scrub and grass, still showing green. Birds were plentiful, and their songs were carried in a disembodied symphony from high overhead. Indian Robins were very plentiful with Baybacked Shrikes (many in juvenile plumage) and pairs of Brown Rock Chats.

Where the slopes eased to form narrow flat areas between the perpendicular of the Kala Dongar and the horizontal of the Rann, the subsoil water was sweet, and herds of cattle were pastured, their tingling bells bringing to mind the high pastures of the Himalayas.

On our left stretched the Rann—startlingly flat and immense; first a white encrustment of salt shining in the sun, then wet mud dark brown and rich, and beyond this the blue water, rivalling the sky overhead and stretching to the horizon where it shimmered into the sky, and it was difficult to discern where one ended and the other began. Here was a region where desert and water had become one in essence, vast and limitless, a land of silence, ruled by the wind and the sun, a home of dancing mirages, a region forbidding yet fascinating in the starkness of its elemental harshness. Yet it was awe-inspiring to see on this cruel stage life playing its part in the great flocks of pink flamingos feeding in the shallows, flying in skeins low over the water or writhing high above the mirages and merging into their unreality like ethereal beings, frail phantoms epitomizing life, fragile yet all-conquering. A little further we came across packed flocks of Blackwinged Stilts, Stints, Ruffs and Reeves, Marsh Sandpipers, Whimbrel, Sand Plovers, and some Blacktailed Godwits. Gullbilled Terns, Blackheaded Gulls, and Brownheaded Gulls flew buoyantly over the water. Tired but happy, we rode along slowly, drinking in the sights of this improbable land. Here was grandeur and beauty rivalled by few other creations of Nature.

We made to Nir late in the evening as the setting sun cast a warm glow on the wonderful landscape. The mirages subsided and far out on the northern horizon we saw lines in pink and white, which Jamal said were Flamingo City. The birds were still breeding. A memorable day had ended well, and tired and happy we lay under the brilliant stars, worn out but contented.

SHIVRAJKUMAR (OF JASDAN), R.M. NAIK, and K.S. LAVKUMAR,
Journal of the Bombay Natural History Society, 1960

Malcolm MacDonald (1901–81) was United Kingdom High Commissioner in India from 1955 to 1960. A keen birdwatcher, he wrote a number of books on Indian birds while in residence in Delhi, notably Birds in the Sun and Birds in My Indian Garden. In the piece below, he writes about the love life of parakeets which inhabited his garden in Delhi.

Green Parakeets in a Delhi Garden

[T]he first birds whom I observed mating in my garden this year were a pair of Green Parakeets. Like most of their kind, they were aristocratic-looking creatures with superlatively elegant figures. Measuring about sixteen inches from end to end, they were trimly streamlined, having well-rounded heads, smoothly tapering bodies and long sharp tails. Their colouring was radiant, the male being especially presentable. His prevailing hue was bright green with washes of bluish-grey on the head, a black band curling round his face, a rose collar circling his neck, and splashes of blue and yellow on his tail. The female was more uniformly green. To complete their charms, both birds had pale yellow eyes, cherry-red beaks, and the solemn, slightly supercilious demeanour of all parrots.

I had been watching the couple for more than six months, ever since they became regular inmates of the garden. The hen bird first flew into my life one morning in the previous July when I caught sight of her inspecting a hole in a jacaranda tree trunk which at the time was occupied by a family of Brahminy Mynahs. During the next two weeks she often visited the spot, gazing covetously at the hole; and when the parent Mynahs were absent she occasionally flew to the opening, perched just below it and chiselled with her beak at the wood, trying to widen the entrance. Clearly she fancied it as a likely residence for herself.

When the mynahs finally departed she at once took vacant possession of the property and began in earnest to enlarge its inner chamber for her own use. For several days she occupied herself with the work of excavation, sitting inside the cavity, nibbling at its walls with her bill, and periodically hurling a mouthful of wood-chips through the doorway.

Often, as she worked, a cock Green Parakeet perched on a branch outside the hole and eyed her approvingly. Thus I became acquainted with her lord. He never lifted a feather to help her in her labour of home-making; but he lent her moral support by the hour. She seemed grateful for his admiration, and sometimes took time off to settle companionably beside him on his sentinel bough. Possibly these were not their first meetings; the pair may well have been a middle-aged married couple who had lived together in previous seasons. Male and female Green Parakeets seem to remain such constant comrades through the year that I should not be surprised to learn that they pair for life.

Before long the hole in the tree was sufficiently enlarged to suit their requirements; and after that the hen bird spent a lot of time in it. Often during the day I saw her peering like a jack-in-the-box from its entrance, and always at nights she slept there. The male did not share her bedroom, but flew at dusk to a lodging elsewhere in Delhi—I knew not where. Shortly before sunrise the next morning he would arrive in the jacaranda tree to bid her good day. She then issued from her sleeping apartment, and they flew away together for breakfast somewhere in the neighbourhood.

Often during those days I saw them standing side by side in the jacaranda tree near their nest-to-be. So their friendship remained, faithful but platonic, from the middle of July until the end of January.

II

Then came their initial intimacy on the first day of February. I have already described the brief encounter: her enticement of him, his initial indecisive response, her further encouragement, his mounting on her back, and then his gradual working up to a vigorous, passionate attempt at union with her. I write 'attempt' deliberately, though I am not sure that the word is justified. His experiment at love-making may have been successful; but I doubt it. The affair was too short—lasting only half a minute—compared with their later couplings, and her abrupt ending of it by pecking at his face showed too much displeasure for the incident to be convincing. However, whether they physically mated or not on that occasion, their partnership was emotionally consummated.

As it happened, later the same day I saw another pair of Green Parakeets go through similar motions of mating. They owned a nest-hole elsewhere in the same jacaranda tree, and had likewise established proprietary rights to it several months earlier. An odd circumstance characterized them; the hen had only one foot, her second leg being a broken stump. Perhaps by chance, their nest had also been occupied in the previous year by a pair of Green Parakeets of whom the female had a solitary leg; but I do not believe this was a coincidence. I think a hen often returns to the same cradle for her offspring each nesting season.

That hen's lameness sometimes led to a certain awkwardness in the pair's marital relations. If the male was too clumsy when he mounted her, she could not maintain her grip with only one foot on the bough where they embraced. So she lost her balance. Several times I saw that happen in the early days of their mating, before he had learned the correct gentleness towards her. Let me describe one such occasion. The two birds were perching side by side on a branch near the entrance to their nest-hole. The hen crouched invitingly with her head tilted upwards, as was usual when she wanted to be ravished. The male sidled up to her and rubbed his beak affectionately in the plumage at the nape of her neck; but after a few moments he stopped, raised his head and looked around nervously, as if he were fearful of being observed. At least ten times he tickled his consort's topknot, and then hesitated and gazed apprehensively round him, extremely undecided. But she did not change her seductive squat, continuing to solicit him.

At last he was sufficiently stirred, made his resolve, and climbed on her back. She crouched lower with anticipatory contentment, and he promptly thrust his beak into the back of her neck. He must have done it too suddenly or passionately, not working up gradually to his frenzy, for she was taken by surprise at the force of his jab, lost her balance and collapsed on their branch. He at once jumped off her, and she tumbled from their perch and hung in mid air by her solitary foot. With difficulty she struggled back to an upright position and recomposed herself. He remained at her side, and for the next ten minutes they stood so, both preening their feathers and performing other nonchalant, unemotional actions. Then she lowered her body into a suggestive posture once more and tilted her head in a fresh invitation to him to pleasure her—which he did with proper skill and apparent mutual satisfaction this time.

The males of both pairs betrayed various signs of inexperience in the earliest days of mating. Thus, a cock's clumsiness when mounting his partner sometimes so discomfited her that she pecked him protestingly, forcing him temporarily to dismount. That never happened when he became more experienced. Again, on several early occasions I saw a male climb on a female's back, attempt coition, and then hop off again three or four times in quick succession, obviously getting no results except frustration from the exercise. Moreover, in his first amorous approaches to his mate a male invariably engaged in protracted, indecisive preliminary gestures such as fondling her neck and head or gazing around to make sure no critical eyes were watching; whereas later he dispensed with these irrelevancies and proceeded with only the most cursory introductory motions to the serious business of coupling. Perhaps another indication of growing expertness was the fact that as the days passed a pair appeared to achieve successful coition more rapidly than in their initial essays, the male remaining on the female's back less than two minutes instead of the three or four minutes which they took before.

None of this was surprising, for the operation was somewhat complicated. It was difficult for a cock to keep his equilibrium on a hen's back while at one and the same time thrusting his beak vigorously up and down at her neck and jerking his posterior energetically to and fro at her rear. Nor was it easy for his lower quarters to make exactly the right contact with hers. To do this he had to twist and curl his body round her in a most awkward angular curve, and the trickiness of the contortion was increased by the existence of their long tails, which were apt to get in their way.

So I am inclined to think their first attempts to accomplish these sexual acrobatics were less than fully successful, and that they only gradually acquired the necessary skill. Certainly the spasmodic jerks of the male's rump in the beginning appeared ineffective compared with the obviously more regular, forceful and satisfying thrusts of his love-making in later days.

III

Every morning I watched the pair at the more accessible nest, hoping to detect a sign that they had reached the next stage in their family affairs by producing eggs. Thus I became quite familiar with their characters and customs. Sometimes the hen bird took the initiative in inviting him to pleasure her, by squatting seductively beside him; and at other times the cock made the first suggestion by sidling up to her and scratching tenderly at her head-feathers.

They were not in the least embarrassed if other parakeets showed keen curiosity about their love-making. Birds are, of course, all uninhibited children of Nature. Frequently half a dozen fellow members of their clan were perched near by when the cock mounted the hen, and as their copulation proceeded one or two of these would move step by step closer to the pair, gazing inquisitively at their odd goings-on, until they were only a foot away. Completely unperturbed, the couple continued their passionate liaison until it reached a climax, when their ardour cooled and they separated. If the fascinated bystander were a male, he then sometimes tried to continue the sport. More than once I saw such a stranger lean towards the hen as soon as her lover left her and attempt to fondle her—only to receive from her a swift, vicious little peck of instant reproof.

The couple were more disturbed by the presence of other species of birds. In particular, if one of the Green Barbets who owned a neighbouring nest appeared while they happened to be mating, they at once desisted, assumed innocent stances and postponed their philandering until the intruder departed. This was common prudence, for they knew from experience that they must beware of physical assault by those short-tempered barbets.

After they started regular matings they themselves become more sensitive about strange birds approaching their nest. Their sense of landlordism seemed to grow, and they drove away Common Mynahs, Brahminy Mynahs and other trespassers who encroached too close. Another indication of their new preoccupation was that the hen parakeet became even more closely attached than before to their hole-in-a-tree. Each day during the previous months she had absented herself from the vicinity for considerable periods;

but now she tended to become an almost permanent squatter either beside the nest-hole or actually inside it. She still left the jacaranda tree to fetch food, but otherwise was invariably there. Usually she sat invisible indoors, but frequently she poked her head through the entrance to survey with rather detached interest the world which she had temporarily renounced. There was something nun-like about her withdrawal from that green and pleasant prospect—except that she was engaged in a pastime in which no respectable nun would indulge.

The cock parakeet was very much in love with her. He spent a lot of time in slavish attendance on her, waiting as a sentinel just outside the nest when she was inside, and flying after her wherever she went as son as she emerged. Impelled blindly by the strongest of all natural urges—the impulse to mate with her and so make his individual contribution to the preservation of their species—he sought at almost every opportunity to satisfy his craving for her. Nor did she object. Green Parakeets are ardent lovers in the breeding season. I watched carefully three pairs who nested in my garden, and they all enjoyed frequent sexual intercourse. It was invariably their first act on rising from bed each morning, and after that the couples mated about once every hour—and sometimes at shorter intervals—as long as daylight lasted. Possibly their ardour flagged a little in the sunniest, hottest noontide; but I doubt, it, for sometimes I saw scraps of irrefutable evidence to the contrary.

For about a fortnight after they began to mate the hen continued to fly periodically from the tree, usually in the cock's company, to gather food at some favourite foraging spot. They were among the regular customers at the Fantail Pigeons' breakfast ground. However, the cock began to feed his partner also by regurgitation. I first noticed this extraordinary performance on the afternoon of February 4th, the third day following their initial mating. Whether it had occurred between them before I do not know; but apparently the male's feeding of the female started at about the time when their sexual relations began to prepare her for laying eggs, incubating them, hatching chicks and rearing a young family—in fact, for spending many weeks almost constantly occupied with domestic duties inside a nest-hole. Soon afterwards she completely stopped fetching food for herself, and became entirely dependent on her lord for nourishment. Thereafter for many weeks—until her chicks were growing quite large—she never left the immediate neighbourhood of the nest, and only emerged from the nest-hole itself for brief spells of fresh air, feather-preening, love-making and being fed.

That last operation was an astonishing spectacle. The hen bird would stand on a bough with an expectant air as the cock approached her. A pace away he would halt and stare at her with an inexplicable sort of curiosity, as if he had never seen her before. Then he seemed to take a decision, drew himself stiffly to his full height and bent his head towards her as he might do if he were about to exercise his husbandly right of possessing her. But instead of tickling her neck, he touched her beak with his beak, leaving their mouths in contact for a moment as if he were tenderly kissing her. Then he suddenly broke off the contact, jerking his head abruptly backwards as he raised his body again to its full height, and holding that pose for a while with an air of pompous dignity. All the time he kept his eyes on the hen, scrutinizing her appraisingly. Then he slowly raised one foot with its toes extended towards her as if he would like either to stroke or to scratch her; but suddenly he put the foot down again, shot his head towards her and once more touched her beak with his. This time his bill lingered for a while on hers and opened slightly to let her bill enter his. When they broke off contact again, the hen made small nibbling motions with her mandibles, chewing bits of food that he had passed her. Meanwhile the cock once more assumed a dignified, erect pose, holding himself with a stiff sedateness which was grotesque to the point of being comic. Occasionally he added to the ludicrous effect by puffing out his chest importantly—and that action revealed the real cause of all his odd gesticulations. Every time that he withdrew his beak from the female's he must regurgitate more food to give her at the next meeting of their mouths; and this process seemed rather difficult. To achieve it he had to contort his neck awkwardly and engage in other straining motions, including (apparently) raising one leg, to assist in the upward passage from his throat to his beak of a further titbit.

The meal always continued for some time, the male executing repeatedly the succession of strange antics which ended each time in his bending towards her, gripping her beak in his, and passing a morsel of food to her. To human eyes the operation looked like nothing so much as a protracted bout of passionate kissing between infatuated lovers.

Usually, though not invariably, this feeding of the hen by the cock was associated with mating. It followed immediately after the act of copulation. As soon as the male parakeet jumped off the female's back, he assumed the stance for feeding her, eyeing her to see whether she looked suitably hungry, and then drawing himself up stiffly for the first effort at regurgitation.

This combined act of mating and feeding was the bird's first preoccupation every morning when they woke. Often I watched the ceremony at the nest-hole in the jacaranda tree. As I have already written, the hen parakeet slept each night in the nest while the cock occupied some other dormitory at a distance. About half an hour before sunrise she would poke her head through her bedroom doorway for a preliminary glance at the reviving world. Then she withdrew out of sight indoors again. A few minutes later the cock would fly across the lawn, shrieking joyfully as he came and making straight for the jacaranda tree. He would perch on a branch beside the nest-hole and at once call impetuously and impatiently. If he received no quick response, he changed his tone to one more gentle and persuasive.

No doubt he was venturing to remark to the occupant inside the hole: 'Good morning, my dear; have you slept well? A new day has dawned. Come out, you lazy bones! Oh, do come out.'

Sometimes she responded immediately, and at other times she kept him waiting for several minutes; but always before long she emerged. Flying straight past him, she made a short flight to exercise her wings, and then alighted in a neighbouring tree. When she appeared he instantly followed her, as close as a shadow and as eager as something much more substantial than a shadow. He landed near her, and after a barely decent interval he sidled up to her. She crouched compliantly, he climbed on her back, and they mated. Then he regurgitated her breakfast for her—after which she promptly took wing again and returned to the nest.

These rites were performed so regularly at almost precisely the same minute every morning that I could have set my watch by them. Nor—as I have already written—was the ritual a service celebrated only at mating but every twenty-four hours. The love parts of it at least were repeated every hour throughout the day, day after day; and the feeding no doubt followed as often as the hen felt hungry.

At the beginning of February I assumed that these passionate exercises would not continue long. After several days, I thought, the hen would lay eggs and the parents would then switch their interest to the responsibilities of incubation. But I was wrong. The intimacies between the pair continued with unflagging zest for more than six weeks, until I thought the couple would never tire of their favourite pastime. For a period at the end of February she sometimes seemed reluctant to submit to his caresses; but that only provoked him to greater ardour, and he became so lecherous that several times he apparently tried to rape her. Later she recovered her amorous mood, and they maintained their matings several times each day until March 18th, when the practice suddenly ceased.

I could not discover when their honeymoon first yielded a material result in the form of an egg, because I could not peer into their nest. But I felt a strong intuition that the hen had borne her first youngster about the middle of February, on the day when she ceased to fly away to fetch food for herself, became entirely dependent on her partner for nourishment, and settled down to almost perpetual brooding indoors. However, I could not prove that.

Evidence mentioned in Hume's *Nests and Eggs of Indian Birds* suggests that an interval of several days can elapse between the successive laying of each of a Green Parakeet's three or four eggs, which no doubt accounts for the protracted continuation of a couple's mating after their first egg has been produced.

IV

The pair of parakeets in my garden stopped love-making on March 18th; so for nearly seven weeks they had mated every day, several times a day.

After that they continued to observe their customary ritual each morning, except for its solemn, complicated climax of copulation. The male bird arrived in the garden shortly before sunrise and flew to the nest to call his spouse. She promptly emerged, made a short sortie through the air to exercise her wings, and then alighted in some nearby tree. He followed and perched beside her. Occasionally in the next few days he tickled her head-feathers with his beak; but this opening gambit of their love-play was apparently only a reminiscent, fleeting, perhaps nostalgic, reversion to an earlier habit, and was never pursued further. Instead he fed her with all the absurd formal posturings required by regurgitation, and immediately afterwards she sped back to their home tree and disappeared into the nest. Evidently she was eager to return quickly to her clutch of eggs. Most of the rest of the day she stayed indoors, no longer reappearing at hourly intervals for romantic dallyings with her mate.

From time to time he visited her, flying to the nest-entrance, perching just below it, and peering solicitously within. Periodically she fluttered out to receive refreshments from him, but always as soon as she had fed she hurried back into the nest. Her whole being seemed concentrated on the business of hatching eggs, which seemed to be almost a full-time job. Some writers on Indian birds have declared that both Green Parakeets share in the labour of incubation, and no doubt their statements are based on careful observations; but I did not myself often see male birds enter their nest-holes at that period. Certainly their share of the work was very much lighter than that of the hen.

On 28th March I noticed an incident at the main nest which might indicate that one or more youngsters had hatched that morning. The female parakeet seemed a trifle agitated because a number of other parakeets showed an inordinate interest in her home. They kept fluttering inquisitively around its doorway, trying to alight there and peer inside. Several times she dashed at them and drove them away; but for some mysterious reason they were always attracted back again. I had not observed any such episode before. Did it mean that the mother bird was sensitive about strangers approaching her residence because it housed newly-born chicks, and that the group of other parakeets were aware of the birth and anxious to inspect the babies? I do not think this is too human an explanation; but I cannot be sure. I could no more tell when the eggs hatched than I had been able to learn when they were laid, for the whole early life of the youngsters was hidden from view deep in the jacaranda tree trunk.

That situation continued with no apparent change for several weeks. The hen bird spent most of her time in the nest, her departures from it being rare and her absences short. At some moment towards the end of March or early in April her progeny must have hatched, and her maternal duties changed from incubating eggs to coddling chicks. After that the cock bird's task as the bread-winner for the family became more exacting, for now he had more than his wife's mouth to feed. The authors whom I have already quoted state in their books that, just as both birds help in incubating eggs, so also they partake in the work of feeding nestlings. I do not know whether these writers intended to convey that the cock bird feeds the youngsters directly. For myself, I can only record that my observations rarely yielded evidence of that. It seemed to me that the male parakeets continued to feed their mates by regurgitation, that they gave them enough rations to appease not only their own hunger but also that of their chicks, and that the hens then promptly flew into the nest to pass the extra provisions to their offspring. Only occasionally did a male fly into the nest-hole to feed the family himself.

As the weeks passed the mother birds began to absent themselves more frequently from their nurseries. Presumably the nestlings needed less constant attention as they grew up, and the parents were freer to divert themselves elsewhere. So they took more protracted flights out of doors to exercise wings and bodies which must have become out of practice for such physical vigour during long weeks of cramped incubation. Until towards the end of April, however, the hen at the nest which I watched most closely did not (so far as I could see) interrupt her vigils at home to the extent of resuming the fetching of her own food. For nourishment she remained entirely dependent on her mate. The first occasion when it looked as if she might have gone on a foraging expedition on her own was on 23rd April; but if that was the case, she also continued to be partly fed by him for a few weeks afterwards. Throughout the occupation of the nest-hole by her youngsters she slept in the tunnel every night with them.

V

One morning, as she sat taking the air alone on a bough in a nearby tree, a strange bird darted at her. With a startled cry she leaped from her perch and flew towards the nest; but the attacker gave swift, wicked chase, and to avoid being struck she had to swerve away. Weaving a skilful zigzagging course in and out among the branches of a row of trees, she fled in panic with the aggressor almost on her tail all the way. Then I saw that the stranger was a Shikra, a hawk so sporting that falconers train it to capture various birds on the wing. The chase was very exciting; several times the bird of prey seemed on the point of catching up with its intended victim; and my heart was in my mouth, for I feared it would deliver a mortal blow and bring an end to my tale of family life among parakeets. But the parakeet showed remarkable resource in evading its pursuer, and the Shikra soon wearied of the hunt. After a tense couple of minutes it suddenly veered away in another direction. The parakeet sped back helter-skelter to its nest-hole, and hastened inside.

Five minutes later the hawk returned to frighten and scatter a group of half a dozen other Green Parakeets clustered in a neighbouring tree. They screamed as they fled in all directions, and the Shikra whimsically settled on the very spot where they had perched. It was evidently in a sportive, playful mood.

On a later morning another kind of intruder disturbed the peace surrounding the parakeet's nest. A Brahminy Mynah arrived on the scene with a strip of lavatory-paper gripped in its beak. Brahminy Mynahs and Common Mynahs are fond of stuffing odd bits of rubbish like the cellophane wrapper off a cigar, silver paper torn from chocolate boxes, or scraps of coloured rags, into their nests as decorations; and this Brahminy Mynah was apparently intent on adorning its home with toilet-paper. Moreover, it seemed resolved to make that home in the very nest-hole already occupied by the Green Parakeets, for it alighted at the entrance and peered hopefully indoors. The hen promptly flew out of the nest in protest and shoo'd the vulgar visitor away. But the mynah was not to be easily discouraged. Often it returned to the tunnel and tried to carry its booty inside; and every time the parakeet intercepted it and drove it off.

Eventually the mynah gave up, retreated to a bough, and let the paper drop from its beak. As it fluttered towards the ground a Common Mynah caught sight of it, felt attracted by its decorative quality, dashed at it, retrieved it and carried it to another nest-hole elsewhere in the same tree. As the bird alighted on the bark, however, some-thing made it open its beak to emit a squawk—and the slip of paper once more descended earthwards. At that the Brahminy Mynah could not resist the temptation to make another attempt at incorporating the flimsy sheet in a nest. Leaping from its perch, it caught the trophy in mid air and carried it again to the doorway of the Green Parakeets' home. But again the parakeet confronted the trespasser and chased it away. Disconsolately, the mynah let the toilet-paper drop.

Then the Common Mynah returned to the charge, dashing at the scrap as it settled on the ground, picking it up and flying with it triumphantly to its own nest-hole. But the triumph was short-lived. A Jungle Babbler apparently questioned its good taste in wishing to adorn a nest with such an article, for it flew so viciously at the mynah that the bird took fright, released the paper and took refuge, empty-beaked, in a clump of foliage.

The game of catch-as-catch-can with the sheet of paper was over. Like a forgotten shuttlecock, it fluttered to the ground, and the various players dispersed. The Common Mynah hopped into its nest-hole, the Brahminy Mynah hastened away, the Green Parakeet returned to its nursery, and only the Jungle Babbler stayed on a bough, eyeing the bit of paper with seeming disapproval ... But I cannot help thinking that one of the mynahs returned surreptitiously to the spot later, reclaimed the prize lying on the grass and carried it at last safely into a nest. If that was so, the repository was not the Green Parakeets' home. Several times in the next few weeks I saw a Brahminy Mynah trying to effect an entry into that desirable residence; but always in vain. The parakeet landlady invariably appeared and made effective landladyish protests, driving the would-be gate-crasher away.

VI

Not until 7th May did a young Green Parakeet reveal itself in the nest-tunnel. That morning I saw a nestling eyeing me dubiously from inside the shaft. After long hesitation it plucked up courage to poke its head through the entrance to catch a glimpse of the outside world; but something must have frightened it, for at once it withdrew indoors again and disappeared from view. I did not see it again for several hours.

In the succeeding days it grew steadily bolder, craning its head ever further out of the nest-entrance to enjoy some sightseeing in the strange, fascinating out-of-doors. It was a charming-looking youngster with a sleek, softly plumaged head, large, innocent dark eyes, and an unblemished pale red beak. Always it displayed curiosity, and sometimes astonishment, at what it saw in the landscape. Silently and cautiously it would stare for hours from its window, no doubt having been warned by its mother that the world is a dangerous place to be regarded with all due prudence.

I never saw a second youngster in that nest, and I suppose this chick was the solitary fruit of all those long months of passionate mating, patient incubation and devoted care by its parents. They continued to nourish and tend it there for another three weeks; and then, on 27th May, it ventured from the nest, flying away gaily with them.

That night the tunnel in the jacaranda tree was empty for the first time for many months. More than ten moons had waxed and waned since I first saw the hen parakeet chipping at its entrance as a signal that she intended to rear a family there. Hers had been a remarkably protracted occupation.

VII

On the day that the parakeets departed I watched a Common Mynah and a Brahminy Mynah both enter the nest-hole several times on inspection visits; and for the next few days they quarrelled over possession of the place. Their contest was vicious; but in the end the Common Mynah and its mate won a decisive victory—and not long afterwards the hen of the pair was sitting on a clutch of attractive blue Common Mynahs' eggs.

VIII

Two other pairs of Green Parakeets nested in the garden. The first was the couple with the one-legged female. Her deformity was no bar to the production of a lusty family, for she bore, hatched and reared three chicks. I first noticed them peering shyly from their residence on 25th April, nearly two weeks before the youngster in the other nest made its similar debut. They left their birthplace on 15th May, also almost a fortnight before it followed suit.

The tale of the third pair of nesting Green Parakeets had a different beginning, a different middle and a different end from the stories of the other couples. As for the beginning, whereas each hen bird of those pairs occupied her nest-to-be as sleeping quarters for several months before she began laying eggs—and so established unchallengeable proprietary rights there—the female in this third case took no such precaution. On the contrary, she and her male partner made their first reconnaissance of the hole-in-a-tree which they favoured only on 22nd February, a considerable time after the breeding season of the others had already begun. They then disappeared for many days, and I did not notice them at the site again until 8th March. In the meantime a pair of Common Mynahs had taken a fancy to the place and made elaborate preparations to occupy it. These mynahs therefore resented the reappearance of the parakeets as potential residents, and a violent scuffle occurred then and there between the rival claimants. The Green Parakeets won this first round, presumably because they were heavier-weight contestants; but what the mynahs lacked in power they made up for in obstinacy. Again and again that day they returned to challenge the parakeets' right to ownership. In fact, they kept renewing the fight every day for the next several weeks.

At the middle of the story, therefore, the parakeets' occupancy of their home was never peaceful. That they established their tenancy was beyond doubt, for each day the pair mated frequently on boughs near the nest, the hen bird spent hours at a stretch squatting within, and she and her mate shoo'd away all intruders, exactly as other members of their tribe did on their respective private properties. Yet the two Common Mynahs never became reconciled to defeat; every morning they bitterly challenged their rivals by standing outside the nest and shrieking at the hen inside, dashing hostilely at the cock whenever he appeared and making themselves a public nuisance in other ways.

As for the end of the story, either the relentless opposition of the mynahs or else some other circumstance caused the parakeets to desert the nest before their eggs could have hatched. Both birds were there (the cock outside and the hen inside the nest-hole) on 15th April—and both were absent all day on the 16th. In their places that morning the Common Mynahs kept flying in and out of the doorway, carrying bits and pieces of all the bric-à-brac which goes to make a mynahs' nest. They lived there happily ever after—or, to be exact, until their youngsters flew from the apartment many weeks later.

I never saw that pair of Green Parakeets again.

IX

During the nesting season Green Parakeets disperse in separate family parties, although I believe some families choose to breed in considerable colonies. Afterwards all the birds revert to their more customary sociability, gathering, travelling, feeding and for ever chattering in large flocks.

These flocks began to assemble in the tree-tops in the latter half of April. Among them were perhaps the most precocious of the new generation of youngsters just emerged from nests. Certainly by the early days of May these adolescents accompanied their elders. More than once I saw small squadrons of parakeets flying overhead at a much slower pace than is usual with those fast-travelling birds. Their wing-beats were leisurely, their movements seemed halting and their whole progress was hesitant, as if they were troops performing a slow march. I presumed they were family groups including fledglings fresh from nurseries, now venturing into the outer world and having their first lessons in flying.

Gradually the flocks grew larger, until companies of fifty, sixty, or more parakeets were a common sight. They were especially active in the early mornings, when they flew from their beds to breakfast off crops in the fields outside Delhi, and again in the evenings when they returned to their roosting places. Every night multitudes of them congregated in various suitable spots in the city, such as Lodi Gardens, where many trees became noisy parakeets' dormitories.

At dawn and dusk, therefore, countless flocks of parakeets sped across the sky. Every few minutes a fresh party of anything between a dozen and a hundred birds would appear, all hastening in the same direction like a ceaseless fly-past of squadrons of aircraft in an air pageant. A flight of Green Parakeets is a beautiful sight. The birds move swiftly and usually in a direct, unswerving line, like arrows shot from bows; but every now and then some whim makes them abruptly change their line of advance, and they turn sharply this way or that, rising and tumbling, twisting and veering in a nervous series of quick, zigzagging, acrobatic manoeuvres. They seem to ricochet from point to point in the air with carefree abandon. Their speed never alters; always it is swift and purposeful. An acquaintance of mine who made a habit of pacing squadrons of Green Parakeets in his motor-car as they sped along the edges of India's straight roads told me that the birds overtook him when he was driving at forty miles an hour, and that he overtook them when he accelerated to forty-five miles an hour. So their normal cruising speed must be about forty-three miles an hour. No doubt if pressed they can distinctly increase this haste. Their speed and capacity for sudden alterations in their line of flight explains why my Green Parakeet was able to evade the hot pursuit of a Shikra.

The number of Green Parakeets who inhabit Delhi must be legion. On several evenings in July and August I have counted the birds flying above my garden in the last hour before darkness, when they were travelling to their nightly lodgings. Each time between 1,200 and 1,500 of them passed overhead. That is an astonishing throng to journey in sight of one confined lawn. If every other citizen of Delhi were to take a similar census, the grand total of parakeets must amount to tens of thousands.

No doubt the birds are pests to farmers, for they guzzle ripening corn and fruit. Nevertheless, such beauty in such masses is one of Delhi's loveliest natural treasures.

MALCOLM MACDONALD, *Birds in My Indian Garden*, 1960

Edward Pritchard Gee (1904–68), was a tea-planter and an amateur naturalist in Assam. He is famous for his discovery of a langur species, named after him—Gee's Golden Langur. He was an active contributor to the wildlife protection policy of independent India, and was a member of the Indian Board for Wildlife. His publications include the well-known book The Wildlife of India, *and this interesting piece about Bharatpur has been taken from there.*

The Breeding Birds of Bharatpur

In recent years there has been a very slight but perceptible increase in 'wild life awareness' in India, and one of the results of this has been the discovery of several more colonies of breeding water birds, each a wonderful spectacle of many species living and breeding together in harmony.

For instance it has recently been found that there is a huge colony of nesting water birds in the Sunderbans, seventy miles south of Calcutta. This place, called Sajnekhali, is now a sanctuary, and is said to have about 15,000 birds in the breeding season, which is June, July and August.

But the most famous water bird sanctuary of India is two miles from Bharatpur, which is about 100 miles south of Delhi, in Rajasthan. Its name is Keoladeo Ghana. Those who know it well refer to it as 'the Ghana', but in print this looks like a certain country of Africa to those not familiar with India!

This place used to be the private wildfowl shooting preserve of the rulers of the old princely state of Bharatpur. Fantastic numbers of duck and geese used to be shot here by parties containing all the famous people of India, from the Viceroys downwards.

In a shoot for Lord Hardinge in December 1914, 4062 birds were killed by forty-nine guns. In a shoot in November 1916 for Lord Chelmsford, no less than 4206 birds were shot by fifty guns. In November 1938 when Lord Linlithgow was present, 4273 birds fell to forty-one guns. This latter occasion seems to have been the record.

The Maharaja still retains the right of holding shoots in the area, where fifteen or more species of migratory duck and two species of migratory geese arrive to feed during the winter months. Pelicans and Siberian cranes also come here for the winter.

But during the other months of the year, that is from July to October, the Ghana is a breeding water bird sanctuary, for the non-migratory water birds which come here from various parts of India.

The success or partial failure of the breeding of these water birds each year depends on the amount of rain water, and on the amount of impounded water let into the area from the local irrigation works. Thanks to the untiring efforts of Sálim Ali, the Maharaja and the Rajasthan Forest Department, the amount of water in recent years has been sufficient to enable scores of thousands of birds to come and breed.

I always couple the Ghana in my mind with the name of Sálim Ali. For not only is he India's leading ornithologist, but also he has done so much for the protection and success of this particular sanctuary. And also, the three times that I have been there, in September of 1957, 1961 and 1963, I had the privilege of being there with him.

In 1957 we, together with Loke Wan Tho, the bird photographer and ornithologist of Singapore, were the guests of the kind and very hospitable Maharaja. In 1961 and 1963 I stayed in the sanctuary Rest House, which is a gift of the Maharaja to the State Forest Department, in order to be nearer to the nesting sites of the birds.

On the 1957 visit I listened intently to everything that Loke had to say about cameras and bird photography. For he is undoubtedly one of the world's best bird photographers, and much of his work has been done in India, in company with Sálim Ali. I remember him saying that a larger sized camera than 35 mm. was preferable, because a larger negative was bound to give a better and bigger enlargement.

I use both a 35 mm. camera, with 135 mm. and 200 mm. tele lenses, and also a $2^1/_4 \times 2^1/_4$ in. one, with a 250 mm. tele lens. I find that there is a lot to be said for both 35 mm. and the larger size, depending on what sort of photography one is doing.

I also remember him repeating 'Every feather! Every feather!' That appears to be his motto, and his results usually bear this out.

When I was with him, he was trying to get exhibition pictures of white ibis. For this purpose every detail had to be right. For three mornings and three afternoons he waited in his hide close to the birds on their nest—without taking a single photograph.

On the fourth day, however, everything went well and he made thirty or forty exposures. What exemplary patience!

Sálim Ali is a slim, elderly man, now heavily disguised behind a neat, white beard. He is a most useful companion to have on an expedition. For when he is there one need not look up his very fine books[1] to see which bird you have seen or for other details. His remarkable memory and long years of experience in every part of India can give you all the answers immediately, and a lot more besides.

In addition to being India's leading ornithologist, Sálim Ali is also her most experienced naturalist in other fields, for he has observed, photographed and filmed most of the important mammals of India. He has also served on several field investigations for different States, and is Vice-President of the Bombay Natural History Society.

There are many roads and bunds criss-crossing the sanctuary, and bird-watching can be done from these. But for those who wish to see and photograph the birds and their nests more closely, there are small, flat-bottomed metal boats and boatmen available.

The early morning journey in a boat through the sanctuary is a most unforgettable experience. It is cool then, and everything glitters with jewels of dew in the bright silvery sunlight. The grasses and reeds part to let the boat through.

Now and then you find open sheets of water, with no grass. Here several kinds of water lilies are in full flower, in various shades of white, pink and pale blue. Other tiny flowers of white and yellow, strung together like beads on long green strands, gleam on the surface of the clear water.

Among this fragile and scanty vegetation some resident birds are feeding, and sometimes nesting: moorhen, purple coot (*sic*), dabchick and pheasant-tailed jacana. This last bird has enormous but spidery feet which enable it to walk about in what appears to be the surface of the water: the aquatic plants in the water take the weight of the bird.

---

[1] The best known of Salim Ali's books are *The Book of Indian Birds*, of which the sixth edition came out in 1961, and *Indian Hill Birds* (1949). A companion book of the former, *The Book of Indian Animals* by S.H. Prater, is now out of print but has been revised and is in the process of being published by the Bombay Natural History Society.

The silence of the peaceful scene is only broken by the quiet swish as the boat moves through the water, and by the distant chorus of so many water birds at their nests.

When doing serious close-up photography here, the technique is to take out two of these boats, with two men to assist you. You select the tree and the nests you wish to photograph, study the direction of the sun for morning and afternoon photography and many other things, and then fix your cloth hide in the right position on one of the boats.

Bamboo poles driven into the ground below the three or four feet of water serve both to anchor the boat as well as for tying the hide in position. You then sit with your equipment and perspire like anything, while the two men depart in the other boat.

You give them instructions how far away to go, and what signals you will give for them to return to collect you.

I used to have two pieces of cloth, used for wrapping up camera accessories. If I hung the red cloth up at the back of the hide, unseen by the nesting birds but clearly visible to the men, they had to take their boat farther away. If I fixed the green cloth, that was the signal to bring their boat up to mine. No cloth at all meant 'Stay put'.

This system worked very well. Sometimes if a bird was in the wrong position, or started to shut its eyes and sleep, I would hoist the green cloth at the back. The men would start to bring their boat towards me, and that would make the bird instantly alert. I would take its picture and remove the green cloth. The boat would stay put. A showing of the red cloth would make the boat go farther away, if necessary.

I had two men with me. One was a 'handyman' sort of person I had brought with me from Assam; and, as could be expected of a man who lived in the plains of such a wet region, he was an expert boatman. The other man was a local labourer detailed to accompany me as 'boat-man' by the Range Officer. I don't think he had ever been in a boat before, because he hadn't a clue what to do with the paddles and poles and water. But he was a willing and cheerful worker.

I don't think he knew much of things on the land either, for he had recently fallen off a bullock cart and hurt his knee. One day while returning from photography in the boat, the following conversation took place in simple Hindi:

'My knee is hurting. It is not getting better,' said the 'boatman' who knew nothing about boats, baring his leg.

'Then you should ask the Doctor Sahib for some medicine,' said my handyman, who was expert with boats.

'Which Doctor Sahib?' I asked.

'The Sálim Ali Doctor Sahib,' replied the handyman.

'The Sálim Ali Sahib is a doctor of birds, not of medicine,' I explained. Then I continued, reflecting on my thirty-odd years on tea estates, in charge of estate doctors, hospitals and the like, 'Possibly even I know more about medicines. You come to me afterwards and I'll give you some medicine for that leg.'

For the next four days outside my room there were hot water, antiseptic, bandages, and the 'boatman'. The results were nothing short of miraculous. The delight on his face was a pleasure to behold, and the faith he had in me was astonishing. On the fifth day he joyfully declared that he was cured; and I think he was, fifty per cent actually and fifty per cent psychologically. I, too, felt better after all this.

I usually fixed my boat and the hide in position each evening, after an afternoon's photography, so that the birds had all night and early morning to get used to it. By doing this, the birds used to return to their nests very quickly every time my two men left in their boat.

The chief discomfort was the heat and humidity while sitting in the hide in the metal boat. I used to do photography for a minute, and then for the next minute or more I had to dry myself with a towel—so great was the perspiration.

It was always a relief to come out of the hide into the fresh and cooler air. There was usually a slight breeze blowing outside.

But any unpleasantness was offset by the immense enjoyment I derived from seeing the birds going about their business of raising a family at very close quarters, quite unaware of my presence.

Usually I could see four or five different species from the tiny windows in the front and sides of my hide. Though my camera on its tripod was probably trained on to one particular nest, yet with my other camera I could 'shoot' other birds at other nests in the vicinity.

And all the time, in addition to getting pictures in colour and in black-and-white of each species of bird, I could also relax every now and then and just enjoy watching them and their ways.

I was amused, for example, at a pair of those magnificently coloured and neatly tailored painted storks. One of them, possibly the gentleman, stood in stately fashion at the edge of the nest, doing not a stroke of work beyond occasionally preening its feathers. The other, its mate, was busy for a full hour arranging and repairing the nest—and then sat down to incubate (Col. ol. 2).

I have not seen these beautiful painted storks at Vedanthangal. But a few pairs have been going to the New Delhi Zoological Park for breeding in recent years.

A pair of spoonbills (Pl. 16) stood silently in their empty nest, with hardly a movement. They looked as though they were waiting for ideas, or for something to happen. Something did happen, for next day there was an egg in the nest.

Another spoonbill was sitting on its nest, incubating. Three feet below it, on the same tree, was a little egret on its nest. Each time one of the birds moved at all, the other one would become agitated and raise its crest, looking in all directions. And a spoonbill with its crest raised looks really amusing.

Yet another spoonbill, on alighting on its nest, took great trouble to screen the sun off its two eggs by stretching out its wings. Sometimes it met with no success at all, and the eggs gleamed brightly in the sun-light.

Little egrets in their breeding plumage of crest and 'aigrette' feathers made a wonderful sight in the bright sunlight, when one of them relieved its mate while incubating. There seemed to be a sort of ceremonial greeting, with all their plumage raised and lowered.

Grey herons are about the most graceful of all birds while flying and alighting at their nests. Purple herons, more colourful but lacking crests, were much more wary. Night herons, comparatively small birds, have very long white crests and blood-red eyes.

Indian darters, or snake-birds, have long and slender S-shaped necks, narrow heads and dagger-like bills—all streamlined to enable them to swim underwater and spear their fishy prey. They feed their young in the same manner as do pelicans and cormorants: the chicks thrust their heads right down the necks of the parent birds to eat the half-digested, regurgitated fish.

Openbill storks, on the other hand, regurgitate the half-digested fish, and the young ones gobble up their food avidly from the bottom of the nest.

Incidentally, no one has yet explained why openbill storks have curved, open bills. They feed largely on snails, and in some Indian languages they are called 'snail breakers'; but other birds also feed on these snails, and do not have curved, open beaks.

Each time that I left a site and took the hide and the boats away, I would stop some distance away and look back to watch the birds return. It was always a relief to know that my photographic activities had not unduly disturbed the birds. I don't think there were any desertions of nests due to my taking of these photographs.

The first time I was at the Ghana with Sálim Ali, I went out with him when he did some ringing of baby openbill storks. One of the forest guards was helping, and his name was Lookhia. Apparently in the old days he had been on big shoots and some very important person who knew only English had addressed him several times with the words 'Look here!' This 'name' stuck.

One of the openbills ringed in September 1947 was recovered four months later in Uttar Pradesh, about 350 miles away. Another was recovered sixteen months later in Bihar, about 510 miles away.

During my 1961 visit to the Ghana, Sálim Ali was leading a team to do mist-netting and banding of small migratory birds. This was part of a regular spring and autumn programme, conducted for the World Health Organisation and the Bombay Natural History Society, to try and discover if migrating birds are responsible for carrying virus diseases.

A virus disease has been diagnosed in a forest in Mysore State, very similar to one prevalent in a certain part of Russia. The Mysore virus disease can be fatal to both human beings and to wild monkeys. A few captured birds have been found carrying ticks: these are extracted and sent to Poona for examination.

One of the finest thrills of bird watching in India is the fantastic courtship dance of Sarus cranes. Huge wings outstretched, they bow and prance round each other—a beautiful and exciting thing to watch.

Sarus cranes pair for life, and the birds are devoted to each other. If one of a pair is killed, the survivor will haunt the scene of the tragedy for weeks, crying distractedly. There is a belief that a surviving mate will pine away and die of grief.

Symbolic of a happy marriage, they are protected in many parts of India by popular sentiment. Consequently they have become confiding and unafraid of villagers and passers-by.

There are usually several pairs of these spectacular birds breeding in the Ghana and its environs. Large piles of grass and rubbish keep their eggs, usually two, above the level of the water.

I was able to photograph them, both in 1957 and 1961. The latter year was more successful for me, because I had by then become better equipped with hides and experience.

Standing on an embankment road, I watched a pair of them at their nest. One of the birds was sitting, while the other was feeding nearby. It was time, apparently, for the one to relieve the other at incubating, and the feeding one stalked up to the nest.

The one, which had been sitting, stood up. One egg, pinkish white in colour, gleamed in the sunshine. The other egg had just become a chick.

The golden-yellow chick roamed around the nest, exploring its new world. Than it jumped into the water and swam to a patch of tall grass nearby.

Some warning note from its parents made the chick freeze and lie hidden—at least it thought it was hidden. I could see it clearly from the road, because I had seen it go there. On a casual wade through the water and grass another person would not have noticed it, because of complete lack of movement.

The relieving bird settled down to brood, and the other walked away. Later it called the chick, which rose and followed its parent on a feeding expedition, for the first time.

Next day I put my hide about forty feet, and then about thirty feet, from the nest. I stood in it, in about one foot of water, to take photographs. It was hot and steamy, but this was not the main difficulty here. It was leeches.

Now I am used to leeches. They are fairly harmless, and only suck your blood. And in any case, people in the old days used to pay large sums of money for leeches to reduce their blood. The technique is to knock or pull them off, or better still to put salt on them—which immediately kills them. Then when you get home you dab some antiseptic liquid on the places where they have bitten, and everything is all right.

These were water leeches which swim well. They are about the size of a rather large fountain pen—that is before they get swelled up with your blood. So I spent one minute in pulling them off and throwing them away, and then one minute in focusing my camera and trying to take a picture. And then again a minute in pulling off the leeches, which lost no time in returning to the feast.

To make things easier, I called up one of the men, and he crouched down and pulled off and kept away the leeches from both of us, while I concentrated on photography. The water was clear and warm, and the object of my endeavours so worthwhile, that I never bothered about the minor discomforts down below.

It is interesting to watch cormorants feeding. They float on the water, but much lower than ducks do, with only their necks and the tops of their backs showing. Very accomplished divers and swimmers, they do all their fishing below the surface, catching small crabs, tadpoles and frogs as well as fish. The little cormorant is the most common in India, the large cormorant and the intermediate cormorant (or Indian shag) being rarer.

Peafowl are everywhere in this area, and cock birds do not let you forget their existence, with their loud screams of may-awe in the evenings and early mornings.

These birds are protected in most parts of west, central and north India by legislation and, more significantly, by popular and religious sentiment. For the peacock is the vehicle of Saraswati (goddess of learning), Kartikeya (god of war) and Subrahmanya (god of yogic powers).

Consequently these spectacular, and to Westerners, exotic, birds have become quite common in these parts, often proudly wandering and even nesting in villages totally unafraid of man. In northeast India the peacock is rarer and hitherto not protected, but it is hoped that legislation will be introduced in these Sates to give it full sanctity.

It is recorded in history that Alexander the Great took back with him from India to Greece two hundred peafowl, and from Greece the birds spread to other countries of western Asia, north Africa, Europe and eventually to America. The Moghul emperors were greatly attracted by the beauty of this bird, and Shah Jehan's famous peacock throne was designed 'its pillars of emerald being surmounted by the figures of two peacocks, ablaze with precious stones'.

Incidentally, the splendid ocellated 'train' of the peacock is not really its tail, but its upper tail-coverts enormously lengthened. For display these are erected and fanned out before admiring hens. Its crest feathers have fan-shaped tips, while those of the Burmese subspecies have pointed tips.

Peafowl, as well as langurs, are well-known as being among the first wild creatures to notice the approach of a tiger or leopard in the jungle and to sound their call of alarm, warning their fellow creatures that a predator is on the prowl.

Peacocks shed all their tail feathers each year and grow new ones. The old shed feathers are picked up and made into fans and widely sold in bazaars and elsewhere.

'The gorgeous peacock is the glory of God' says a Sanskrit verse, and in a country of pageantry and colour it is only fitting that the peacock has been officially proclaimed as the national bird of India.

An added attraction of the Ghana sanctuary is the presence of a few black-buck, chital and other animals that roam about in the drier parts.

It was here, 'where deer and antelope roam and play', and where she had 'taken pictures of deer and a newly born fawn' that Ylla Koffler was cremated in March 1955. She was killed in an accident near Bharatpur while photographing a country fair. The story of her camera shikar in India is well told by Suresh Vaidya in his book *The Jungle Lies Ahead*.

She was one of the world's best photographers of animals, and I am proud to have assisted her in arranging her tour in India, and of advising her of the various difficulties which beset the wild life photographer in this country.

She stayed with me in Assam in the month before she met with the accident, and I introduced her to the rhino, wild buffalo, wild elephants and other inhabitants of Kaziranga. I admired greatly her courage and vivacity, and her ability to select and compose what would make a really good picture, showing the typical, representative character of the subject.

But she found making her book *Animals in India* vastly more difficult than her *Animals in Africa*. For, as I had told her, wild life in Africa is more numerous, largely diurnal and inhabits open, sunny places. Whereas wild animals in India are rarer, mainly nocturnal and their habitat dense.

She wrote to me from Mysore in the early days of her stay in this country, saying 'the grass is very high and the forest dense, and I have not been successful in wild life photography. I sometimes wonder how I will ever do in my book.' She correctly anticipated having to include pictures of tame animals, such as Brahminy bulls, domestic buffalo, temple monkeys, circus animals and so on in her book.

She sleeps in the forest at Bharatpur, near the wild animals and birds she loved so much. It is a pity she did not live to see the Ghana and its birds in all the glory of the breeding season, a sight which would have given her so much opportunity to exercise her consummate photographic skill and which would have appealed so much to her imagination and kindly nature.

E.P. GEE, *The Wildlife of India*, 1964

R.A. Stewart Melluish worked as Editor of Oxford University Press in Madras and Bombay, and played an important part in the publication of Sálim Ali and S. Dillon Ripley's ten-volume Handbook of the Birds of India and Pakistan. *A knowledgeable birdwatcher, he contributed many articles to the* Newsletter for Birdwatchers.

Notes from Madras

The birdwatcher who visits the city of Madras in winter will not regret spending an hour or two at the sanctuary of Vedanthangal. But he should arrange to stay long enough to go further afield, and not rest content with what is little more than armchair watching: for though the birds there are entirely free, and there is probably nowhere else locally where the larger waterbirds like spoonbill and ibis can be seen at their nests so conveniently, Vedanthangal is not unlike a zoo, and if any more municipal cannas and carpark notices are planted and garden seats and observation towers erected many birdwatchers will be driven away. There are plenty of other interesting places to visit, and those who prefer birdwatching when it involves some physical effort can exercise themselves well. The mudflats to the west of Point Calimere, for example—mile upon mile of glutinous ooze—are a severe challenge to any enthusiast's stamina.

The moment you leave Vedaranniyam you are amongst the birds. In winter, at least, and according to the foresters all the year round, the western horizon over what they call the swamp is fringed with a pink line of flamingoes. This is not one of the world's great flamingo feeding-grounds: the numbers are relatively few—from the most realistic accounts, not more than five thousand *roseus* in winter—but it is probably the best Southern India can offer. The numbers at Pulicat are not, as far as I know, ever as great as this. If any reader knows of similar or larger concentrations of these birds in the South, I hope he will publish the fact, because the seasonal movements and habits of this species away from their known breeding ground in India seem to have been little documented, in spite of its conspicuous and interesting appearance.

If you are new to flamingoes, you set off after them on foot and begin your day-long plod through the mud and water. Flamingoes' feet are a better shape than yours, though, for mud, and they can walk faster than you can, so once they realise you are anxious to watch them or photograph them, and not simply catch shrimps like a local fisherman, they wander nonchalantly away. They seem to prefer to taunt you in this way rather than take to their wings, and so, perhaps, give you the opportunity you may be seeking to admire them, in flight. But the effort of tramping through the mud, slow though the progress is, and however foolish the flamingoes make you feel, is well worth while; indeed, it is essential if you are to see much else, because although you can engage a local boat it will hinder as much as help you, and anyway it can only go where there's water.

If you do wade out, and there is mud and water in the right quantities and the time of year is satisfactory, you will see a great deal. Sálim Ali, in 1962, undertook a trial catch of waders with a local fowler's device consisting of a row of nooses, 'strung out at random along the mudflats', and so in a very short time collected, of the *Charadriidae*, Lesser Sand Plover, Redshank, Marsh Sandpiper, Wood Sandpiper, Little Stint, Kentish Plover, Ruff, and a single Rednecked Phalarope. Of these only the last can be regarded as unusual, though I would not call Ruff a common winter visitor to these coasts either. The other species caught are to be seen at any suitable spot in Madras at the right time of year, but not in such huge numbers as at Calimere. If the visitor there is lucky in his timing, he will find the mud on the landward side of the shore, one scurrying, fidgeting, chittering, fluttering mass of small waders, frenziedly poking about in the slime in their hunt for food. He will see Terek Sandpipers in sizeable flocks of fifty or more, quantities of Little Ringed Plover and Greenshank; also Stilt, Curlew-Sandpiper, Large Sand Plover and Turnstone. He may even spot, among the Stints, a group of larger chubbier birds with downcurved beaks which, when flushed, do not show the tell-tale white upper tail coverts of the Curlew-Sandpiper: these are probably Dunlin. I saw four of these birds at Calimere on 12 January 1964, but I have not been able to confirm the record yet; they are not, evidently, one of the common wintering birds that reach the south regularly. And who knows how many Temminck's, Broadbilled and Longtoed Stints, Sanderlings and other such tiny snippets pass the field observer by unnoticed in the mass of confused movement and hasty flight?

The larger, more sedate birds are there too: rows of plump Golden Plover stand in the shallows, all facing the wind; a Bar-tailed Godwit probes about in the banks of a creek; Whimbrels hasten overhead whistling their seven whistles; a party of Grey and White Plover, *Squatarola*, beat upwind with a neat and precise motion of their wings, their black axillaries rhythmically flashing; a curlew calls, and a number rise languidly from their feeding, disturbed, perhaps by the impetuous fighting of nervous stints and plover zigzagging between them. These, together with the usual egrets, herons and storks, and the terns, which fill the air with their squawks and buoyant flight (mostly Caspian, Whiskered, Gullbilled, and Lessercrested) make up the bulk of the great, seemingly limitless, concentration of birds which, in winter, dominate the mud.

Engrossed in all this, the birdwatcher may well neglect the shore itself. For if he turns away from the mud and all the activity and looks towards the strait, the world is immediately empty—except for some dauntless butterfly fluttering off towards Ceylon, an inch or two above the waves, or a Brownheaded Gull. The transformation is astonishing. The sea, for all its fidgeting waves and the shimmering facets of its surface, is relatively lifeless, and its shore vacant. It is curious how dull tropical shores can be, and how fruitless a watch on one so often proves. If one sits on the edge of the Baltic, or spends an afternoon on a headland in Norfolk, and scans the waves, something or other is sure to turn up. Geese will fly purposefully along the coast, a Fulmar will wheel over the crests of distant waves, a raft of Scoter or Merganser will appear, or some diver-like Blob will attract one's attention a mile or more out to sea, unidentifiable, baffling, but hypnotic and fascinating for hours. This just doesn't seem to happen on the southern coasts of India; at least, all my shore watches on the Madras coast have been most disappointing.

To dismiss the shore, though, is a mistake. There is little doubt that the Palk Strait off Calimere offers a lot of excitement in the months when the migrations are on, for Ceylon entertains many visitors from the centre of Asia, and many if not the majority of these must cross the strait. An expedition to Calimere in September or October ought to be most rewarding, and give a new dimension to one's view of the sea and its shores. And the birdwatcher who goes there after the main movements are over, as I have done, should remember that apart from the conspicuous Oyster-catcher there is at least one remarkable shore-dwelling bird which is unlikely to be noticed at all unless one deliberately and diligently examines the tideline: the improbable Crab Plover. This extraordinary bird—so odd that it is classified in a family of its own, *Dromadidae*, all by itself—is thoroughly at home on the remote undisturbed beaches west of the point, and there would seem to be little reason why it should not burrow its quaint tunnels in the sand there, and breed its solitary young. Ripley says it breeds off Ceylon at Adam's Bridge, but does not mention its nesting in India, so I suppose nothing is known about its movements and possible or actual breeding localities here. I spent one afternoon last November watching a party of seven at Calimere. They didn't do anything much, except wash and preen themselves, and then prospect a little along the water's edge. But their heavy bills and pied plumage and generally singular appearance enthralled me, and I sat on the sand and watched them through a telescope for the best part of two hours, and was only roused from my reverie by what seemed to me an abrupt and ill-considered decision of the tide to rise and set me awash. No other birds seem so completely in tune as these with the atmosphere of remote, unpeopled seclusion which prevails over faraway and almost inaccessible shores and the sight of them, justifies any number of barren days spent on empty coasts.

R.A. STEWART MELLUISH, *Newsletter for Birdwatchers*, 1965

Poet, playwright, and art critic Nissim Ezekiel (1924–2004) is considered one of the foremost Indian writers in English. He was Professor of English at Bombay University. He was also the editor of Imprint, Quest, *and the poetry section of* The Illustrated Weekly of India. *'Poet, Lover, Birdwatcher' reprinted below, is one of the most beautiful of Ezekiel's poems. The title, reminiscent of Shakespeare's 'The lunatic, the lover and the poet' (*A Midsummer Night's Dream*), puts the poet, the lover, and the birdwatcher on the same level. While Shakespeare put them together because all of them were 'of imagination compact', Ezekiel finds another common denominator in this group, namely that they all show the same sensitivity to experience and possess the virtue of patience.*

Poet, Lover, Birdwatcher

To force the pace and never to be still
Is not the way of those who study birds
Or women. The best poets wait for words.
The hunt is not an exercise of will
    But patient love relaxing on a hill

To note the movement of a timid wing;
Until the one who knows that she is loved
No longer waits but risks surrendering—
In this the poet finds his moral proved,
  Who never spoke before his spirit moved.
The slow movement seems, somehow, to say much more
To watch the rarer birds, you have to go
Along deserted lanes and where the rivers flow
In silence near the source, or by a shore
  Remote and thorny like the heart's dark floor.
And there the women slowly turn around,
Not only flesh and bone but myths of light
With darkness at the core, and sense is found
By poets lost in crooked, restless flight,
  The deaf can hear, the blind recover sight.

NISSIM EZEKIEL, The Exact Name, 1965

Thomas Gay (also known as Thomas Waterfield in some other writings) joined the Indian Civil Service (ICS) around 1928. A man of wide ranging interests, he started a book reading club in Bombay known as 'Silverfish'. He wrote several articles in the Newsletter for Birdwatchers, *and also published a book on butterflies.*

An Evening at Pashan Lake, Poona

I scanned that lake from end to end. And there they were, scattered over a portion of weedy water swelling here and there into a small mud-flat away beyond a long stretch of tall reeds. My powerful binoculars showed me cotton teal and common pochard for certain, and probably other kinds too; about 100 in all, with dozens of coot paddling and diving busily among them. I walked down to the reeds, leaving my children to start their game of cricket, or get tea ready, as they preferred.

A bay-backed shrike flew to the top of the thorn bush red wattled lapwings, eyeing me warily, moved out of my path on mincing feet. Beside them, a lone blue rock pigeon took off in typical fashion, as though it had just remembered an important engagement. Passing an inlet of the lake, I stood transfixed by the gorgeous colouring of a pair of purple moorhens, trampling the edge of a reed-bed in slow motion, and floodlit by the westering sun. A jungle crow called harshly from a babul tree behind me. The ground fell away towards the long reed patch, which I entered to the protests of a reed warbler and now I could no longer see the open water.

I forced my way through the reeds as silently as I could, noting with relief that the muddy water did not come much above my ankles, until the stems thinned sufficiently to show me the blobs of white, black and brown still well ahead. I raised the binoculars, looked through the last row of stems, and began to count the ducks. There were 112 of them, and whereas the drakes could be identified with certainty the females were a little bewildering. It would not be far out, I thought, to say '50 cotton teal, 45 common pochard and 14 common teal'; about the three magnificent pintail drakes there could be no doubt at all. Among the duck bobbed a dabchick or two, as well as large numbers of coot, which there was no purpose in counting. Away up the lake were egret, but I had eyes only for the duck.

At last I tore myself away, to go on duty as third fielder. Cricket on a tussocky pitch is a lively game, and we were soon hungry enough for tea and sandwiches beneath a gnarled old mango tree on the bank of a dried-up rice-field. Now the air was full of redrumped swallows with an occasional pariah kite floating lazily above. Red-vented bulbuls flitted along a thorny hedge, and a king-crow surveyed his domain as he balanced on the top of a babul. Ring-doves flew down and picked industriously in the rough grass. Large grey babblers shouted 'Creaky-creaky' to each other.

Suddenly I became aware of a bird flying from behind me towards the further side of our field. I recognized the dipping flight of a woodpecker; but when the bird settled on a thin horizontal spray of a babul, exactly like a dove, I thought I must be mistaken.

But I wasn't; the binoculars showed it to be a yellow-fronted pied wood-pecker. And there it sat for several minutes, with a self-conscious look that seemed to say, 'I know that woodpeckers are not supposed to sit like this, but I'm going to do it, all the same.'

There was still one more treat in store. A flock of some forty red Amandavas came over the field like wind-blown leaves, and settled beside some brahminy mynas among the grass—tufts beyond an earthen bank. I stalked them carefully, and got close enough to be thrilled by the cocks' astounding crimson heads and breasts. And in a few minutes the air grew chill; the sun had gone; and suddenly it was an almost birdless world.

THOMAS GAY, Newsletter for Birdwatchers

Peter Jackson was Reuters' Chief Correspondent in India for several years. He accompanied Sálim Ali frequently on his surveys, and played a key role in the declaration of Sultanpur Lake, close to Delhi, as a bird sanctuary. He later joined World Wildlife Fund International in Geneva as Director of Information. Jackson also served as Chairman of the Cat Specialist Group of the International Union for Conservation of Nature (IUCN). His famous piece 'A Day's Worth of Delhi Birds', originally published in Newsletter for Birdwatchers, *illustrates what is known as the 'twitcher's style' of birdwatching—the aggressive pursuit of birds, the surge of adrenalin in which the birder attempts to tick off on his list as many species as possible within a limited span of time, although the author observes that 'birds have more to offer than just being marked on a list'.*

A Day's Worth of Delhi Birds

I am not a 'tick-hunter'—a compiler of lists of birds seen, just for the sake of it. But there were many times when I took friends out around Delhi during the winter when we did tot up our score and we often found that it was more than 100. Inevitably, I began to think of how many birds one could see in a day. From our Delhi checklist I reckoned that it should be possible to top 150. The day was constantly put off until, on 1 March 1970, I awoke with the realisation that I was leaving India in July, the winter was passing, and the chance might be lost for ever.

I got off to a quick start as I left the bedroom—our resident house sparrow sped across the living room to feed its nestlings on top of the almyra. Before I was in the car I had the house crow, green parakeet, common and brahminy mynahs and the pariah kite. As I turned down Prithviraj Road, heading for Mehrauli, blossom-headed parakeets flew alongside. I wished I had taken a portable tape recorder to note birds while driving, but I had to keep the list in my mind for the next stop.

The first halt was at Mehrauli, where the dry, stony landscape, the rocks and the ruins produced some 20 more species, including the rufous-fronted wren-warbler, the blue rock thrush, brown rock chat, yellow-throated sparrow, and dusky crag martin I had relied on—others noted could have been picked up elsewhere.

My route then went on to Gurgaon, noting pale harrier, white-necked stork, white-eyed buzzard and steppe eagle, among others, on the way. From Gurgaon I turned west, taking the old Farrukhnagar road via Dhankot to Sultanpur, a magnificent jheel which the Haryana Government has now farsightedly turned into a Wild Bird Sanctuary. All along the road I had to stop to write down what I had seen whenever the list was getting too long to remember.

All those interested in birds in the Delhi area regretted the draining of the huge Najafgarh jheel. But in fact this probably led to the concentrations at nearby Sultanpur. Although shot over to some extent, its bare margins provided little cover for hunters and the water birds were relatively undisturbed. The flocks of rosy pastors and flamingos were unmistakable and I sat down happily under a shady tree with a telescope to work over the various species of duck, geese, waders, herons, and storks. By lunchtime my list was going well at 87. I would have liked to remain enjoying myself just watching at Sultanpur, but the Yamuna had to be covered.

I went back to Gurgaon by the same road, but this time took the road to Palam, where I turned east and crossed my morning route at Mehrauli, going on past Tughlaqabad to the Mathura Road. From Madanpur a road runs down to the river at a point where the market gardens south of Okhla end. This was a good place to pick up species, such as various terns, streaked weaver, pied mynah, purple gallinule, and as a bonus, a spotted dove, which is very uncommon in the Delhi area.

My plan to follow the cart-track by the market gardens to Okhla, which can be very productive, was foiled by some earthworks in progress and I had to speed round by the Mathura Road. I was now conscious that the pied kingfisher, which I had reckoned one of the easiest to find, was not on my list. Surely, at Okhla ... no, not for me today. A look at the ridge as the sun set yielded little, except one 'chuck', immediately noted as a nightjar. But no more 'chucks' followed, and I reluctantly crossed it off my list as uncertain. It was dark passing through Chanakyapuri, but I paused at the crossroads near the American Embassy. I shone my spotlight into a tree—one of my spotted owlet friends did not let me down. He was at his nightly post.

I was home. My score totalled 128—reduced 127 when the nighjar was rejected.

And now a confession. My leap from bed that morning had not been as the first light of day spread over Delhi. I have to confess that it had been 9 a.m. when I set forth—a shameful time for a birdwatcher. I claimed the right to a full 24 hours for my count, giving me to 9 a.m. next morning. This time I was up with the dawn, with a prepared list of species to get, and raced to the zoo. Not cheating! Delhi zoo attracts a fine selection of wild birds, and I was able to add several, including stone curlew, large pied wagtail, night heron, coucal, and whitebreasted waterhen. I was up to 141. A visit to the Ridge again produced the golden-backed woodpecker, woodshrike and white-cheeked bulbul—144. The minutes were ticking away. I was desperate. My garden still should have something. It did—I ticked off the magpie robin, and as the second-hand nudged up to nine a.m. I closed with the red-whiskered bulbul.

I had failed-four short of my minimum 150 target. I cannot say I sank back into depression. It had been an exhilarating chase, and I don't think anyone would say 146 represents a bad day's birding anywhere in the world.

In retrospect it seemed ridiculous not to have seen such common Delhi birds as the darter, pied kingfisher and white-eye, and even the coppersmith, which regularly nested in my garden. On the other hand, species such as the spotted dove and pale harrier were not to be expected, and so it probably worked out about right in the end. Even so, I am sure that the 150 mark could be reached reasonably easily, and perhaps pushed above 160, especially with more than one pair of eyes working together. Was it my most exciting bird watching day while in India? No, not really. I have had so many blissful days birding in India. I could not really select one occasion, but I felt it worth recording, as a mark for others to challenge. But I hope that 'tick-hunting' in some will not become in India the obsession it has become in some part of the world birds have more to offer than just being marks on a list.

PETER JACKSON, Newsletter for Birdwatchers, 1971

Philip Kahl is well known for his studies on the ecology, behaviour, and systematics of storks, notably the American Wood Stork. His interest in storks brought him to India during the 1970s. As he writes in the piece reprinted here, 'legends about the white stork's ability to supply human babies have been passed on for generations, sustained as cute and convenient ways of avoiding early sex education, relatively little has been written about how storks themselves propagate.'

The Courtships of Storks

Many people know the stork mainly as a decoration on greeting cards and from cartoons depicting it in flight with a human baby suspended from its bill. This image, associated with the white stork, derives from the European folk belief that the stork is an omen of good luck, portending many children, particularly boys.

Although many legends about the white stork's ability to supply human babies have been passed on for generations, sustained as cute and convenient ways of avoiding early sex education, relatively little has been written about how storks themselves propagate. These mostly tropical birds have evolved complex patterns of breeding behaviour.

Of the 17 stork species, only two, the white stork and the black stork of Eurasia, regularly leave the tropics in large numbers to nest in temperate regions. They breed only in the spring and summer and migrate back to the tropics, or beyond, to pass the winter. Resident tropical species, however, can nest at any time of the year, depending upon local ecological conditions.

For many tropical residents, food availability seems to be the environmental trigger that determines the timing of their breeding season. With the exception of the insectivorous Abdinm's stork, they do not undertake a true seasonal migration; instead, they may wander widely within a general region in search of suitable feeding conditions, then return to their ancestral nesting grounds to raise a family.

The wood storks of Florida, for example, scatter throughout the southeast, mostly along the Atlantic and Gulf coasts, during late summer when water levels are high and their fish food is widely dispersed. At the beginning of the dry season in early winter, the wood storks return to their southern Florida nesting sites and commence to breed. Egg laying may be earlier or later in a given year, depending on when the rains end and the water levels start to fall. As the marshes dry up, fish populations become highly concentrated, furnishing abundant fare for the birds. It is apparently this increase in food availability that triggers the wood storks' reproductive cycle, for in winters following inadequate summer rains, when few fish are produced, the birds do not congregate at nesting areas and may fail to breed entirely.

Similar environmental triggers are at work for the yellowbilled stork of Africa and the painted stork of India. These birds are similar to the wood stork in their ecological requirements, and their breeding seasons also seem to be triggered by an increase in available food. Owing to the different topography of their habitats, however, the fish upon which these birds feed are most available during the rainy season. Southern Florida is extremely flat, with few lakes, rivers, or other bodies of permanent water; fish are therefore widely dispersed in the wet season and highly concentrated during the dry season. The availability of fish for the yellowbilled and painted storks, however, is governed by the lakes and rivers of their habitats: the fish retreat to these deeper bodies of water at the onset of the dry season, making it difficult for the birds to reach them. Only with the first flooding of the rainy season do the fish enter shallower marshes to spawn and become available to the birds. Thus, these three similar stork species respond to the same environmental trigger of food availability, but because of differing ecological conditions, they do so at different stages in the wet-dry annual cycle.

Conversely, marabou storks nesting in western Kenya only a few miles from yellowbilled stork colonies usually have large young in the nest, almost ready to fly, before the yellowbills have even laid eggs. The marabous, being largely scavengers, nest during the local dry season when their food is most concentrated around drying water holes and at grass fires.

Whenever their respective breeding seasons arrive, all storks enter a similar process of courtship and pair formation. Some, such as the solitary nesting saddlebill, blacknecked, and jabiru storks, probably mate for life; thus, the process of pair-formation does not take place every year. The colonial species, however, seem to choose a new mate each season and, in some, courtship and pair-formation is a prolonged and elaborate affair.

Courtship in storks can perhaps be best exemplified by choosing a 'model' stork and following the actions of a typical pair from arrival at the breeding site until egg laying. For such a model, I have chosen the painted stork, which I studied at Bharatpur, India, for two seasons.

In August of most years, when the monsoon rains have flooded the area around the nest trees, flocks of painted storks begin arriving at the colony site from all points of the compass. Some may have spent the dry season as far as several hundred miles away in search of favorable feeding areas. Now, as they arrive back at Bharatpur, most have acquired the bright, new black, white, and pink plumage of their nuptial dress. Their naked heads and necks are a brilliant reddish orange and their legs a deeper magenta.

'Bachelor parties' of unmated males and females gather in the low acacia trees, which were filled with nests in the previous season. Weathering over the past year, plus the ravages of other birds pilfering sticks for their own nests, has reduced last year's flimsy nest platforms to mere remnants. Immediately after a bachelor party lands in a tree, males begin to jockey for nest sites—fighting, flapping, and supplanting each other. Any nest foundations that do remain from the previous season are quickly appropriated by the more aggressive males.

At first the established males behave aggressively to all other storks—male and female alike—attacking and driving away those that approach too closely. Other, non-established males often fight back; sometimes they succeed in driving off the original male. Each male that acquires a nest site begins to 'advertise' his status almost at once by repeating two ritualized and stereotyped displays. At first glance, these movements do not appear to be displays: it looks as if the bird is just arranging its plumage and testing the stability of nest sticks or nearby twigs. If one watches closely, however, it can be seen that the movements are formalized and tensely executed, and often do not complete the action they appear to serve.

For example, the male repeatedly preens the primary feathers of one wing by stripping them down from the front with his bill. If you glance around the tree, you will see other males performing the same 'display preening' operation, and each is making the downward movements at precisely the same rate, keeping pace with nearby birds. A closer look will show that often the bird is not actually touching the primaries with its bill, but is only 'pretending' to preen them. After a long series of display preens on one side, the male shifts to the other wing, repeating the process there. Should a female land nearby or approach, the male's display intensifies, with the amplitude of the downstrokes becoming greater and the movements becoming even more tense and stiff.

Between bouts of display preening, the male bends forward and slightly to one side of center and grasps lightly at a nest stick or nearby twig. He then releases his hold, pivots slowly to the other side, the grasps another stick. The process may continue for several minutes, and when performed with irregular intervals between movements it is difficult for the observer to realize that he is witnessing a true display. Motion pictures viewed at a faster than normal speed, however, clearly reveal the nature of this behaviour. With the exception of the saddlebilled and blacknecked storks, male and female storks are essentially alike in appearance, and the birds probably use courting behaviour as a clue to sexual identity. When one or more females have been attracted to performing male, they alight on nearby limbs. As she draws nearer to the male, the female stands with her head held low, wings widely spread, and bill gaping. A short time after landing or moving along a limb, the female may close her wings, but if she remains near a courting male, she continues gaping for long periods. Should the female edge too close, too quickly, the male will probably drive her away with a vicious grab. Courting females do not fight back, however. They approach slowly, usually in a ritualized submissive posture; if attacked by the male, they merely retreat to a nearby perch, wait a short time, and then attempt to approach the male again. Such females often receive harsh treatment from their respective mates, but they 'won't take no far an answer!' They return again and again.

Some males are more aggressive than others, and I have witnessed males driving courting females away for days with vicious attacks. Feathers may fly and blood may even flow, but rarely, if ever, does the female fight back to defend herself. She merely flies way, waits patiently, and later tries again.

At last the female's patience is rewarded and the male allows her to step into the nest. This is a critical movement in the formation of the pair. Initially, both birds appear tense and on edge, and any sudden movement or miscalculation by the female may precipitate another attack by the male.

Frequently, midway through this process of pair-formation, the birds are disturbed by the intrusion of another stork; or the entire bachelor party may leave, a few birds at a time, and form in another, nearby tree. In such cases, the chain of events is broken and the process must start anew. But the persistent, repeated, and gradual approaches of the female allow her to rejoin the male on the new nest site. Each pair-formation, consequently, usually spans several hours or even days.

When fighting stops and a compatible pair is finally formed, they begin mutual greeting displays and copulation becomes frequent as often as three or four times per hour whenever both birds are on the nest. With the passage of time, and numerous copulations, the female gains the male's complete acceptance. The male then shifts to another phase of the nesting sequence: he begins to gather sticks for the nest.

A few sticks may have been added to the platform before a female was accepted, but nest building starts in earnest only after a mate is acquired. At this point, the male leaves the female alone on the nest between copulations and makes repeated expeditions to gather sticks. As the male returns to the nest with a stick, the birds greet each other with a mutual display, which in most storks includes snapping or clattering of the bill and raising and lowering of the head. Owing to the head movements, I have termed the homologous, or closely related, forms of this greeting ritual in the various species the up-down display. The form of the up-down shows wide diversity among the 17 species of storks, but there are enough similarities to indicate that it is, indeed, a homologous display that has undergone species-specific changes through evolution.

The up-down is shown in its simplest and probably most primitive form by the two openbills and the four wood storks. In these six species the display consists mainly of lifting the head and gaping the bill skyward, uttering a series of short and simple vocalizations, and then lowering the head until the bill reaches almost to the floor of the next. In the openbills and the American wood stork, no bill snapping or clattering is heard during the up-down. The yellowbilled stork incorporates single or double snaps of the bill between vocalizations as the bill and head are lowered. And in the painted and milky storks these snaps become multiplied into short bill rattles or bursts of bill clattering.

Bill clattering during the up-down in shown to a greater or lesser extent by all other storks, and reaches its peak of development in the prolonged series of bill clatterings heard from the white stork on European rooftops. In this display, which may last up to ten seconds, the birds throw their heads up and back until the crown is resting on the back feathers, then the head is thrown forwards again to the normal position, with loud bill clattering continuing all the while.

Another bizarre up-down is given by the black-necked stork of Asia. In this species the head remains approximately in the normal position throughout the clattering, but the wings are widely spread to the sides and flutter rapidly during the display. Since the blacknecked is one of the tallest of all the storks and has pure white primary and secondary wing feathers the display is especially spectacular.

The saddlebill stork, the close African cousin of the blacknecked, will probably be found to possess a similar up-down. At the present time, however, the saddlebill remains the only species of stork for which the up-down has not been described.

A few days after copulation begins, the female starts to lay eggs, which appear singly at intervals of about two days until the completed clutch of three or four eggs is laid. The pair continues to engage in some displays during this period, and even the young, one day after hatching, exchange displays with their parents. The up-down is the primary display retained by the adults during the period of egg laying, incubation, and rearing of the young; the function is probably to help maintain and reinforce the pair-bond.

While the prolonged and complicated pairing ceremony may appear to be an unnecessary waste of time, it serves several useful biological functions. Not only does it help to synchronize the physiological states of the participating birds but it also prevents 'mistakes' in a mixed colony of closely related species. Since each stork had its own species-specific set of genetically influenced courtship displays, the formation of hybrid pairs is effectively prevented. The male of one species has the 'key' that will fit only the 'lock' of a female of the same species.

It is often possible to speculate, with a fair degree of accuracy, on the evolutionary pathways by which each display has developed. In the painted stork the display-preening and twig-grasping ceremonies of the male probably evolved from true preening and nest-building movements. The gaping and open wing postures of the female may have evolved from panting and balancing movements. Each movement or posture has taken on the new function of social communication in the behavioural repertoire of the species and is largely, if not fully, emancipated from its original context and function. For instance, gaping which probably originated from thermoregulatory panting, now occurs during courtship even when the weather is cool and the bird is not overheated.

Since much of this highly ritualized, stereotyped, and species-specific behaviour is genetically influenced, it is often more helpful than morphology in determining the relationships and differences between the various species. Based upon newly acquired behavioural evidence, for example, several changes in the previously accepted taxonomic classification of the storks would now seem to be in order.

The most recent, widely accepted classification of the storks was that of Peter's *Check List of Birds of the World*, published in 1931. Peter's treatments were based largely on morphological evidence, for little was known at the time about most stork species in the wild. Consequently, some anatomical differences were overemphasized and some similarities were ignored, without reference to their functional significance in the birds' lives or the rapidity with which they might have changed through evolution.

My studies of the comparative behaviour of storks have shed light on some taxonomic relationships within the family that were obscure in the past, when only museum specimens were studied. For example, all four wood storks the milky, painted, and yellowbilled storks and the American wood stork share extremely similar courtship displays. This is particularly true of their up-downs in which all the major sequences are shared, with only minor quantitative differences in bill snapping and rattling. All four species, now divided between the two genera *Ibis* and *Mycteria*, should be combined into one, *Mycteria*.

Likewise, the closely related saddlebill and blacknecked storks of Africa and Asia have previously been classified into two monotypic genera, *Ephippiorhynchus* and *Xenorhynchus* respectively. But in addition to certain obvious morphological similarities (such as sexual dimorphism of eye colour), they share a 'flap-dash' display unique to them and to the jabiru stork of South America. In this display a male, and occasionally a female, runs away from its mate when both are feeding in a marsh away from the nest. The bird runs for several yards, taking huge steps and flapping its outstretched wings as if about to take off, then runs back to the other bird, stopping short a few feet away. The saddlebill and blacknecked storks are solitary nesters and both species probably mate for life. Neither associates closely with any species other than its own, and they also share a feeding behaviour featuring a slow walk while probing the water for food. Since these storks are alike in so many important respects, both morphologically and behaviourally. I recommend that they be combined into one genus, *Ephippiorhynchus*. (The jabiru stork cannot be included because of other basic differences from the saddlebill and blacknecked storks.)

Other examples of similar behavioural relationships are now known, and further illustrate the importance of field studies of behaviour, which, when combined with our knowledge of the bird's morphology, give us a much better understanding of how present-day storks are related and how they evolved.

PHILIP KAHL, Natural History, 1971

Madhav Gadgil (b. 1942) did his Ph.D. in mathematical ecology from Harvard University. His paper 'Life Historical Consequences of Natural Selection' is regarded as a classic in evolutionary biology. Since 1973 he has worked at the Indian Institute of Science, Bangalore. He served as a member of the Science Advisory Council to the Prime Minister of India during 1986–90. A recipient of several awards, he received the Padma Bhushan in 2006. His article 'Ornithology in Bandipur' reprinted here focuses on the type of questions which can be asked about birds and their behaviour in the context of modern evolutionary biology. Written in the setting of the Bandipur Tiger Reserve in Karnataka, it reads like a travelogue or recordings from a

naturalist's diary. It is a brilliant introduction to behavioural ecology and about asking interesting questions, by observing birds in the field.

Ornithology in Bandipur

BIOLOGY IS FOR THE BIRDS!

Bandipur lies at an altitude of one thousand meters just where the Mysore plateau joins the Nilgiri hills. Perhaps some two hundred years ago, this area was a plain under cultivation as witness the numerous irrigation tanks strewn over the reserve tell us. Depopulated, probably during Tipu's wars, the land was reclaimed by dry deciduous forest which has remained rather open with plenty of undergrowth and grass. The numerous tanks provided excellent water sources for wild animals and the whole region became a rich wildlife area. The Mysore Maharajahs made it their hunting reserve and protected the forest well till it became a wildlife sanctuary and then a Project Tiger Reserve. The low open forest with its juxtaposition of woods, grassy glades and ponds with their bamboo-fringed banks is an ideal habitat for elephants, gaur, chital, wild dog, panther, and of course tiger. It is also a great habitat for birds of deciduous forests and one is guaranteed at least fifty species in a couple of days of birdwatching at Bandipur.

OF WOODPECKERS

We spent quite a lot of time watching birds as a part of our scientific programme at Bandipur. A most striking component of this bird fauna is the Woodpecker guild. Bandipur can boast of no less than seven species of woodpeckers ranging in size from the pygmy to the majestic Great Black Woodpecker. Their drumming is as much a part of Bandipur as the '*kuk-kuk*' calls of chital and the *miouwing* of the peafowl. The commonest species of woodpecker is the handsome Goldenbacked. One has to look rather closely at this one, for there are three very similar species in Bandipur and its neighbourhood—the Goldenbacked (*Dinopium banghalense*) the Goldenbacked Threetoed (*Dinopium javanense*) and the Larger

Goldenbacked (*Chrysocolaptes lucidus*). A close look confirms that the Bandipur species is the Goldenbacked Threetoed woodpecker. Sálim Ali's *Handbook of Indian Birds* tells us that the Goldenbacked affects lighter forests, plantations and gardens while the Goldenbacked Threetoed and Larger Goldenbacked both affect moist deciduous and evergreen forest.

We are quite familiar with the phenomenon of two similar species belonging to the same genus distributing themselves in somewhat different habitats along an environmental gradient. Thus while Bandipur harbours the Crimsonbreasted Barbet (*Megalaima haemacephala*), the moister forests of Mudumalai harbour the very similar Crimsonthroated (*Megalaima rubricapilla*) species. The distribution of *Dinopium benghalense* in open forests and of *Dinopium javanense* in moister forests therefore falls within a well-known pattern. We believe that such pairs of species derived from a common ancestral stock, and on becoming adapted to divergent ecological conditions, diverged a little in appearance as well. The convergence in appearance of the Goldenbacked Threetoed (*Dinopium javanense*) and the Larger Goldenbacked (*Chrysocolaptes lucidus*) is a much more remarkable and less well-known phenomenon. Here the two species belong to different genera, and hence to different ancestral stocks. In the course of evolution they have come to occupy very similar habitat, often overlapping in range, and have evolved to converge in appearance, starting out from ancestors which no doubt differed far more in their appearance.

THE FRUIT-EATERS

Another remarkably diverse group of birds at Bandipur is that of the frugivores. Our list shows no less than eighteen species of regular fruit-eaters: Green Pigeon, three species of parakeets, lorikeet, koel, Grey Hornbill, Small Green Barbet, Coppersmith, five species of mynas, Large Cuckoo-Shrike, Goldfronted Chloropsis and two species of bulbuls. Many of these, such as the Green Pigeon and the barbets, feed almost exclusively on fruit. This whole guild of fruit-eaters is essentially a tropical phenomenon. In the colder latitudes with harsh winters there cannot be a year-round supply of fruit, and hence no room for confirmed fruit-addicts. Our barbets, parakeets and flying foxes (fruit-bats) are therefore phenomena of much scientific interest.

The fruit-producing plants and fruit-eating birds have evolved in step, and the greatest of the bird-fruit genus of plants is the *Ficus*, to which belong the peepal and the banyan. The *Ficus* seed can happily germinate on the trunk of another tree or on a rocky ledge, and it is fruit-eating birds which serve to disperse the seeds through their droppings. At Bandipur we watched a number of *Ficus* trees fruit and watched the fruit being eagerly devoured by an army of birds. In the early months of January–February, a number of such trees came into fruit and we noted how the fruit crop grew and attracted birds maximally when the fruit was fully ripe. This is when the seed would be mature, and the birds reciprocate for the food provided by dispersing the ripe seed—a nicely attuned process of natural benefit. And then in the last days of our stay we saw the process get out of gear. In mid-March a single *Ficus* tree fruited profusely on the campus next to the Vaneshree lodge. There were no other fruiting trees in the vicinity and the whole greedy army of birds camped on this single tree for a period four days polishing off the entire crop leaving not a single fruit to ripen. All the energy invested by the tree in the fruit-crop was thus a total waste. At the same time, if the birds had waited prudently allowing the fruit to ripen, they would have got much more out of it. But the blind, selfish competition amongst the birds would not allow this to happen.

We were able to devote but little time to this fascinating phenomenon. The energetics of fruit production and fruit consumption, and the co-evolution of fruit trees and fruit birds is a subject worthy of a detailed study, and India with its huge population of *Ficus* tree is the best place to undertake it.

THEY SIP NECTAR AND FIGHT

Just as plants use birds to disperse seeds, they use them to pollinate flowers. The nectar secreted in such flowers serves as the attractant for the pollinators. Plants exercise great economy in the production of such nectar, seemingly secreting not an iota more than absolutely necessary to attract the pollinators to the flowers. The amount necessary to attract the birds would depend on alternative sources of food available to the birds. It is therefore advantageous for bird-flowers to bloom in a season in which other food sources are at a low level. This is precisely what non-specialized bird flower such as the Silk Cotton and Flame of the Forest do. They bloom in the dry season from December to March when the insect food of their major pollinators—the mynas—is at its minimum. They further enhance their attraction by advertising the blossom in a burst of mass flowering when the leaves are shed. And the birds do come to them in large flocks.

Our stay at Bandipur coincided with the blooming times of the Silk Cotton as well as the Flame of the Forest, and we spent long hours watching the birds at these trees. The major consumers of nectar were the White Headed Myna, Jungle Myna, Indian Myna, Small Green- and Crimsonbreasted Barbets, Redvented and Redwhiskered Bulbuls, the Goldenfronted Chloropsis and the Maharatta and Golden backed Threetoed Woodpeckers. Even the Black Drongo appeared to sip the nectar on occasion, though it mostly concentrated on pursuing flying insects. It was obvious that nectar was in short supply and there was continuous squabbling amongst all the birds for sips at the nectar. The squabbling never ended up in real blow, but took the form of visual and vocal threats followed by surrender of the disputed position on the tree by the submissive party.

Most of the interactions are amongst the three commonest visitors—Jungle Myna, Indian Myna and White-Headed Myna.

Intraspecific interactions are only a fourth as common as interspecific interactions. The Indian Myna, in particular, is a highly querulous bird and is responsible for the bulk of interspecific aggression as it vents its aggression almost exclusively on other species. The Jungle and Whiteheaded Myna do squabble a great deal amongst themselves and most of the intraspecific disputes are restricted to these species. What struck us as most unusual, however, was the amount of flack that the Black Drongo took from the Indian Myna. The Black Drongo is famous for its pugnacity, which is exhibited particularly in the vicinity of its nests. But on the Silk Cotton tree, the Indian Myna reigns supreme. The drongo rarely sips nectar, but comes to the tree primarily to hawk insects, particularly the honeybees. Its motivation for defending a particular perch is therefore unlikely to be as great as that of an Indian Myna. It is perhaps this factor which tilts the scales in favour of the myna—which is by no means a mean fighter on its own merit.

THE BANDITS

Plants have evolved a mutually beneficial relationship as pollinators and seed dispersers with many birds—bulbuls, barbets, sunbirds and mynas. These birds utilize the unlimited amount of pulp nectar made available by the plants and serve to disperse the seed or affect pollination. But there is another set of birds who are not willing to rest content with what the plants provide; rather they want to exact tribute from the plants without rendering any service in return. They have developed a most formidable weapon in a strong beak in order to achieve this end. These bandits are of course the green jewels of the tropical forests—the parakeets. With their strong beaks, the parakeets are capable not only of feeding on the soft pulp which the plants provide as an inducement for the birds to disperse seeds, but can break through and eat the seeds as well. Similarly, at a silk cotton tree or on a flame of the forest, the parakeets not only go for the nectar but also feed on buds and petals as well, destroying thousands of flowers in the process. The plants seem to have no effective defense against these bandits and have to suffer their depredations. And parakeets are a most successful group of birds at Bandipur with large populations of three species—Roseringed, Blossomheaded and Bluewinged.

BIRDS OF A FEATHER

We have mentioned a number of birds so far, and each one of them is characterized by some specific social habits. The Goldenbacked, Threetoed Woodpeckers largely occur in pairs, as do the Indian and Jungle Mynas. The Whiteheaded Myna occurs in small flocks and so do all the three species of parakeets. Each of these social habits must have an adaptive significance posing yet another challenging problem for the biologist. We ourselves made some interesting observations on the flocking behaviour of two other species—the Jungle Crow and the Spotted Dove.

The Jungle Crows do not sleep at Bandipur—they have a roost outside the Tiger Reserve in a nearby village. From there they fly into Bandipur very early in the morning, in loose straggling flocks. Then they start looking for food, loosely spread out over the forest. It is evident that they keep in touch with each other all the while. One of their choicest foods at Bandipur is pieces of meat at a kill made by predators such as wild dog. The Jungle Crows keep a sharp lookout for the predators, and invariably follow them if the predators are out on a prowl. They also search out the kills and having found a kill fly and call around it in a most characteristic fashion. The tribals use crows as infallible guides in locating the kills, and no doubt other crows can do the same. A flock of crows therefore builds up very rapidly once one or more crows discover a kill. They feed on the morsels available till these are exhausted, and they disperse once more. The flock is thus not a permanent entity at all, but builds up and breaks down opportunistically. Such flocking behaviour seems geared towards proper utilization of resources which occur patchily in an unpredictable fashion.

The Spotted Dove was another bird we looked at closely. We noticed that in their flocks there were two types of birds—those with the bright white and black chessboard on the neck, and those without. Observation of the courting and mating behaviour suggests the possibility that the birds lacking the chessboard, somewhat smaller in size, are the females. This is something which needs to be investigated further. While feeding, the doves often formed large flocks on the road. These were circular in shape with the presumptive females largely inside the circle. The flock moved in a constant direction. Now, it is known from studies of Wood Pigeons that not all the birds in a flock can feed equally well, but rather those in certain positions get pushed around continually and get less time to feed. It was found that in these spotted doves the peripheral individuals made an average of 77 pecks per minute, while the central areas made 82 pecks per minute.

DEATH COMES EARLY IN THE TROPICS

We also had a look at several Spotted and Ring Dove nests. The nest would be made the eggs laid and we would expectantly watch for further developments. And, alas, the eggs would be gobbled up by some predator. Not a single dove nest that we saw could succeed. This falls in line with the accumulating evidence that mortality of eggs and chicks is considerably higher in the tropics than in temperate zones. On the other hand, the adults in the tropics seem much less susceptible to predation than are the adult birds that breed in the colder latitudes, whether they later migrate or not. This generalization, if substantiated, has many important implications for our understanding of ecological processes. It may tell us why tropical birds have smaller clutches, whether the adults are expected to be in more acute competition with each other, whether the ecological requirements of tropical birds are expected to be narrower and so on. There is hardly any data today on mortality in the egg and nesting stages of Indian birds, and no data at all on the mortality of adults. Only a long term study based on individually identifiable colour-ringed birds will tell us more.

In contrast to the Spotted Dove, the House Sparrows of Bandipur had excellent nesting success. We observed three nests of the house sparrow from beginning to fledgling. One of these had a clutch of 4 and the other two clutches of three each, and the entire broods were raised successfully. These house sparrows bred inside a hut in rather protected situations, and their nesting success may be rather atypical, but this needs to be examined further. A very interesting point that emerged out of the sparrow study was that although the sparrows are monogamous, the males and the females do not share the burden of rearing the chicks equally. This was consistently so in all the three nests investigated. In all cases females made roughly 55 per cent of the trips to the nest, while the male made about 45 per cent. In addition, the female came to the nest with food in 90 per cent of her trips. The male, on the other hand brought food only in about 80 per cent of the trips. If we assume that the male and female bring about the same average quantity of food, then the male is sharing only about 40 per cent of the burden of feeding the chicks. In addition, the male was always much more cautious of his each visit to the nest.

This is another very interesting finding in view of the recent theoretical developments in sociobiology. We now realize that the relations between members of a mated pair or parents and offspring are made up of elements of both co-operation and conflict. At the same time that the male and female are co-operating and fulfilling the shared goal, each partner may be trying to get the maximum out of the other, and trying to give out the minimum feasible. An analysis of the relationship of the male and female at such a game has proved most illuminating. We now believe that in most cases, even of monogamy, the males come out of the game making a smaller contribution to the task of rearing the offspring as they have the female at a disadvantage because of her initial heavier investment in reproduction. Our sparrow results are therefore worthy of further pursuit.

THE LONG TAIL

So we come to the last part of our narrative which shall deal, appropriately enough, with tails-and rather long ones at that. The longest tail of all—or rather the longest train made up of upper tail coverts—belongs to the peacock. Surprisingly enough we know little about the breeding behaviour of our national bird. It is however clear that the peacock practices polygamy—he has many wives and he apparently does nothing for his own chicks— not even the 40 per cent that the male house sparrow puts in. In this polygynous system, not all males can mate, and this introduces tremendous competition amongst the males for the acquisition of a harem. The success of the males in this endeavor apparently rests on their being able to dazzle the females by showing off their great trains. We spent a lot of time observing peacocks in different stages of the growth of trains and it is evident that the long train of a peacock is a serious disadvantage to it in flight. A long tailed male cannot take off the ground quickly, nor can he maneuver properly once up in air. At least on the surface, it appears as if the long tailed males are severely handicapped. Why should the females then bestow their favours on these handicapped creatures? We speculated that a male which is alive and flying in spite of his tremendous handicap must obviously be a vigorous individual and therefore should rank high in the female's view. Just as social pressures for showing off have led to the practice of spending exorbitant sums on wedding ceremonies which leave the bride's parents in debt for years to come, the pressures for showing off before the females have forced the male peacocks into developing an enormous train which in every way is a serious drain on their resources.

This notion of selection for a handicap is another new and interesting development in sociobiology. It would be very much worthwhile testing this out carefully in the field by an investigation of the mortality rates and reproductive success in peacocks with different lengths of trains—and following this over a few generations.

The drongos are another group of longtailed birds, but the long tail here is persistent round the year and common to both sexes. It does not seem to serve a display function and is probably not a handicap, but rather an asset in the drongo's normal business of life. Professor R. Narasimha, our aeronautical engineer who visited us at Bandipur spent a long time watching Black Drongos hawk insects from a telegraph wire near the Bandipur tank. These sallies are remarkable for their twists and turns in mid-air, and Professor Narasimha speculated that the long, bifurcated tail may play a critical role in enhancing manoeuverability. The point to be stressed here is that by simply looking at the bird, we could not have guessed at the function of its tail. This was possible only after watching it in the field.

THE MORAL

The moral of this long tale is that as students of biology, we could learn a tremendous amount from birds. Birds have played a critical role in the development of much of modern biology at the level of organism as a whole. Many giants of modern biology—Ernst Mayr, Konrad Lorenz, Niko Tinbergen, David Lack and Robert Mac Arthur, to name a few, have been ornithologists. It is through their studies of birds that we have gained most of our understanding of evolutionary, ecological and behavioural processes. Sadly enough, our biology curricula pay scant attention to these exciting fields, concentrating instead on a dull, repetitious study of dead structure. Our zoology students spend endless hours dissecting dead rats, cockroaches and earthworms—and from there they graduate to dissecting more dead rats and cockroaches and earthworms but now with a dead lizard or a scorpion or a pigeon thrown in. They are never encouraged to look outside the laboratory and observe the rich drama of tropical life going on all around us.

This is a great pity. Half of a modern biology curriculum should deal with the molecular and physiological level and the laboratory exercises should deal with interesting experiments illuminating these fields. The other half should deal with ecology, behaviour and evolution—and the practical training should almost entirely concern itself with field work. To understand their proper context, ecological and behavioural observations should be made under natural conditions—conditions under which the organism has evolved. And birds, along with squirrels, monkeys, bats and lizards provide the richest and most easily accessible material for Indian students of biology for conducting their field work.

Our own major emphasis at Bandipur was not on birds. Nevertheless, we did spent some time watching the birds, and out of these observations emerged a number of interesting ideas. We have presented above a selection of these covering a spectrum of biology ranging over biogeography evolutionary theory, community and population ecology, behaviour and sociobiology. Hopefully, we have made a convincing enough case for the claim that we can teach, learn, and do serious research in modern biology by just watching the birds in their natural setting.

Most of the readers of this Newsletter would not be professional biologists—which is wonderful, in that amateur bird watchers, besides enjoying themselves have made remarkable contributions to science. It is also a tragedy in that so few of our professional biologists are interested in learning about, and exploiting the rich teaching and research potential of our bird fauna. Our course at Bandipur was perhaps one hopeful sign that this apathy is slowly dissolving, and that our scientific community is waking up to the challenge of the winged bipeds. Very soon, let us hope, it will be ringing with cries of biology is for the birds!

MADHAV GADGIL, *Newsletter for Birdwatchers*, 1978

(Late) Hamida Saiduzzafar was Professor of Ophthalmology at the Gandhi Eye Hospital, Aligarh Muslim University. She contributed many articles to the Newsletter for Birdwatchers, *and also wrote a piece on bird vision in the* Encyclopedia of Natural History *(published by BNHS and OUP). Her article on the Blue Jay, reprinted here, represents a logical way to piece together a story, and is remarkable for the simplicity of its style and an approach based on evidence and deduction.*

Some Observations on the Apparent Decrease in Numbers of the Blue Jay

Over the last twenty years I have been travelling by car from Aligarh to Delhi and back by the Grand Trunk Road (G.T. Road) fairly often, and usually in the mornings between 6 to 10 a.m. Apart, from crows and mynahs, the parakeets, drongoes, doves, white-breasted kingfishers, and blue jays are the birds most commonly seen and identified, sitting on telegraph poles or wires running parallel to the road for about 85 miles. From among these the Blue Jays (*Coracius benghalensis*) are by far the most colourful and spectacular in flight, and catch the eye as they fly up or down to catch their morning breakfast! Around 1965–6, to amuse my young nephew (who was getting interested in bird-watching) he and I used to count these blue jays and I remember noting down our tally, which varied between 60 to 80 birds between Aligarh and Delhi.

About a decade later around 1976, I once idly did a count again and was surprised that I couldn't see more than 40. I reported this to Sálim Ali the same year, but he didn't give it much importance, because he said their numbers would vary depending on the seasons and the time of day, and had I taken these factors into account? The time of day was mostly the same, but I had not paid attention to the seasons and couldn't find my old notes at that time, so I thought that my observations were probably fallacious and that I was just chasing a 'red herring' rather than a blue jay!

Around 1980–2 I again travelled many times to Delhi by car, and this time I did the counts again, carefully noting the seasons as well as time of day. The count remained between 25 to 35, so I began to look around more carefully also on other routes, e.g. to Narora and back. The impression persisted that these birds were on the decrease. June of 1982 brought a devastating tornado-like storm in this region, which swept across the belt between Khurja and Ghaziabad, bringing down a great number of old and very big trees. The absence of these big old trees have left large gaps on the roadside and many trees have dried up. All new roadside plantation by the Forest Department is of Eucalyptus only, and I cannot see any of the shady trees being replaced as such.

It is also a fact that the density and noise of traffic on this highway has increased at least three or four fold since 1965. So, this year, 1984, when I again remembered to count my blue jays, travelling to Delhi on three occasions within two weeks I was quite shocked to find my count was down to 6, 10, and 12, respectively, on these three morning journeys.

Whatever other reasons there may be, I am now quite convinced that this bird is decreasing in numbers in this region. However, this impression needs systematic and accurate verification, because the Blue Jay is said to be a useful bird from the agriculturist's point of view on account of the large number of insect-pests which it consumes. I sincerely hope that my idea is incorrect, and that the birds have simply changed their perches, and gone further afield, away from the roadside noise and telegraph wires.

HAMIDA SAIDUZZAFAR, *Newsletter for Birdwatchers*, 1984

Prakash Gole is founder-director of the Ecological Society in Pune, and has been editor of the society's journal since its inception. He was head of the Crane Working Group of the Indian Subcontinent, and an active worker of the International Crane Foundation. He is best known for his pioneering study on the status of the Sarus Crane across India, and the piece reprinted here discusses the problems of this magnificent bird.

The Pair Beside the Lake

It happened long ago. A poet who lived by a lake saw a strange sight—two great birds dancing. The male courted the female: he danced around her, pirouetted, jumped in the air, and made chivalrous bows. The coy female accepted with gentle nods.

Both were so entranced that they forgot the world around them. Then the poet saw to his horror that a hunter was approaching. A swift movement, and an arrow struck the male. The female saw her mate fall, but she did not run away. She tried to rouse him with her beak and doggedly stood her ground. The hunter closed in and the female, too, fell beside her mate.

When the poet saw the tragic end of the couple, a poem took shape in his mind. This was the birth of the great Indian epic, the Ramayana, which still enthralls millions in India, Burma, Thailand, and Indonesia.

The cranes that inspired this epic still grace the Indian countryside. But the hunter, now in a different garb, still stalks the cranes and may yet put an end to their lives. So while there is still time, let me tell you the full story of the great birds since that epochal event beside the lake.

PEOPLE AND CRANES COEXIST

Millions of years ago, a great sea surged where the plains of north India sprawl today. As the Himalayas rose, the sea retreated, leaving behind a great trough which was filled with silt brought down from the mountains by India's great rivers, the Ganga and the Yamuna. A great alluvial plain formed with innumerable streams, lakes, and marshlands.

Then man came and saw with delight the great variety of animals and birds that belonged to the fertile plains. Though he killed some of them for food, he admired many others for the nobility of their character.

One of the largest birds always appeared in pairs, and he called it Sarus for its strident, bugle-like call. He admired the bird's graceful looks, its ardent courtship and the great devotion the paired birds showed each other. For dwellers of the plains, Sarus became the symbol of conjugal love. People felt sorry for what happened on that fateful day beside the lake, and vowed never to repeat it. The Sarus were protected, almost lost their fear of man, and came to live beside the village.

During the tumultuous period when the fortunes of dynasties waxed and waned, India's countryside and village life changed but little. People drew water from wells and the village pond was bordered by tall reeds; and in these reeds the Sarus pair built their nest and laid two eggs.

The marsh around the pond provided food for their chicks, and the pair often visited fields to glean fallen grain after the harvest. No one bothered the pair or their chicks. They were a part of the village scene, as was the stork nesting in the tall banyan tree or the monkey living on the temple tower. The land teemed with ponds and marshes, and Sarus were everywhere. Their clarion-like calls marked the beginning of another day for many a village family.

THE END OF TRANQUILITY

It seemed that the hunter had been permanently banished. But no! The tranquility of village life was shattered by events that took shape in faraway capitals. Cities continued to grow, and then started dictating what the villagers should sow and reap. They sent to the village new seeds, fertilizers and water from canals. Many village ponds soon went into disuse, filled with village waste, and dried up. Machines roared into fields and drove the birds away. Roads and highways cut through the countryside, dividing fields and bisecting the Sarus territories. Fields that once grew grain changed to cash crops like sugar cane and tobacco, and marshes were drained for expanding agriculture.

Finally, industry made its appearance in the village. Labor from distant lands poured in for construction and factory work; the new people set up shacks, and weeds spread where once the lotus bloomed and the lily blossomed. Fertilizers and insecticides washed from fields into the ponds and killed freshwater plants and animals. The village ponds that remained became cesspools of filth.

I clearly saw all this when I traveled several hundred miles through the Sarus country. I traveled across Orissa, the region where rice is said to have its origin. I knew that in the absence of marshes, Sarus often resorted to flooded rice fields. But there was no Sarus in Orissa. From Orissa, I entered south Bihar and then stepped into the great alluvial plain of the Ganga and the Yamuna rivers.

In vain did I look for Sarus till I reached the very center of the Ganga-Yamuna flood plain. I asked the villagers, and showed them coloured pictures of Sarus ... but everyone told me: 'Sarus used to be in our village some years ago. But now it is gone.'

LUCK AT LAST!

When at last I met my first Sarus pair, I had already crossed half the region of the Sarus range as described by the ornithological texts.

I jerked my vehicle to a stop when I saw the pair quietly gleaning grain in a harvested field. Not far from the pair, a farmer and his family worked. It was the same tranquil scene that I had envisioned many times. A thrill ran through me, probably the same feeling the poet had experienced on that fateful day.

I asked the farmer to show me the village pond. He took me to a small lake surrounded by stately mango trees. In one corner was the familiar reedbed—the home of the Sarus. 'Two or three families nest here,' he told me. I then visited another lake not far away. This was overgrown with aquatic vegetation. But there I saw a Sarus pair with two juveniles—gray and cinnamon-brown, small replicas of their tall parents.

I continued to meet Sarus as I traveled. I saw them in the fields gleaning fallen grain, even attacking standing crops of rice and wheat. I asked the farmers how they felt about the theft of their crops. 'It's not much,' they carelessly remarked. I saw Sarus with juveniles foraging in shallow ponds and marshes. Once as I was watching Sarus in a marsh, a village boy came and told me that a road was being planned through a part of the marsh. 'Do you like it?' I asked him. 'No, I don't want to lose all these birds,' he replied.

I saw Sarus coming together for rest under trees, and I witnessed their evening social gatherings when they collected in flocks to play, run, chase one another, dance and jump in the most delightful manner. As the sun set, I saw them winging their way to a roost in a wide and shallow basin of the river.

Gradually, I came to understand how the Sarus spent their days. Principally, their time passed in feeding and resting. While pairs with chicks fed chiefly in marshlands, pairs without chicks fed in agricultural fields. During the warmer part of the day, they rested under shady trees, in large pools or on riverbanks. A number of pairs gathered at these resting places, where they preened vigorously to keep their feathers clean and trim.

The evening social time was important too. It was then that the newly recruited adults probably found their mates, and already mated pairs strengthened their bonds. These social gatherings took place in fallow fields and what we called 'waste' lands. The habitat combination that Sarus appeared to favor best was marshes, ponds, fallow land, and cultivation—in that order.

CAN THE SARUS SURVIVE PROGRESS?

By now, I had crisscrossed a lot of Sarus country. I had observed over 1,200 Sarus in different regions. It was gratifying to see that Sarus was still protected. Except for a few pockets inhabited by hunter-gatherer tribes who killed with bow and arrow anything that was moving, Sarus was not hunted.

The great birds even prospered in certain areas. They had taken over strips bordering irrigation canals where water seeping through canals had created wetlands, and here the Sarus nested throughout the year. The year-round availability of shallow water and marsh probably triggered this unusual nesting behaviour.

It was wonderful to find the hunter vanished from these Sarus areas. But when I probed deeper, certain disturbing trends became visible. Even in regions where Sarus were numerous, not many pairs could breed successfully. For many of them, sufficient breeding habitat was just not available. Marshes and ponds had gone under the plough or had been reclaimed for other uses. If the Sarus is not able to breed with success in its area of greatest concentration, declining population will result.

In regions where agriculture was mechanized and dependent on heavy doses of fertilizers and insecticides, no Sarus could be found. Sarus appear to have retreated to the so-called 'backward' areas, where agriculture is still traditional, where harvests are low because very little fertilizer and insecticide are used, and where human population densities and urbanization lag behind the rest of India.

What if these so-called backward regions also come under the spell of economic progress? Where will the Sarus go then? Are great birds like the Sarus and the stork incompatible with economic progress? Which future do you prefer, dear reader? The pair beside the lake, or the car and the supermarket?

It is not the bow or the gun that threatens the Sarus in India today. The 'hunter' has been transformed into something more wily and perhaps more dangerous, for he now comes in the garb of Technological Man!

PRAKASH GOLE, *The ICF Bugle*, 1989

Asad Rafi Rahmani, currently Director of the Bombay Natural History Society, is well known for his work on the conservation of the Great Indian Bustard. In this piece, he draws an interesting link between civic cleanliness and the population of the Adjutant Stork.

The Greater Adjutant Stork

Storks are some of the largest birds in the world. The Greater Adjutant Stork *Leptoptilos dubius* provides another example to the axiom that bigger animals face more dangers to their survival than the smaller species mainly because bigger animals need more space to live. With a height of 130 to 150 cm, the Greater Adjutant Stork (hereafter called 'Adjutant' in this article) is a tall bird. It is slaty-back, grey and white, with nearly bald head and neck and a massive wedge-shaped bill. A large naked, pinkish pouch 25–30 cm long hangs from the front of the neck. It was named 'adjutant' by British ornithologists because its deliberate, measured gait reminded them of military adjutants who walk in a similar manner. It is mainly a carrion eater and sometimes eats quite large bones, hence its Assamese and Bengali name 'Hargilla' or 'bone-swallower'.

The overall distribution of the Adjutant extends from northwest India (Gujarat and Rajasthan) to Vietnam and possibly up to Borneo although some ornithologists now consider the Borneo record as an error, referring instead to Lesser Adjutant *Leptoptilos javanicus*. In our country it was seen in Gujarat, Rajasthan, and Gangetic plains, the northeast and sporadically in central India. There was no record from Tamil Nadu or Kerala. However, William Harvey of the British Council wrote to me that he saw eight Adjutants on 21 December 1980 on the backwaters at Mahabalipuram near Madras. This is a range extension of this species in south India.

Although ringing data are not available, the Adjutant is supposed to be nomadic, moving to the plains of India during monsoons from its main breeding grounds in the northeast India and Burma. It is found in jheels, marshes, paddyfields, and drying puddles. Its main food is carrion, fish, frogs, reptiles, insects and dead or dying birds, or any small animal matter which comes within the range of its massive bill. Its African cousin, the Marabou Stork *Leptoptilos crumeniferus* has been reported to be very destructive to flamingo colonies.

Owing to its proclivity to eat carrion, the Adjutant is considered unclean and thus generally not shot for food. It is even tolerated by local people and even now, in the Brahmaputra Valley in Assam, it can be found inside towns. In spite of this fact, there has been a drastic decline in the number of Adjutants. Fifty to sixty years ago, it was not an uncommon bird in the north and northeast of India. Vast numbers were seen on the garbage dumps outside Calcutta and other towns. It played an important role in keeping the environment clean. With improvement of sanitary conditions and urbanization, it disappeared from Calcutta. However, we cannot explain its disappearance from rural areas where sanitary conditions have not improved and may even have deteriorated much to the advantage of this species. The population of livestock has gone up and thus there are more carcasses to feed on. Studies by my colleagues in the Bombay Natural History Society indicate that the population of Whitebacked and Longbilled vultures have increased due to the increase in food supply. Similarly, the Marabou Stork of Africa, which is ecologically similar to the Adjutant, is increasing and is commonly seen on garbage dumps around towns and villages and like in our country, tolerated by Africans.

My own surveys and the information collected from fellow ornithologists, show that, in all over its distributional range the Adjutant has become rare. As far as I know there is no recent record from Gujarat, Madhya Pradesh, Andhra Pradesh, Orissa and Uttar Pradesh. Only in Keoladeo National Park in Rajasthan, some Greater Adjutants are regularly seen during summer months but every year less and less come. In 1983, Dr Vibhu Prakash of the BNHS who did studies on the raptors of Keoladeo, saw 7 to 10 Adjutants and 150 to 220 Lesser Adjutants but in 1989 only one Greater and three to four Lesser Adjutants were seen. In Bihar, I saw seven in April 1988 from the main road. In West Bengal where hundreds were seen earlier around towns and villages, drew a blank when my colleagues Goutam Narayan and Lima Rosalind did a roadside survey in June 1989. Only in Assam we could see some Greater Adjutants. In May 1989, we found 57 right in the middle of Tezpur town and a few more were seen earlier around a fish market in Guwahati town. Between 29 April and 9 May 1989 we counted 80 storks in 10 spots in five districts of Assam. During this roadside count, we saw five carcasses and every carcass had a few Greater Adjutants along with vultures and kites. Inquiries from local people revealed that this species is still seen around slaughter houses and garbage dumps in many towns of Assam.

What could be the causes for the rarity of Adjutant? A literature survey revealed that there are very few nesting records of the Adjutants from our country. Nearly 100 years ago, few nests were found in Gorakhpur district in Uttar Pradesh and a small breeding colony in the Sunderbans in West Bengal and one or two records from Orissa. In the 1960s, five to six nests were located in Kaziranga National Park in Assam. However, this does not mean that these were the only nesting areas of Adjutants in India but it indicates that even in the olden days, most of the adjutants seen in India were coming from outside. The nearest country where they were seen in large breeding colonies was Burma. Owing to paucity of data, I have not been able to chronicle the decline of the Adjutant but it is generally and rightly surmised that this decline started with the destruction of breeding colonies in Burma.

However, this could not be the only reason. Why has the resident population of the Adjutant, which was present in India, not increased (like the Marabou Stork) to fill the vacancy created by the decrease of the Burmese population? Food of Adjutant in the form of carcass has not decreased, so what could be the reason for its sharp decline? The Adjutant, unlike other storks such as Blacknecked, is not very sensitive to human disturbances but still it has fared badly. I have seen Adjutants right inside Tezpur and Guwahati towns around slaughter houses or fish markets, so why should this species become uncommon when it is not molested and it can get all the food it wants, thanks to our dirty sanitary habits? Is Adjutant's dependence on human beings the major reason for its fall? Is it suffering from pesticidal poisoning and is its rarity only a symptom of environmental degradation? Unless we do a detailed study on this interesting species, it is difficult to answer these questions.

Asad Rafi Rahmani, Newsletter for Birdwatchers, 1989

Otto Pfister is an amateur ornithologist and photographer. While working as an official in the Swiss embassy in Delhi, he indulged in birdwatching and photography and travelled across India. His book Birds and Mammals of Ladakh *published in 2004 was widely acclaimed.*

Cranes of Hanley

For many years I have been anxious to find that observe the rare Black-necked Crane. Fortunately, I received permission in the summer of 1995 to visit the remote corner of eastern Ladakh, which is reported to host one or two breeding pairs of *Grus nigricollis*.

The Black-necked Crane *Grus nigricollis* is a tall grey bird. The male, almost the size of a Sarus Crane, is about 145 cm tall while the female is lightly smaller. Its head (except the red 'cap' and the more or less visible small white patch behind the yellow eye), its neck as well as the curving tail feathers and wing-primaries and secondaries are black; the coverts are white-grey. The beak shows a grey-green colouration, turning reddish at its root. The couple intercommunicates in a discreet *gorr-kro, gorr-kro*, whereas the shrill trumpet call is used mainly to mark and manifest territorial claims and courtship.

In addition to the Hanley Valley in eastern Ladakh, further records of breeding places of these scarce birds are the marshes of Tchuchur, near Pangong Lake, and the borders of the waters in southern Tibet, all at altitudes between 4000 and 4500m. Observations of feeding birds in Ladakh are reported from Joje-Tso, Tsokar and Puga; even the Shey marshes have been visited in earlier years but there are no recent sightings because of the disturbance caused by building human settlements and the drainage of part of the wetlands to be converted into grazing land for the numerous donkeys, horses and cattle. In Phyang Monastery close to Leh, an 'air-dried' specimen, hanging in the Mahakhala Temple for many decades, proves Black-necked Cranes were frequently seen and enjoyed a wider distribution towards the west till the area around Leh. It is now estimated that just about 200 birds survive though no exact records are available. The largest accumulation of Black-necked Cranes is currently observed in the remote valley of Phobjika in central Bhutan, where about 100 birds pass the winter months from November till March. Other valleys like Bhumtang, east of Phobjika, and some isolated valleys in Arunachal Pradesh also host wintering birds.

TOWARDS HANLEY

I left Leh on 13 June 1995. It was a beautiful clear morning, with blue skies and a fresh breeze. As we followed the Indus river eastwards, the topographic surroundings changed constantly: narrow valley passages, steep loose gravel slopes and colourful rock formations were followed by wide open high altitude plains, grasslands and even sand dunes after Mahe. After nearly two days of hard driving we reached Loma, where our route on the well maintained blacktopped road ended abruptly. We tuned right, crossed the Indus and entered the Hanley Valley. An unexpectedly hostile landscape greeted us—barren arid hills and valley floor, interrupted only by a narrow green strip of grass bordering the Hanley river. The valley widened as we drove on, the river winding down in a number of branches, creating marshy pools here and there. Just about half way up to Hanley, while we were passing a swampy area called Lalpari, I suddenly spotted two big birds wading through the reedy grass—I could not believe it, but looking closer, I saw a pair of Black-necked Cranes feeding. The driver guided the vehicle gently down the slope until we reached a mound which provided sufficient cover. Our presence did not seem to bother the birds so I had plenty of time to view these beautiful rare creatures. When the slightly smaller bird, the female, came stalking through the water towards me and started to lift dead aquatic plants onto a heap situated about 10m. off the nearest bank, I realized to my delight that I had a breeding pair in front of me.

I pitched my tent on a stony plain about 250m. away from the spot where the cranes were building there nest and decided to spend time on observation. When I returned to the swampy pool to have a closer look through my binoculars at the flat mound on which the female was still standing, I noticed one creamy coloured, olive green-brown splotched egg, about 10 cm. long, lying among the twigs. She sat down on the egg, watching me carefully, even though the hide made me invisible. I slowly approached some 30m. along a dry strip of grass stretching like a tongue into the water. Still, the bird felt no disturbance. Half an hour later, she suddenly rose and I immediately saw that a second egg had been added (15.6.95, 14:30 hours)! Without wasting time she changed position and continued to incubate the first egg, ignoring the new one beside her. I was confused—why did this female lay an egg but not breed over it afterwards? From reports, I gathered that two eggs are normally laid but only one chick is reared.

Thus I concluded that the parents probably do the selection at this early stage. After two hours, the female left the nest to join the male, who was still feeding in an adjoining pool. The latter returned to the nest after while, but he too ignored the second egg. The male's patience on the nest was exhausted after 30 minutes. It rose and walked away, leaving the nest unattended. I then realized that the light had changed and the sun was setting. Just as the red disc disappeared behind the snow-covered mountains, the female crane returned to the nest and I left my observation spot.

I was back in position at sunrise the following morning, installing myself for the day. The female was still (or again) incubating and continued to do so till after ten a.m., but the second egg was no longer there! As she got up, I discovered that it had been included into the clutch. My earlier confusion was cleared—

the freshly laid egg with its still soft shell would be ruined while being sat upon, so it is exposed to wind and sun to allow it to dry and harden. The female called the male with a repeated, soft *gorr-kro gorr-kro* and the usual handing over ceremony started—the departing bird added nesting material while the one taking over turned the eggs carefully before settling down. The female walked off, crossing the grassland towards the river which was some distance away.

About an hour later, the male suddenly became nervous. A shadow glided over my hide and I spotted another pair of Black-necked Cranes landing just 50m. away from the breeder, who instantly got up from the nest while his partner returned in a straight flight from her distant feeding place. Both resident birds walked up and down, passing the next, flinging their heads into the air, bending their necks slightly backwards and with beaks wide open trumpeting their territory call (very similar to the well known call of the Sarus but somehow higher in hurl; the male, however, does not half-open the wings to impress like the Sarus does). The intruder couple pretended to be occupied with grooming their feathers, while the territory calls of the local pair echoed unceasingly through the valley; yet no aggression was exchanged. After 15 minutes the interlopers took off. Excitement prevailed as the residents continued to confirm their territorial claim at regular intervals. Even while seated on the eggs the female would answer the male's calls. Gradually calm and peace returned to the marshland, the female incubating while the male remained close by, feeding (mainly on tubers and shoots, but he also dug further into the mud to extract other eatables which looked to me like small snails). The crane's dark grey-green coloured legs turned black while in contact with the water and the normally distinct white patch behind the eye was hardly visible.

The rest of the day as well as the next passed without further excitement for the cranes. They changed shifts at more or less regular intervals (the male would remain on the eggs for about half an hour to one hour, whereas the female would breed longer, namely three to four hours) and remained silent apart from the routine *gorr-kro* conversation while taking over. Interestingly the birds mainly faced the sun while incubating. As I still wanted to get to Hanley which was another few hours away, to the south, I reluctantly decided to move on.

THE HANLEY PLAINS

After a bumpy, backbreaking three-hour ride we passed the dominating monastery which guards the entrance into the vast Hanley plains. We were at an altitude of 4,400 m., the soil was very marshy, saline and covered with sturdy grass—the ideal habitat for more Black-necked Cranes. However, there is a large concentration of nomads whose animals have taken over the plateau. After talking to local farmers (mainly of Tibetan origin, who live here on the gentle slopes of this high altitude valley throughout the year) I learnt that around the middle of May, three cranes had migrated in and a week later two more had joined them—a total of five birds. One of the farmers was kind enough to accompany my driver and myself to the spot where the *Cha-Thung-Thungs* (as the locals call these birds, which means: the 'tall bird') were last seen that same morning. It was a very helpful gesture, since looking for those birds in that enormous expanse, amongst all those yaks, sheep, goats and horses, would have been a lengthy procedure.

After a short while we found the five cranes (probably three males and two females) heavily involved in territorial fights: three birds constantly jumped into the air, leaping at each other, trying to knock the rival down, trumpeting loudly, supported by the two females standing near the scene. Finally, one fighter took off at full speed, leaving the other two to continue the combat in an even more persistent and aggressive manner. They now started chasing each other over the prairie, running with wings widely spread out, their beaks wide open in exhaustion; then they flew off, followed by one of the females, only to land a few hundred metres away and start the whole fight again. This went on for at least another hour.

We drove on to find a suitable camping spot in the vicinity. From a hilltop, I later looked for the birds through my binoculars and found them at three different places, feeding amongst yaks and goats—two couples and the single one. Whenever the single crane tried to approach one or the other couple, it was chased away with a loud trumpeting and beating of wings. In the late afternoon, one pair approached a waterbody not too far from our camping site. One bird lay down while the other one foraged among the reeds. But not for long. A group of horses waded through the water in search of their preferred aquatic plants and the cranes had to vacate the spot. I thus witnessed the reason why these birds were still desperately trying to find a suitable nesting place whereas the other couple down the valley had located a site, built a nest and were already incubating. No wonder only one or two pairs of this endangered bird species manages to breed in Ladakh in a season (according to *High Altitude Wildlife*, D.B. Sharma).

OTTO PFISTER, Sanctuary Magazine, 1995

Mark Cocker is a British journalist, birder, and author. He contributes to the 'Country Diary' column in The Guardian, *besides writing for* The Times. *The piece reprinted below provides an interesting description of the quest for the Satyr Tragopan in the mountains of Nepal.*

Rare Bird of the Mountains

A loose translation of the name satyr tragopan, might be 'horned god of the woods', and for once I feel the early naturalists found a title to match the creature itself. It's a type of pheasant, but any attempt to describe the bird in terms of the hand-reared fowl that blunder daily into British car windscreens is like trying to compare an Apache warrior to a balding overweight businessman.

Imagine a bird the size of a really big cockerel with an electric blue face, black feather horns that he can erect when excited, and a body plumage of deep blood red. Overlaying this magical colour are hundreds of white ocelli, each encircled by a crisp black margin and so bright they seem almost luminous.

Most western ornithologists agree that it's the ultimate species on any birdwatching trip to the Nepalese Himalayas. These mountains comprise most of the tragopan's world distribution, although the Nepalese bird atlas shows that the species has been seen in just 10 of the 81 tetrads covering the country. Moreover, its highly restricted range on paper only hints at the exertions involved in finding it on the ground.

Tragopans are mainly recorded between 8500 and 12,000 feet and are creatures of dense oak and rhododendron forest with thickets of bamboo. To add spice to the challenge they favour extremely steep slopes. My previous quest lasted about a fortnight in an area sandwiched between the mountains of Annapurna and Dhaulagiri.

Every morning we would gaze up at these five-mile-high giants looming on either side of the Kali Gandaki valley, the deepest in the world, and reflect on how they seemed the perfect setting for the ultimate Himalayan bird.

Each day then resolved into long exhausting ascents, frequent halts as our lungs began to panic in the thin mountain air, followed by jaded descents during mid-afternoon. I came to understand why so few ornithologists have made the effort to see more than a handful of tragopans, and why some have settled for the sight of a female, a subtle blend of grey and brown sprinkled with dull spots.

It is probably because of these fruitless memories that it seemed incredible, during a recent visit to the Himalayas, that I could be listening to this mythic creature just a stone's throw away through the forest. So typical of the bird, the male's dawn call is an unearthly and un-avian wail, usually transliterated as '*W-a-a-a-a-a*'. Both this and its other main call—a repeated '*Ka-ka-ka-ka-ka-ka-ka*'—have a quite definite mocking quality. We crept towards the direction of these sounds until it was so close it seemed equally incredible that we still couldn't actually see it.

There is one further dimension to my obsession with tragopans, which concerns the person who accompanied me 13 years ago during my previous search. After I had left Nepal he made a final mountain trek and reached a place called Tharepati in the Langtang National Park, to the north of the capital, Kathmandu. Ignoring the strains of the climb to this mist-shrouded spot, he went out in search of the tragopans he could hear calling, and has never been seen since.

Suddenly, almost casually, our tragopan wandered into view. For one, perhaps two seconds I watched it as it descended the tree from which it had been calling. I could make out its large, full-chested shape, the dark face and brilliant red plumage.

The circumstances of my friend's disappearance made this individual bird one of the most beautiful and haunting I have seen in my life.

MARK COCKER, Guardian Weekly, 16 June 1996

S. Theodore Baskaran, well-known natural history writer and birdwatcher, worked as a senior officer in the Indian Postal Service. A regular contributor to The Hindu, *his book* The Dance of the Sarus, *a collection of articles on natural history and conservation, was published recently. The following piece discusses the long forgotten story about the mysterious disappearance of birds in Haflong.*

The Haflong Phenomenon

Ever since E.P. Gee wrote about the bird mystery of Haflong in his book *The Wildlife of India*, for every birdwatcher a visit to Haflong has been a dream. The phenomenon occurs in October and lasts only a few days, depending on certain weather conditions. So, as soon as we got the news of its occurrence, we hurriedly got organized and the very next morning, were proceeding along the mud road connecting Shillong and Haflong, across the North Cachar hills.

Jatinga, very near Haflong in Assam, is a tiny village atop a ridge, at a height of 1,066 meters. Every year in October hundreds of birds of different varieties come diving down to Jatinga at night, attracted by the lights put up by the villagers and are collected for the spot.

At the beginning of this century a group of Nagas settled down at this spot, because it was a secure location. On a winter night they noticed that birds were flying into their houses. Taking them for evil spirits, they pushed them out with long poles, too scared even to touch them. The birds continued to crash in. Worried at the tenacity of the 'evil spirits' the Nagas abandoned the village. But it was a good site. So, U.L. Suchiang, a Jaintia tribal headman, bought the site in 1905 and settled 100 Jaintia families. (Suchiang's grandson, who is the headman of the village now, told us this story.)

On a winter night, a Jaintia was walking across to another house in this village with a lit torch in hand and he noticed some birds falling around him. Realizing that the birds were attracted by light, the villagers began taking advantage of this and collected the birds. By 1926, the operation had the sanction of tradition and collecting these birds became a community affair.

This phenomenon has been baffling ornithologists. Though some travelled up to Haflong, none had been able to actually observe it, the timing being so uncertain. It was only this year, during the puja holidays, that first two scientists of the Zoological Survey of India and then our group—which included three scientists from the Life Sciences Department of the North Eastern Hill University—that a first ever authentic observation has been made.

Leaving Shillong early in the morning, passing through about 150 kilometers of primeval forest and after crossing two rivers by ferry, we reached Haflong after nightfall. Spotting a number of rare birds, including the shy Red-headed Trogon, took the edge off the fatigue of travel. On arrival at Haflong we were told that on the previous night a large number of birds had come. The tenth day moon was still shining brightly; we waited till it disappeared behind the hills and then set out to Jatinga, just three kilometres away. At the village we saw three petromax lamps kept at three different elevated spots. Half the portion of the lamp was covered with cardboard and the beam was directed over the valley.

At 11.30 p.m. we took our position near one of the lights. It took only a few minutes to befriend the group of Jaintia young men sitting around. Cigarettes passed around and our purpose explained, we became part of the group. Myriads of moths were fluttering in the beam of the light. A Jaintia youth, who stood with a long, thin flexible bamboo pole that he used as a whip to bring the birds down, was desultorily practising his aim at the moths. A ghostly fog, in seemingly great masses, swept across the beam of the light and the surrounding area was lit up by the reflection, creating strange lighting effects as on a modern stage. There was a promising wind from the south.

Suddenly out of the fog, a bird shot out and crashed near the lamp. It was a White-breasted Waterhen. We sat transfixed at this incredible sight; two more birds flew right into the verandah and fell near us—Indian Water Rail, a bird that most birdwatchers read about but rarely get to see. As the birds fell, they were too stunned to take off again and were easily collected. The bamboo whip brought those that just flew across the light beam down.

It was well past midnight and we had logged eleven species. A slight drizzle started and the air once again was filled with fog. The boy with the bamboo whip grew taut and peered into the fog. We could hear the distant calls of some kind of duck. There were many of them. Very soon the air around us was reverberating with their frantic, high-pitched calls. Two ducks dived and fell into a thicket near the light, and were soon picked up—Lesser Whistling Teals. I recall that their call was so different from the low whistle that they usually make while in flight. We kept vigil till 4 a.m. and when we picked our way towards the jeep, dawn was already cracking over the valley. Our bags were full of precious cargo, for observation and identification. We had listed sixteen species—ranging from the tiny Paradise Fly catcher to Wedge-tailed Green Pigeon. And at our spot sixty birds had been picked up.

What causes this phenomenon, the like of which has not been recorded anywhere else in the world? Most of the birds that dived to their doom that night were diurnal ones, not a single nocturnal bird, though there were plenty of night jars and owls around.

Also no bird of prey showed up, though later we spotted Shikra, Kestrel, and eagles in the area. Nor do the lights attract any of the common birds—mynas, crows, and babblers—that can be seen around Haflong town.

Almost all the birds that come are residents and only one species was migratory—the Indian Water Rail. They too must have come from the dense forest around Jatinga where they had taken temporary winter residence. (Earlier it was thought by many that only birds on their migratory route got attracted by the lights.) One thing seems to be clear. The birds are disturbed from their nightly roosts in the thick jungles that clothe the slopes around Jatinga, by some factor and take off. Once in the air, they are attracted by the lights. Nocturnal birds, used to seeing street and car lights, are not attracted by the lights of the villagers. What is it that disturbs them and provokes them to get air borne in the night?

As I write this, the Hooded Pitta, one of the most colourful birds of the Indian forest, which I had saved at Jatinga and brought with me, is crouching in one corner of the room, pumping its rump up and down, in typical Pitta fashion, as if to mock at my amateurish efforts to understand its mysterious ways.

S. THEODORE BASKARAN, *The Dance of the Sarus*, 1999

Zai Whitaker (b. 1954) grew up in Mumbai. She moved to Chennai after her marriage to Romulus Whitaker, the renowned snake expert, and for the next twenty years they worked on various reptile conservation projects together, including the Madras Snake Park and Madras Crocodile Bank. Zai has written several books for children including Andaman's Boy *and* Kali and the Rat Snake. *Known for her witty and humorous writing style, the piece reprinted here presents an interesting take on birdwatching.*

Misty Binoculars and Other Strategies for Survival among Birdwatchers

In our family, it was taken for granted that a normal person had to be interested—preferably passionately—in birds. While normal families went to the movies and had Sunday barbecues, we went to Karnala or Borivli hoping to see and hear the calls of the shama and other rare birds. The night before the birding trip, binoculars were spit-polished, bird-books thumbed, small notebooks located and pencils sharpened. (For a true birder makes his/her List with a pencil, not a pen). Much to my secret chagrin, food was the last priority. 'Make some sandwiches, Paul, for tomorrow morning', my mother would tell our cook. Paul believed that six people could survive on ten sandwiches ... but only I seemed to notice that we were being starved to death. The others were too preoccupied with calls and crests, vents and whiskers, bills and backs. If I were a bird I'd mind terribly, being taken apart like this, but I guess our avian friends have no choice.

Well, neither did I. I grew resigned to my fate. But going along was one thing—it could be done, at a pinch; the real difficulty lay in hiding one's ignorance. Often, a friend of my parents'—bird-friendly of course—would be along and I simply had to pretend a certain proficiency I did not possess. There was also our uncle Sálim, who got very upset when you didn't recognize a call or identify a small dark silhouette about a thousand metres overhead. Another complicating factor for me was that my sister was a good birder. As a small kid she astonished a bird-group by identifying a black-winged kite. This was bad enough; but over the years my brother, who used to joke about blue-bottomed nitwits, defected. And the outside world assumed I was just as bird-savvy as the rest of them. A difficult situation .... Over the years however I developed certain survival tactics which I feel it is my duty to share, in case there are other non-birders like me who have the misfortune to be born into a birding community.

Number One, don't fight it (i.e. this whole birdwatching thing). It doesn't work. Birdwatchers are simply unable to understand that someone can be uninterested in seeing the nest of a blue-eyed caterwaul. Or walking ten miles in the (slim) hope of seeing an immature mud-slider. I don't think it's anything to do with intelligence because when you steer them away from birds (once in a while) they seem fairly okay. But when it comes to this, forget it. Just go with the flow.

Two: have your binoculars round your neck at all times. This is a useful talisman: it immediately makes you part of the club, establishes your identity as a birder. As the bird walk starts, twiddle the focus thing with a knowing, slightly puzzled air. The reason for this will be divulged later in this paper.

Three: If there's food along—and there usually is, for many birders like to eat—don't indicate any interest in it at all. Pretend you just don't care. Once in a while, even suggest it be left behind in the car. This is tough I know but it does create a good effect. There is bound to be someone who loudly vetoes the idea, probably the best birder in the group who doesn't have to resort to strategies like these. In fact my most vivid memory of uncle Sálim is of his dismay when the cold coffee had been forgotten at home.

Four: Soon enough, the dreaded moment will come when a sorry brown object flies across the path or croaks from a tree, and the birders stop mesmerized and ask 'What was that?' Binoculars go up, mouths fall open. Someone or the other will single you out for an opinion. *Don't give it*, because like truth, ignorance will out. You can blame your misty binocs—this is where the focus-twiddling is useful—or say you just got a flash, or glimpse. The untrained ignorant will immediately shout some name—some bird not found here or off migrating somewhere—and his grave is dug.

Five: Every now and then, snatch your binocs and glare piercingly at a tree. Squint, standing absolutely still for at least forty or fifty seconds. Then look dejected, and say 'Oh, it's only a babbler.' Or parakeet, or other common species. This has a two-pronged effect. It tells the other that you know your stuff, and that you're above babblers and bulbuls. If you can use the Latin names instead, all the better. But don't overdo it. Once is plenty.

Six: Avoid going with the same birding group twice. Sooner or later one gets found out, and it isn't pleasant. Play it safe. And never, never go with just another person because then you've had it. I have been on the run from birding groups all my life because as Sálim's grand-niece and our editor's daughter*, people expect me to be an expert.

Seven: Quick thinking is a must. Just the other day I was asked the difference between the Palni and Nilgiri laughing thrush. 'Oh, I'm not one of these nitpicky birdwatchers,' I said. 'But I can look it up for you.' I managed to hide the fact that both thrushes are total strangers to me.

There are other strategies but since I've crossed the editor's word-limit already I'll stop: but only after adding my latest birding victory. Two years ago I was in the States. My hosts were friends of my father's and decided to take me birdwatching 'as a special treat'. They even lent me—ouch—a pair of superb binocs, recently cleaned. 'Shall we take a picnic, David?' asked Mary. David looked apologetically at me. 'Of course not. The last thing these bird-wallas think about is food.' I smiled unhappily and off we went. We turned off the highway on to a dirt track which led to a lovely lake. Just then a group of noisy objects flew in and landed clumsily in front of us. 'Isn't it a bit early for the Canada geese?' I asked David. They were both impressed. They hadn't seen the small Wildlife Department signboard we'd passed. I, on the other hand, had. So we come to Number Eight: keep your eyes peeled.

ZAI WHITAKER, *Newsletter for Birdwatchers*, 1999

---

*The author is the daughter of Mr Zafar Futehally, Founder Editor of the *Newsletter for Birdwatchers*.—Ed.

Personalities and Controversies

6

TOP: A younger Sálim Ali (sans the trademark white beard) arriving at the 1950 International Ornithological Congress in Uppsala on a motorcycle. Also in the photograph are David and Elizabeth Lack
BOTTOM: Edward Pritchard Gee (left, with dog in hand) and Colonel Richard Meinertzhagen (smoking a pipe) in Doyang Tea Estate, Assam, 1952. In the centre is Theresa Clay, Meinertzhagen's niece, and an acclaimed entomologist
[Source: Sálim Ali, *The Fall of a Sparrow* (1985)]

Remembering Sálim Ali

Zafar Futehally (b. 1920) was Honorary Secretary of the Bombay Natural History Society from 1962 till 1973, before he moved to Bangalore. As Founder Trustee of the World Wildlife Fund, India and Vice President of International Union for Conservation of Nature (IUCN) he was instrumental in initiating Project Tiger. He edited the Newsletter for Birdwatchers from its inception in 1959 till 2004. This piece is an intimate portrait of Sálim Ali, the grand old man of Indian ornithology.

Remembering Sálim Ali

My first recollection of Sálim Ali was during my school days in 1937. Sálim was a great friend of my father's and their friendship was based on the fact that both of them had a tremendous sense of humor. I never heard any serious conversation between the two of them—it was all banter and laughter. I soon found that Sálim had come to stay with us because he had to have a mastoid operation. This was a serious affair and could have been fatal, but there were no signs of fear in Sálim's attitude. His wife was away in Japan at the time, but the operation was urgent; and fortunately it proved successful. However, as a result of the operation one ear went out of commission, and it is amazing that in spite of this handicap Sálim was so good at identifying bird calls. He once told me, for example, that the ioras of Pali Hill had a special tune of their own, different from the ioras of other localities. In spite of being so acutely discriminating about bird calls, he did not, strangely, have a good ear for music.

Later, in 1939, I visited his home in Dehra Dun. It was a lovely little house made beautiful and charming by his wife Tehmina, who was well known for her good taste and her ability to make a rupee go very far. Since Sálim was not a man of means at the time, this was a great asset. The landlord, Ugra Sen, a great admirer of Sálim, contributed to his tenant's survival by charging a negligible rent. Sálim, in spite of his shortage of cash, was never short of company because of his charm and humour. I recall him telling me that the most serious occupation in UP was, '*Guppe Haankna*', that is, telling of tales. The landed gentry of those days had the time and raw material for this pastime and had made it into a fine art. At such gatherings Sálim's contribution was to stoke the fires of conversation by his incisive wit. The Census Department of the Home Ministry must have been confused by what Sálim's wrote under the heading '*Profession*'. 'writer of books and teller of stories'.

The 1940s was still the decade when the Englishman was held in some awe by the natives and anyone who stood up to his arrogance was greatly admired by his congeners. I remember an occasion when one of the *sahibs* in the party made a thoughtless, if characteristic remark about the unsuitability of 'natives' for some position in government. Sálim turned on him with the confident though unexpected remark 'The best thing for people of your way of thinking is to leave the country and go home'. Silence reigned supreme for the rest of the evening.

The death of his wife in 1939 shattered him, and the fact that he was able to recover rather quickly was due largely to the generosity of his sister (my mother-in-law) and her husband, Hassan Ali. Soon after Tehmina's death, Sálim wrote to them demanding accommodation in their house. For nearly 45 years Sálim lived with them in their house at 33 Pali Hill, Bandra, Bombay.

It was fortunate for Sálim that during the pre-independence period many maharajas and nawabs were keen sportsmen, and though most of them preferred to have a duck on the table rather than in the bird room of the Bombay Natural History Society (BNHS), they were very willing to support Sálim's ornithological surveys. These surveys provided the basic data for his *Handbook*, and so much has been written about these expeditions that I need not mention them here. The fact that the princes were supporting his surveys, and he often enjoyed their hospitality, was never allowed to inhibit his forthrightness. The late Maharaja of Bharatpur once said, ruefully, 'that every time Sálim Sahib comes to Bharatpur, he makes me cross off another species of duck from my menu'.

My wife and I were invited to join his camp in Palanpur, Gujarat, in 1944. The camp was sited in Bolaram, a delightful jungle, where the roar of a tiger and the sawing of the panther could still be heard. It was at this camp that I came to know Sálim intimately. It was fun to see the way the maharajas ate out of his hand, though it was they who provided him with the luxuries of life. During a state dinner hosted by the Nawab of Palanpur, there was a glittering assembly of royalty, and throughout the day there were sounds of cannons, announcing the arrival of some prince. From the number of booms, experts could identify the status of the guest. During the party, Sálim whispered in my ear that her 'white' Highness (the wife of HH Palanpur) was a 'negotiable document' as she had been married before to a prince and was now transferred to our host.

I was with Sálim in several camps, in Kutch, Bharatpur, Bhutan, and others. In all these places there were severe inconveniences and dangers, which were rather worrying. Saw-scaled vipers were common in Kutch and the sand-flies which swarmed over your eyes made any purposeful activity difficult. But Sálim was 'least worried' as they say, and continued with removing birds from the mist nets, measuring, weighing, de-ticking and recording his findings. The ticks were collected for Teresa Clay, Meinertzhagen's niece and the well-known entomologist. Ticks often help to determine the sub-species of the host. During collecting trips of Bhutan, the stinging nettles protecting the plants were a forbidding experience, and while I often avoided any responsibility by looking the other way, Sálim walked through the bushes and recovered the birds he had shot for the reference collection.

Physically he was both tough and fearless. He told me once that it was a great joy to him to have developed himself from a puny, sickly boy into a strong physique. One discipline to which he subjected himself was never to drink water during the day. This enabled him to continue working in the 45°C heat in his camps in Kutch and other places. In fact, he often reacted unnecessarily harshly, I thought, to anyone demanding a drink or sustenance during 'office' hours in his camps. 'Office' hours, it may be noted, lasted from pre-dawn to midnight.

He was a wonderful companion when he was in the right mood, and for some reason he was extraordinarily generous to me. To give one example: he had lent me his Exacta camera, and

I lost it in a taxi in Calcutta. When I went in a trembling state to report the loss, he tapped me one the shoulder and said, 'we are all human and mistakes will happen'. On the other hand, he could be savagely critical about minor offences.

I need hardly stress that being out with him in the field was always an education. I remember one walk with him in Borivli National Park—the Park as it was in the 1940s without the ugly artefacts which have now been put up. What a delight it was to walk though this country from Ghodbunder Road up to Kanheri Caves. Nothing could be more beautiful than the deciduous trees glossy with new foliage, and the streams tumbling beneath them. As we walked along we heard the call of the Malabar Trogan. It is not often that we can hear this bird but Sálim stopped in his tracks and said that it was not the trogan but the drongo imitating the call of the trogan. This was a remarkable identification with only one ear at his disposal.

His eyes, of course, had the sharpness of an eagle. In Kihim one morning, while walking through a scrub jungle destroyed by the activities of the Thull Fertilizer Project, two yellow-wattled lapwings were seen walking furtively. It was obvious, even to me, that there was nest around, but I had no hope of being able to discover it because it was so expertly camouflaged. The eggs match the ground colours, and the nest is just a scrape in the ground. Sálim signalled to me to stand still and he walked on. Within a few seconds he had located the nest and also a single chick some distance away.

My association with Sálim grew closer and closer. I came more and more under his wing, and he was, of course, instrumental in my climb up the conservation ladder. His irritable temper notwithstanding, he had a tremendous capacity to make and hold on to friends. World figures—Sir Landsborough Thompson, Peter Scott, Richard Fitter, Frank Fraser-Darling, and a host of others, were in close touch with him. J.B.S. Haldane, perhaps the most versatile scientist of our age, always extolled Sálim's capacity to do fieldwork with just a pair of eyes and bare hands. No need for sophisticated technology and foreign grants.

My first serious meeting with this international group was in New Delhi in 1965, when the Government of India organized a

PERSONALITIES AND CONTROVERSIES 345

meeting with a group of IUCN delegates who were on their way to Bangkok. The group included Sir Frank Fraser-Darling, Richard Fitter, and Sir Peter Scott. I was Honourable Secretary of the BNHS, and it was during this meeting that I was invited to Lucerne to attend the ninth Triennial General Assembly of the IUCN. And it was at Lucerne that I was made a member of the Executive Board of IUCN, and I cannot deny that my connection with Sálim had much to do with it.

My connection with Sálim developed a formal side during the twelve years that I was Honourable Secretary of the BNHS, while Sálim was the President. In 1962, there was serious disagreement among the members of the Executive Committee of the BNHS about the priorities of the Society. The committee consisted among others, of three retired members of the Indian Civil Service who were, by their training and back-ground, *'Lakir Ke Fakir'*— literal followers of the written word. So when Sálim wanted to use the services of the staff of the bird room and the reference collection for his bird-banding trips, some members of the committee objected on the ground that the Grant from the Government of Maharashtra, which enabled the Society to maintain the staff, could not be diverted for fieldwork. The staff had to stay put in the office of the BNHS, dusting and fumigating the specimens.

Sálim's view was that taxonomical work had been largely accomplished, and what was required was information about the living bird and its ecology. This would, in fact, give 'life' to the dead specimens in the references collection. The dispute became so serious that first Sálim resigned as President, then the Hon. Secretary resigned. After this Sálim requested me to become the Honourable Secretary of the BNHS. My principal task was to pour oil over troubled waters—a task at which I succeeded only partially.

Naturalists everywhere have an egocentric view of the world, and are not willing to abide by the discipline of a committee. Unfortunately Sálim, in spite of his pre-eminence as a naturalist and author, was not a good Chairman, and left the onerous task of conducting the meetings to me. He told me once that attending these meetings were some of the more unhappy experiences of his life.

One problem which Sálim faced when he started his monumental ten volume *Handbook* was to find a suitable typist—someone who would not be bewildered by the scientific names of birds. Several of the probationers were found to be unsuitable. I then suggested to the committee that we loan our librarian and senior steno, J.S. Serrao, to Sálim to help him with his book.

Surprisingly, this created a storm and they insisted that he could not be given to Sálim for his personal work. It was an extraordinary attitude and I pointed out that the completed work would enhance the standing of the Society. Fortunately, as Hon. Secretary I was able to release Serrao for Sálim's work. Serrao with his unfailing memory, meticulous typing, and tremendous admiration for Sálim, did a splendid job. When the last volume was completed, a beaming Serrao came to me and showed me a cheque for Rs 1000 which Sálim had given him in appreciation of his work.

At the end of 1973, my wife and I decided that we had enough of Bombay, its high-rise threats and other unfortunate developments. So we moved to Bangalore and my active connection with the BNHS came to an end.

My personal bonds with Sálim continued and we were happy that after the death of her parents, my wife suggested that he should make his home in her house in Kihim. He had once told her that this was the landscape he loved best in the world. It was the right place for him to end his days.

ZAFAR FUTEHALLY, *Sanctuary Magazine*, 1995

In his later years, as the Director of the BNHS and the principal Investigator of several important research projects, Sálim Ali was feared as a quick-tempered man and a hard task-master, but always with his 'boys', as this amusing piece indicates. Bharat Bhushan, the author, is well-known for his rediscovery of Jerdon's Courser—a bird that was thought to be extinct—in 1986.

The EmPee Saar in Andhra Pradesh

It was a typical hot stuffy morning on 17th January, 1986, at the Renigunta Airport near Tirupati, in Andhra Pradesh. Beyond the airport one could see the grasslands fringing the foothills of the Eastern Ghats. Had it been May, the heat of the afternoon would no doubt have caused a splendid mirage, perhaps, a cheetah chasing a blackbuck.

The lone policeman at the security gate asked me to state my reasons for straying out of the visitors lobby. I informed him politely that I had taken the Airport Manager's permission to receive my boss beyond the security gate and if need be, to approach the aircraft. The policeman looked at me unbelievingly, and asked what was special about *this* boss. I had to explain that he was ninety plus, a frail old man, a very important personality, who was also a member of the Rajya Sabha. At this, the policeman began to get increasingly skeptical. Where was the VIP reception and the retinue of an important person? Which political party did this MP of mine belong to, anyway? I explained that he was not affiliated to any political party and he was not expecting any VIP treatment either. I was the only person receiving him at Tirupati. The policeman now began to look at me as if I was a suspected terrorist or Naxalite. I did look like one, I think, with my longish beard, khaki and olive green clothes, and a stout walking stick that I had brought along. I explained that the 'Rajya Sabha MP' was coming all the way from New Delhi to see a bird in Cuddapah. Immediately, the police constable asked 'Oh? Is he *Daaktar* Sálim Ali? The Bird Man of India?'

To this day, I have never ceased to be surprised at the immediate connection that people make between birds and Dr Sálim Ali, even in remote and unlikely places.

I had been waiting at the Renigunta airport to meet Dr Sálim Ali, to escort him to Siddavatam near Cuddapah in Andhra Pradesh. The Jerdon's or Double-banded Courser had been rediscovered three days previously and had made great ornithological and natural history news all over the world. Dr Sálim Ali, the President of the Bombay Natural History Society, was the Principal Investigator of the Endangered Species Project. I was working on this Project in the Eastern Ghats and had been searching for the Jerdon's Courser since mid-1985.

On his arrival at Renigunta, I went with all the pride that I could muster at being able to talk to Dr Sálim Ali about the rediscovery of the Jerdon's Courser. He had barely walked four steps when he turned to me and said, 'What is your name? Are you with the forest department?'

Bang! That was the end of my pride and ego. I had met Dr Sálim Ali at least five times and presumed that I had impressed him enough for him to remember me. And now he did not even remember my name, or worse, that I was from the BNHS!

I turned helplessly to Mr P.I. Shekar, his Man Friday, who was walking behind us, lugging all sorts of shapeless cabin baggage. Mr Shekar was laughing and enjoying my discomfiture. Realizing that there was no help from that quarter, I patiently explained that I was from the BNHS and had participated in the rediscovery of the Jerdon's Courser. The 'Old Man' did not even smile and in a *very* serious undertone, told me (I remember his words to this day) 'Are you sure it is the Jerdon's Courser? If you are wrong, I will not bother to hang you from the nearest palm tree. You should do it yourself!'

Well, up to that point I had been very certain that the bird I had seen in Cuddapah was the Jerdon's Courser. But the menace in Dr Sálim Ali's voice made me very very uncertain. Was the bird indeed the Jerdon's Courser? What if it wasn't? And now this 'rediscovery' thing! What if I was wrong? I did not speak for an hour. Just escorted him mutely, until Mr Shekar, now shaking with silent and uncontrollable laughter, told me to relax, to watch the Old Man's eyes to see how he was enjoying himself at my discomfiture.

Dr Sálim Ali, guessing at the exchange, smiled and said, 'So? That bird is the Jerdon's Courser? Congratulations. Are you going to give me a party? Chilly Chicken?'

BHARAT BHUSHAN, *Hornbill*, 1995

Sálim Ali on Others

During his childhood, spent in the shadow of the British Raj, Sálim Ali imbibed the traditions and values of the British, but at the same time remained deeply conscious of his 'Indian-ness'. His encounters with Englishmen and members of the Indian royalty display his unique sense of humour and wit, as the following excerpts from his autobiography, The Fall of a Sparrow, *reveal. Colonel Meinertzhagen's name figures again in a piece included later in this section.*

Colonel Meinertzhagen

Richard Meinertzhagen was one of the most colourful, original and, in many ways, likeable characters that ornithology introduced me to. He had a passing acquaintance with India, where he had done temporary stints of service and convalescence during the First World War. When asked if he had managed to pick up any Hindi or Urdu while in India he replied that he had only learnt the useful term *sooar ka bacha* (swine), which he had encoded as 'SKB' and frequently used in private conversation when referring to 'opposition' folk.

Meinertzhagen seemed to be as indifferent to physical pain as to personal danger. While we were collecting in a reedy marsh near Kabul, he, wearing khaki shorts with legs uncovered, accidentally stepped on a barb-pointed reed which broke off, leaving about three-quarters of an inch of its length within his flesh. Regardless of this, he continued splodging through the marsh while his blood flowed freely. Finally, after some persuasion, he agreed to return. As he was limping back to the car to get back and have the barb removed by the embassy doctor, he noticed a Bearded Vulture—a wanted species—some 300 yards away in a different direction. Ignoring the projecting reed and the flowing blood he limped up to the bird and shot it before getting back to the car.

Following our first meeting at Bombay where he arrived on 11 February 1937, Meinertzhagen's meticulously kept diary, which he kindly permitted me to read 'At your risk(!)' shortly before his death, when our earlier shaky contact had ripened into abiding friendship, says 'I then went on to the Bombay Natural History Society where I met Prater and Sálim Ali. I was favourably impressed by the latter and liked what I saw of him. It is as well if we have to travel together for the next few months. He seemed intelligent, but is hideously ugly, not unlike Gandhi.'

Meinertzhagen had asked me to look out for a suitable botanist to take along with us on the Afghan expedition. A competent young student, K.N. Kaul, was recommended by my friend Birbal Sahni, F.R.S., then Professor of Botany at Lucknow University. Meinertzhagen interviewed Kaul, was well impressed but finally turned him down, I couldn't understand why. I know the reason now. The relevant entry in his diary says: 'Lucknow 8.3.1937. A young Hindu student, Kailash Nath Kaul, geologist [*sic*] wanting to accompany me to Afghanistan. He is a young man, nice mannered and intelligent, but I am a little doubtful whether I can stomach two seditionists for three months all day and every day. Sálim is a rank seditionist and communist, so is Kaul (a brother of Jawaharlal Nehru's wife) and it would probably end in disaster.'

My experience of the type of Englishman one normally came across in the heyday of the Raj was that he was a bully where one lower in the 'peck order' was concerned. Indians as a rule are too mild and submissive and thus lend themselves readily to being bullied. Another peculiarity of the British character I have found is that if you stand up to the bully and hit back, you command his respect. So it has been with me and so it was with Meinertzhagen. One of his biographers has described him as 'physically a powerful, violent and ruthless man'—a description which, happily, is only partly true. Though possessed also of many admirable qualities, he had a distinct streak of the bully in his make-up and could be unreasonable to the point of brutality at times. Due to his excessive meekness, Pillai was a perfect foil for a bully and he lived in obvious terror of Meinertzhagen because, in the latter's estimation, Pillai could do nothing right or in the way it should be done—neither could I, as I later discovered from his diary—and I had often to intervene when the hectoring got too far.

Revealing Excerpts from the Diaries of Colonel Richard Meinertzhagen, D.S.O.

14th April 1937 Ghorbund Valley, 6500 ft., Afghanistan. 'Pitching camp was a long and tedious business as Sálim is quite useless at anything of that sort and none of the servants knew anything about tents or camping. Sálim is so accustomed to be waited on by an army of servants that he is impotent when he has to do something himself, and yet he advocates that his class is capable of governing India. They must first lean to govern themselves.'

15.4.1937 'I'm not enjoying this at all. I find Sálim trying. He is inefficient and cannot bear being told how to do anything and must always do everything in his own way, which is often wrong. His ignorance of camp life and his helplessness in camp are pathetic. He tells me he has never had to fend for himself in camp and has always had masses of servants before. The pleasure of camp life so much depends on ones companions.'

30.4.1937 'I am disappointed in Sálim. He is quite useless at anything but collecting. He cannot skin a bird, nor cook, nor do anything connected with camp life, packing up or chopping wood. He writes interminable notes about something—perhaps me. ... Even collecting he never does on his own initiative. Like all Indians he is incredibly incompetent at anything he does; if there is a wrong way of doing things he will do it, and he is quite incapable of thinking ahead.'

20.5.1937 'Sálim is the personification of the educated Indian and interests me a great deal. He is excellent at his own theoretical subject, but has no practical ability, and at everyday little problems is hopelessly inefficient, yet he is quite sure he is right in every case. His views are astounding. He is prepared to turn the British out of India tomorrow and govern the country himself. I have repeatedly told him that the British Government have no intention of handing over millions of uneducated Indians to the mercy of such men as Sálim: that no Englishman would tolerate men being governed by rats.'

9 Jan. 1952 Sikkim (post-Independence). 'I find Sálim very touchy about India and Indians. He resents any trace of criticism and is extremely bitter about South Africans' treatment of the Indian question. My experience of Indians both in Kenya and S. Africa is that they introduced disease, dishonesty and sedition.' [Then follow violent views on Gandhi, 'the rat Gandhi', and present conditions in India of inefficiency, dishonesty and squalor, as compared with under British rule!]

Sálim Ali, The Fall of a Sparrow, 1985

Edward Pritchard Gee, who made important contributions to the conservation of Indian wildlife, was a close associate of Sálim Ali. In this short extract from The Fall of a Sparrow, *Sálim Ali remembers his friend Gee, who like him was also hard of hearing. In the subsequent extract Sálim Ali takes a dig at Indian royalty.*

Edward Pritchard Gee

After retirement from a long innings of tea planting in Assam, E.P. Gee—a 'chronic' bachelor—settled down in Shillong where he assembled one of the finest private orchid collections in Assam, mostly taken in the wild by himself. As a young planter he had been an exceedingly keen sportsman-naturalist and an inveterate fisherman, which he remained to the end. By about 1948, when I first met Gee on the BNHS's survey of the rhinoceros population in Kaziranga Wildlife Sanctuary at the invitation of the Governor of Assam (Sir Akbar Hydari Jr.), he had given up shooting and taken to wildlife photography with a vengeance—a hobby in which he soon came to excel. He joined hands with Colonel Burton and R.C. Morris, who had also turned to conservation with missionary zeal. After meeting him at Bharatpur in 1957, Loke's diary describes Gee as

a fairly heavily built man, balding, and wears tortoise-shell covered spectacles. Like Browning's thrush he repeats everything twice over, the second phrase tumbling out after the first, 'peeneka pani hai, peeneka pani; he got fed up, he got fed up, so he shot himself, so he shot himself.' Gee is rather hard of hearing, and this may be the reason for the trick of repetition. Sálim, too is pretty deaf, and when he and Gee talk to each other, the one in his high piercing voice and the other in his lower monotone, the world does not have to strain its ears to hear what they are saying!

During World War II Gee had volunteered for service and, since he was used to handling a large plantation labour force, he was assigned to the Pioneer Corps to supervise the building of one section of the famous Burma Road. As it happened, and unbeknown to either, there was another man of the same name supervising a different section of the road a few miles further on, and the superior officers were constantly getting confused between the two Gees. So, since our Gee was rather fond of talking, they aptly dubbed him 'Chatter-Gee'!

SÁLIM ALI, The Fall of a Sparrow, 1985

The Maharaja of Kutch

Among the Indian princes and princelings whom I had opportunities of knowing a little more intimately than others—chiefly on a naturalist's plane—was Maharao Vijayarajji of Kutch. He was over sixty when he came to the *gaddi*, having been on a patient and seemingly unending probation as Yuvraj for forty years or more, thanks to the robust good health of his father, Maharao Khengarji, who had come to be regarded by two generations of his loyal subjects as an ancient, indestructible monument of Kutch. Both father and son were keen sportsmen and knowledgeable naturalists, the former as a hunter of big game, the latter particularly interested in birds—game as well as in general. Besides being an excellent shot with gun and rifle, Vijayarajji was an accomplished tennis player in his younger days and a 'habitual' entrant in all-India tournaments, which he frequently won since many of the renowned players of the day were only too happy to partner him in the doubles.

I first became acquainted with Maharao Vijayarajji in 1942, soon after he, at long last, ascended the *gaddi*. By then he had lost some of his youthful vigour and assumed a comfortable, portly shape, abetted by lack of exercise forced by an injury to his knee. Though having to cut down on shikar jaunts needing physical exertion and mobility, he still retained an enviable expertise in small-game shooting and a lively interest in watching birds, especially of his own state. Thus it was at his invitation and under his generous sponsorship that I undertook a field survey of the bird life of his fascinating state with a view to producing for him an illustrated book on the birds of Kutch on the lines of my *Book of Indian Birds*, which had caught his fancy. In 1943, World War II was still very much on and petrol was severely rationed in India, bringing private transport virtually to a standstill. As a special sop to the ruling princes, however, an extra quota of petrol was allotted to them which, in the case of Kutch, enabled freer movement for the bird survey and visit to out-of-the-way places otherwise difficult to reach.

In between camps my party usually spent a couple of days in Bhuj for re-fitting, and each time I was in, the Maharao would invite me to accompany him on his evening drives to some scenic point in the environs of the town and 'take the air'. He was usually alone, attended only by a flunkey armed with a thermos, a bottle of 'pege' and a supply of pistachios and almonds and things of that sort for His Highness to while away his time pleasantly, while listening to or discussing my report on the progress of the survey. One of the things that struck me as singularly odd at the time—especially coming from a man normally so courteous and considerate—was that never in all these outings did he even once offer me any of the things he was stolidly munching away while the replenishing flunkey stood attentively at his elbow. That it should never have occurred to him to do so seems queer and inconceivable, yet there it was.

I am reminded by a note in my diary of that time of a crude manifestation of the anachronistic feudalism that still persisted in Kutch. I felt outraged, while responding to the Maharao's request to meet him at the palace for some discussion, to discover too late that it was mandatory for 'natives' to alight from their vehicles—whether car or horse carriage—at the main palace gate and cover the fifty yards or so up the drive to the entrance porch on foot. This mandate applied uniformly to all Indians, of whatever status, whether residents or visitors, official or non-official. The enormity of the *diktat* was that even the Indian *dewan* (chief minister) of the state visiting His Highness on official business had to 'crawl' in this fashion, while *any* European or Anglo-Indian of howsoever dubious a quality could drive straight up to the porch without let or hindrance, and perhaps even with a welcoming salute from the armed sentry at the gate. The 'reigning' dewan at the time of my survey, a highly respected senior Indian civilian, had to submit to this perverse indignity, while the lowly Anglo-Indian Customs Inspector could drive right up to the porch. I got a shock when ordered to alight at the main entrance, created a scene, and later protested to the Maharao in no uncertain terms about this insulting iniquity. I hope it had some effect, but I never had occasion to visit the palace a second time.

A peculiar oddity that amused me greatly when observing the intra- and interspecific habits and behaviour of that now-extinct genus—the maharaos, maharajas and nawabs—in the course of my bird surveys of the various Indian states, was the comic ostentation with which the rulers addressed each other, back and forth, as 'Your Highness' in tête-à-tête conversation even though they might be old friends and contemporaries or close relations. When talking to one of 'lesser breed' some of them took good care, when referring to a brother Highness, to slip in inconsequentially—as though in parenthesis—such vital information as 'He is 13 guns, you known, I am 17' and thereafter run on with the discourse.

Sálim Ali, The Fall of a Sparrow, 1985

Controversies

Conservation in India remains a complex issue. There are different perceptions about the co-existence of wildlife with humans, and also inevitably questions about why preserve wildlife at all? Or, as Madhusudan Katti asks, are the tiny, nondescript birds, such as warblers less important than mega fauna like the tiger, lion, and elephant when it comes to allocating limited funds for conservation. Madhusudan Katti is an alumnus of Wildlife Institute of India, Dehradun and is currently pursuing research on birds in USA.

Are Warblers Less Important than Tigers?

Now what kind of a stupid question is that? Everyone knows that tigers are more important, being large predators, as apex species, at the top of the food chain, flagship species for conservation, ... etc. etc. ... etc! These are arguments I have to face often enough when I tell people I am studying warblers—in Kalakad-Mundanthurai Tiger Reserve! For some reason, studying these tiny, nondescript, common birds is thought to be an entirely trivial, indeed arcane, academic pursuit of little practical or conservation value. 'What can studying little birds tell me about the habitat of large mammals, which are my primary concern?' asks the reserve manager. On the other hand, if we focus on the larger mammals—the apex species philosophy of Project Tiger— and do our best to improve their habitat, other species will also naturally benefit. Given limited funds and manpower for conservation (research and action), is it not better to focus on the mega-fauna and let the mini- and micro-fauna take care of itself? The only small creatures one should worry about then are those that may form part of the food chain leading up to the larger focal species.

Before you accuse me of a biased perspective (which is undoubtedly true, for I make my living watching little warblers!), let me state that, in defending these little creatures, I am also arguing in favour of a broader ecological perspective in conservation, one that goes beyond the charismatic mega-fauna, and starts looking at species more in terms of their ecological role in the system, rather than their appearance/charisma or tourism potential! So what is the ecological role of my favoured little leaf warblers?

Leaf warblers (genus *Phylloscopus*) must surely rank among the least glamorous vertebrates, so utterly lacking in charisma that even many diehard birdwatchers dismiss them lightly, scarcely bothering to try and even identify them to species level. Part of the problem is, of course, the fact that they are all small, dull-green coloured, and highly active in the forest canopy, making identification in the field difficult. It is only rarely—either when one is truly nuts about birds or when the fate of one's PhD thesis hangs on such identification—that one develops the eye for the subtle morphological, auditory and behavioural differences between species. These difficulties in identifying species, however, need not bother our busy manager too much, since the leaf warblers are all pretty similar ecologically as well—their role in the forest is largely independent of their taxonomic status, except insofar as structural aspects of their foraging microhabitat within the forest canopy are concerned.

All 18 species of leaf warblers occurring in the Indian subcontinent are migratory. They breed during the temperate summers from the Himalaya north to the Arctic Circle, and take over the peninsular forests, including those in the Himalayan foothills and much of northeast India, from September through May. While each individual may weigh only seven to 11 grams (this range includes all species, give or take a gram), one may still emphasise the term take-over when describing their relationship to their forest habitats: they number in the billions and form probably the most abundant avian guild in the subcontinental forests during our tropical winter. My study at Mundanthurai (in the southern Western Ghats) records a density of six to eight leaf warblers (of two species) per hectare of forest—usually any given patch of forest may have two to three species, depending on the type of forest. I doubt whether there is any forest habitat in India that does not host at least one species some time of year. Picking a random hectare from my 20 hectare study plot at Mudanthurai, I find six leaf warblers (of two species) making it their home for seven to eight months—for these are territorial individuals that remain on site for much of the winter. And what do they do during this period? Eat insects, mostly. Humdrum as their lives may sound, they spend over 75 per cent of their waking hours foraging for insects (and other arthropods, but insects predominate) in the foliage. Since they are not concerned about finding mates or raising young during this season, and want merely to survive in good shape for the next summer, their other activities—preening and maintaining territories through vocal and visual dialogue with neighbours—does not take much time. Hmm ... a bunch of small, dull birds spending most of their day peering at leaves in search of insects—do I seem to be only weakening the defence? Not really. ...

Consider the fact that each leaf warbler, on average, eats three insects every waking minute (this is averaging over all their activities throughout the day). Since they forage by picking prey off a substrata—mostly leaf, sometimes also twigs and flowers—the prey largely consists of herbivorous insects. In the case of my one hectare on Mundanthurai, it is mostly leaf-eating caterpillars. A single leaf warbler thus eats an average of 180 insects every hour, or about 1,980 per day (assuming an average 11 hour working-day from dawn to dusk). The six individuals on our plot thus rid the plants of almost 12,000 insect pests every day!! Multiply that with the number of days (200 to 250) that they are in residence on that one hectare plot and you may begin to appreciate the service they render to all the plants. Now I ask you to consider removing these warblers from the study plot, since they seem to take away so much research and conservation energy from your more favoured mammals, and picture the forest as it may appear in a few weeks' time ...! The scenario could become even more dramatic if you (in your large-mammal chauvinism) remove all the other insectivorous birds from the plot as well: I estimate each hectare of Mundanthurai's forest has at least 40 insectivorous birds, including other warblers and flycatchers (both resident and migrant), minivets, shrikes, drongos, babblers, etc. The average number of prey may come down to just over two per bird per minute—which gives a total of about 5,000 insects per hour, or 55,000 per day in every hectare of forest! Remove those insectivores: ... and don't be surprised if in a few weeks your plants start to appear ragged with their foliage tattered ... and your endangered langurs become unhappy because so many leaves are now packed with toxic antiherbivore compounds produced in response to caterpillar nibbling ... and the plants make fewer flowers and fruits as they are forced to spend too much energy in self defence ... in turn making the nectarivores and frugivores unhappy ... and regeneration of the forest slows down as fewer seeds get produced and dispersed ... and the ground starts to dry faster because the canopy is thinner and more sunlight gets in ... I leave you to work out the rest of the ecological cascade effects on your own!! For now, I'd be happy if you simply pause to appreciate the job done by the nondescript little green jobs—the leaf warblers—and their insectivore colleagues that travel thousands of kilometres every year to eat all those insects.

Before you start protesting that you will never contemplate removing all those birds, and that I am just another doomsayer, consider the fact that 80 per cent of the warblers (especially the green leaf warbler, which is the most common one here) as well as the next most abundant migrant (Blyth's reed warbler) spending each winter at Mundanthurai come from the forests from the hill regions around the Caspian Sea, from Turkey east through Kashmir, including bits of southern Russia and Afghanistan. Now imagine that these hills—breeding grounds for so many migrant insectivores—are deforested on a large scale (either directly by us or through effects of global climate change) cutting down the bird population by 90 per cent. Such a decline is not very unrealistic, as those studying migrant forest birds in the Americas will tell you—though they worry more about forests in the wintering areas being cut down rather than in the breeding grounds. In fact, over the past two decades, Americans and Europeans are increasingly facing the prospect of another Silent Spring. Not, this time, due to the factors mentioned of Rachel Carson's clarion call in the 1960s—overuse of chemicals in agriculture at the height of the green revolution—

but to a suite of other human activities that have hit the habitat of avian migrants in both their northern breeding grounds and southern wintering grounds. Many species of migrant songbirds, which enliven the northern spring after the dreary and silent winters, have been pushed to the brink of extinction—some, like the Kirtland's warbler, down to a few scores of breeding pairs—over the past two decades, even as my ornithologist comrades in the West are racing against time to figure out the causes of these declines, so that we may try to reverse the process. The culprits are, of course, us humans: deforesting the tropical wintering grounds, fragmenting the temperate forests into suburban woodlots more accessible to human subsidised nest-predators such as domestic cats and other small carnivores (wild or feral) thriving on our garbage and directly subsidising populations of non-migratory nest-parasites like the North American cowbird through backyard bird feeders, enabling them to survive the harsh winter, and fool over 200 gullible species of songbirds into raising their offspring! We seem to be particularly adept at causing damage to the ecological fabric of this planet, even when we mean well—feed them poor little birdies in the winter, or the cute raccoons at night.

Getting back to our continent, where we have no information on population trends of forest birds at all—whether resident or migratory, in tropical South and South-East Asia, or temperate Russia, Mongolia, and Siberia—declines paralleling those on the other continents are very much on the cards, if, indeed, they haven't occurred already! Given the contempt that these migrants have for human geopolitical boundaries, their populations are subject to forces beyond the control of any one national conservation agency, let alone the manager of a single tiger reserve. If their populations are found to be declining as drastically as many New World migrants have over the past several decades, mammal chauvinists may be reduced to haplessly watching the habitats of their favourite creatures getting degraded.

Do you think even the tigers might get worried about such a scenario? Is it worth studying these warblers, trying to figure out what makes their populations tick, and how to save them and ensure they continue to keep all those insects down?

Are warblers less important than tigers? Isn't the question itself meaningless?

MADHUSUDAN KATTI, *In Danger*, 1997

The bird ringing programme of the BNHS undertaken under the stewardship of Sálim Ali had interesting ramifications as some people then thought. The (migratory) birds were coming to India via China from the (then) USSR; they were believed to be carrying a deadly virus. The ringing project was funded by the Americans and those were the days of the 'Cold War'—it may sound a bit like a chapter from an international spy thriller and it is left to readers to discover the rest for themselves.

The Bird Ringing Controversy

Before 1935 I only knew the 'Ghana' of Bharatpur by its reputation as a phenomenal private duck-shooting reserve of the maharajas and their elite VIP guests. That year, at my bidding, my friend Prater, Curator of the BNHS, wrote to Sir Richard Tottenham, ICS, a keen naturalist-sportsman member of the Society, to enquire what facilities would be available from the state for setting up a pilot wildfowl ringing centre at the famous Keoladeo Ghana *jheel*. Sir Richard was the Administrator of the state during the minority of Maharaja Brijendra Singh, and the High Panjandrum whose word was law. That is how I got my earliest introduction to this fabulous wetland which has since become one of my most regular stamping grounds, and in the process developed into the Society's chief centre for the study of bird migration in India. Prior to that time practically no bird ringing had been done in the subcontinent, barring the pioneering experiment in Dhar State (now in Madhya Pradesh) in 1926 by the enterprising ruler,

Maharaja Sir Uday Singh Puar. Aluminium rings bearing the legend 'Inform Maharaja Dhar' had to be specially handcrafted for him by the BNHS (in the proverbial style of the early Rolls Royce engines). It was a slow and laborious job as strips of appropriate sizes had to be cut from the metal sheet, their sharp edges filed smooth, and the legend and serial number manually punched, letter by letter, from steel dies. Under the circumstances the operation could only be on a very limited scale. Despite this the recoveries reported from such distant places as Turkestan, Siberia, and other parts of the USSR were so exciting and encouraging that I was determined to have the BNHS take up bird ringing as one of its major field activities if and when funds and facilities ever made it possible. Using the Dhar success as a lure the Society was able, through the good offices of some of its civilian members in the Sind administration, then a part of Bombay Presidency, to persuade a handful of affluent sporting zamindars to take up the ringing of migratory ducks. But here again the bottleneck was the ring supply. Rings marked 'Inform Bombay Nat. Hist. Society' had still to be manually prepared and were sparingly available. In spite of the modest scale on which ringing was done in Sind the recoveries were gratifying beyond expectation.

It was a long wait, but 'possibility' did drop into our lap fortuitously and unsolicited, via the World Health Organization. Around the year 1957 there erupted an apparently new form of encephalitis in the Kyasanur forest area of Karnataka (South India), affecting villagers and monkeys, sometimes fatally. The tick-borne virus of this encephalitis (known as 'Monkey Disease' or KFD) was reported by the Virus Research Centre, Pune, to be closely related to Omsk Haemorrhagic Fever and the Russian Spring-Summer Encephalitis (RSS) complex. The apparently transcontinental distribution of the virus suggested that it could have reached here through the agency of ticks on birds migrating between USSR and India, and the WHO seriously seized upon the problem. In March 1959 I was invited to a meeting in Geneva of a 'Scientific Group on Research on Birds as Disseminators of Arthropod-borne Viruses', with the project proposal prepared at their request for the study of bird migration in Kutch and north-west India. The proposal was warmly approved, and with financial support from WHO the first-ever organized scheme for bird ringing and migration study in the subcontinent was launched. This is the genesis of the Society's Bird Migration Project, a study of which the seed was well and truly implanted in me during the visit to the Ornithological Observatory at Heligoland in 1929.

Up to the mid-sixties the WHO was itself receiving substantial financial aid from the US government for its wide-ranging public health research programmes, chiefly in underdeveloped countries. After three or four years WHO's funding to the BNHS ceased. Because of its massive involvement and commitments in the Korean war the US was obliged to prune its research grants drastically, which in turn brought the BNHS under the axe. The Society's bird ringing project was seriously jeopardized, and it would certainly have collapsed but for the timely intervention at that point of the Smithsonian Institution in Washington. They considerately took over the funding for the interim period of uncertainty, followed fortuitously by the dropping from the blue of a totally unlooked-for source.

At that time the US Army Medical Research Laboratory, a component of SEATO with headquarters at Bangkok, was conducting a massive bird ringing programme known as MAPS (Migratory Animals Pathological Survey) in South-East Asian countries like Japan, the Philippines, Thailand, Malaya, Indonesia and Taiwan. MAPS was anxious to extend its activities to India, where the excellent performance of the BNHS in bird migration study had been commended to it by both the WHO and the Smithsonian. The overall MAPS programme was directed and co-ordinated by the ebullient and dynamic American ornithologist Dr H. Elliott McClure, an extrovert with whom it was a pleasure and an education to work. I have always been a stickler for the utmost economy in the handling of institutional funds. Therefore I was bucked no end when McClure expressed to me later his amazement and appreciation of the Society's performance in terms of birds ringed to dollar spent, as compared with the other South-East Asian countries MAPS was financing.

The bulk of the funds for the MAPS project came from the US Army Research and Development Group (Far East), Japan.

MAPS had been following the BNHS's ringing work in India with interest and acclaim, and on learning of the possibility of its having to be wound up promptly offered to step in with the funding on condition that our data and results would be made available to it to complete its own information. That seemed fair enough, and I did not at the time suspect that such a useful scientific collaboration could or would land the Society and me personally as Chief Investigator in so much unpleasantness and adverse publicity.

It seems that an enterprising journalist, styling himself 'Scientific Correspondent' of some north-Indian newspaper, had got wind of the Society's ringing collaboration with MAPS. He came out on his own with a highly imaginative alarmist story, imputing that the Society was in this way colluding with the United States to explore the possibility of migratory birds being used in biological warfare for inducting and disseminating deadly viruses and germs in enemy countries. Since our migrant birds came chiefly from the USSR, and possibly also China, this lent credibility to the report in view of the Cold War, which had hotted up considerably about that time. The report caused a furore in Delhi's political circles and generated much heat and noise among our pro-communist, anti-US 'patriots' in the Lok Sabha. The outcry resulted in two successive enquiry committees, both of which fully absolved the Society and the project director Sálim Ali of criminal intent or subversive action. But the MPs remained unconvinced. In order to stop the noise and clear lingering doubts and suspicion once and for all, the government appointed a third so-called 'high power' committee of three top-ranking scientists, one each from the Virus Research Centre (VRC), Pune, the Tata Institute of Fundamental Research (TIFR), Bombay, and the Zoological Survey of India (ZSI), Calcutta, to review the matter thoroughly and *de novo*. It was not till their verdict fully endorsed the earlier findings that the Society regained its credibility and public image. However, during the two years or so that the hullabaloo was on, our migration study had practically come to an end for want of funds, the MAPS project itself having terminated meanwhile. To avoid a repetition of similar awkward incidents the government decided that the entire funding for all collaborative projects approved by it would in future be undertaken by itself on a request from the American collaborating agency from the counterpart rupee funds held in India under the PL 480 scheme.

SÁLIM ALI, *The Fall of a Sparrow*, 1985

The following extract, from an article published in Buceros, *touches upon the controversy of changing names of birds. All the authors are from the BNHS. R. Manakadan, M. Inamdar, and Gayatri Ugra work on the staff. A.R. Rahmani is the current director of the BNHS and J.C. Daniel has had a long association with the institution. He is well-known for his work on the reptiles of India.*

Common English Names of the Birds of the Indian Subcontinent

The newsletter *Buceros* (Vol. 2, No. 4) of the ENVIS (Environmental Information System) Centre at the Bombay Natural History Society (BNHS) had an in-depth discussion on the issue of the recent common English name changes of birds of the Indian subcontinent. Subsequent to that, we had short-listed 25 ornithologists associated with Indian ornithology for their opinions on the list of avian common names proposed by us for the Indian subcontinent.

After a perusal and analysis of their views, we have short-listed what we feel are the most appropriate English avian common names for the Indian subcontinent. This list will now be sent to the International Ornithological Congress committee, which is in the progress of standardizing the list of avian common names in English for the world. On the basis of this document, we anticipate that the views of those who represent the Indian subcontinent, will be heard. To aid those who do not have the copy of *Buceros* mentioned earlier, we give below the important points that were discussed in the issue.

THERE SHOULD BE NO OVERLAPPING OF NAMES

Example: The Mountain Quail in India refers to *Ophrysia superciliosa*, while in the USA it denotes *Oreortyx picta*. The former has now been renamed Himalayan Quail. Similarly, a species should not have different names in different countries, for example, Broad-billed Roller and Dollarbird.

Necessitated when a species or subspecies has been split/demoted/ upgraded to species/subspecies level, or when placed in another taxon after reassessment of its taxonomic status.

Example: The race *japonica* of *Coturnix coturnix* (Common Quail) is now treated as a species, with the name Japanese Quail *C. japonica*. In such cases, there can be no disputes regarding the necessity for name changes. Such changes have been taking place with regularity in bird taxonomy, as and when it was necessary.

To Rectify Wrong, Inappropriate Names or Shorten Lengthy Names

Examples:
Correcting wrong or inaccurate descriptive or regional names of a species. For example, giving the qualifier *Red-headed* to a species when only top of the head is red. In such cases, the names have been changed to *Red-capped* or *Red-crowned*. Or where the earlier known distributional range recorded for a species is valid no more—e.g., a species recorded earlier in the Western Ghats was later recorded from the Eastern Ghats also.

Shortening names such as Prince Henri's Laughingthrush to Henri's Laughingthrush. Deleting *Grey* in Grey Hypocolius, as there is only one Hypocolius species—use of qualifier *Grey* gives the wrong impression that there are other Hypocolius species. Or in case of unnecessary double qualifiers, for example, Australian Black-fronted Dotterel and Small Indian Pratincole. There is only one Black-fronted Dotterel and Small Pratincole, so additional qualifiers *Australian* and *Indian* are clearly unnecessary.

To Ensure that Name is Applicable to all the Races of a Species

Example: The White-rumped Magpie *Pica pica* has a black-rumped race (*bottanensis*) and hence the name is inappropriate. The new name proposed is the Black-billed Magpie—both the races have black bills. By this, the need to give names to subspecies (as was done in Ali and Ripley's *Handbook*) can be totally eliminated.

SUGGESTIONS OFFERED FOR RENAMING

In addition to the above mentioned points, we offer further suggestions for the renaming exercise.

- Traditional names, especially those that find a place in everyday life of literature, should not be meddled with, unless unavoidable. For example, two of the larger species of kingfishers of Australia are known as kookaburras. The kookaburras are so well entrenched in Australian lore, literature and song, that changing their names to kingfishers (the family to which they belong) would be uncalled for.
- Group/family names of birds should ideally accompany the species name as far as possible, e.g., Coppersmith versus Coppersmith Barbet. This makes it easy for birdwatchers especially of a foreign country (many of whom have no inclination to learn scientific names) to easily know what kind of bird it is. For the same reason, shortening of names should not be at the expense of qualifiers useful in differentiating genera or subgroups within a group, e.g. Hill-Partridge, Pygmy-Woodpecker.
- In cases where different names are used for a bird species in different countries, the more widely used name could be selected—unless the less common name is more appropriate. For example, in the case of the Woolly-necked Stork versus White-necked Stork, the latter name is more appropriate—the soft wool-like texture of its neck feathers can only be felt (dead birds!), while the white neck is easily seen in the field.
- Existing alternate names (within a region) should be deleted. Examples are Black Drongo or King Crow, Crow-pheasant or Coucal, Spotted or Dusky Redshank, Cape Pigeon or Cape

Petrel. In such cases, the more appropriate or sometimes popular name should be retained.
- It is improper to change a name that had been named after an ornithologist as it is part of ornithological history, and honours the ornithologist concerned. It may be allowed in cases where another very apt name has been in use for the species elsewhere.
- Assigning region or country based names to birds should be avoided, unless the birds are endemic to a region or largely to that region. Earlier region-wise coined names, which are no more valid after discovery of populations outside the range, should be changed.
- The qualifier *Common* has to be used with caution, as it can be regionally biased. At the most, it could be used for a species that has a widespread distribution over countries or continents (e.g. Common Coot *Fulica atra*). It should not be used for a species that is restricted to only a region or country, with closely allied species in other parts of the world. One example is the Common Grey Hornbill *Ocyceros birostris*, which has now been renamed Indian Grey Hornbill. *Common* could also be used for a family that has very few species and one of which is much more common or abundant than the other(s).
- Uniformity and logic in renaming is advised. For example, the White-throated Munia (*Lonchura malabarica*) is proposed to be renamed as the White-throated Silverbill. The reason given is that a similar African species is called Silverbill. However, except for two species, the other thirty odd species of this genus are called Munias and not Silverbills. It would be more logical to change the African bird's name. For the same reason, we have accepted Hanging-Parrot versus Lorikeet for our two *Loriculus* species, as the birds of these genus are called Hanging-Parrot elsewhere.
- The coining of new names, in most cases, have followed the rules of grammar and syntax, which should be welcomed to some extent as a name is a name only. After a perusal of the different styles of use and non-use of hyphens, we feel that the following pattern is ideal:

a) Hyphens in adjectival cases, e.g., Whitecheeked Bulbul and not White-cheeked Bulbul. Though the use of hyphens in adjectival cases does not really make much of a difference—

unless one wants to adhere to proper grammar—it is necessary when the same alphabets meet together in two words, e.g., White-eyed Babbler versus Whiteeyed Babbler. And, for the sake of uniformity, let hyphens be used in all such adjectival cases.
b) Hyphens are very useful and essential in the case of compound group names, e.g. Green-Pigeon, Night-Heron, Flycatcher-Warbler, Eagle-Owl. Use of hyphens in such cases makes it clear to ornithologists and birdwatchers that these are group names, and not just adjectival names used for birds. For example, by use of the hyphen in Tibetan Eared-Pheasant, one would immediately know that the Eared-Pheasant is a group name, otherwise one could presume it to be a descriptive name as in Asian Brown Flycatcher. In this case, Flycatcher is the group name and *Brown* (or *Asian Brown*) is the descriptive name. Another good example is Red Collared-Dove. If hyphens were not used (i.e., Red Collared Dove), a novice birdwatcher could imagine it as a dove with a red collar, instead of a reddish dove of the Collared-Dove group.

SALIENT FEATURES OF RESPONDENTS' VIEWS

- A significant feedback was from an old and well-established group of ornithologists from Sri Lanka. They were unhappy that the endemic bird species of their country which carried the tag of the island's old name (Ceylon) have now been replaced with Sri Lanka. They had decided to retain the country's old name for their bird names due to historical reasons and the charm of the old name. Respondents from Sri Lanka (and also other countries) questioned, that if Ceylon was to be replaced with Sri Lanka, would birds with the qualifier of say Burma be changed to Myanmar?!
- It appears—our assumption may be wrong-that there is a tendency for Europeans and Americans to pass on the burden of changing names for standardization process to others. How else can one reason their reluctance to change the name of the Woolly-necked Stork to White-necked Stork. One argument for their view is that wool is anyway white! This is not true for India at least—where we have good populations of brown and black sheep!

- The perpetual European-American conflict comes out into the open with regard to bird names also! The Americans prefer the name of Loon, the Europeans Divers; the Americans class the smaller Skuas as Jaegers, the Europeans clump all of them together under Skuas. They also insist on calling their *Icterus* and *Myioborus* species as orioles and redstarts respectively, while the rest of the world uses these terms for birds of the genus *Oriolus* and *Phoenicurus*. The Americans should accept the nomenclature being followed in most parts of the world to make the name standardization exercise a success.
- One of the respondents mentioned that shortening of names is a good idea but expressed the need for caution, saying it should not take away the 'substance' of the name.
- Two respondents came out strongly against the qualifier *Common*, suggesting that *Common* be totally done away with! We do not advocate this, but add that since there is in any case an exercise to standardize names, it matters little if long established names with the qualifier *Common* are given a hard second look.
- A respondent asks the pertinent question of some cases of use of qualifier *Eurasian*, when the species also occurs in Africa. Another did not like its overuse in the bird renaming exercise, and instead suggested the use of more descriptive qualifiers.

A suggestion of a foreign respondent is logical—if a species has its stronghold outside the Indian subcontinent, then the decision on the bird's name should rest the ornithologists from the bird's stronghold area (and vice-versa)—unless our name for the bird is much more appropriate.

R. Manakadan, J.C. Daniel, Asad Rafi Rahmani, M. Inamdar, and Gayatri Ugra, *Buceros*, 1998

Michael Lipske uncovers a plot and outlines the skillful detective work of Dr Pamela Rasmussen, of the Smithsonian Institution, who rediscovered Blewitt's Spotted Owlet in India some years ago. Interestingly, Colonel Meinertzhagen (with whom we are already familiar, from Sálim Ali's piece in this section) figures again prominently; this time in a sinister ploy to destroy evidence. Michael Lipske is a freelance writer, well-known for his writings in National Wildlife *and* International Wildlife.

Forest Owlet Thought to be Extinct is Spotted Anew after 113 Years

Ornithologist Pamela Rasmussen felt both panic and elation one morning in 1997 when she gazed, only half trusting her eyes, at a long-lost species of bird perched in a bare tree in western India. Panic because, *Anthene blewitti* the forest owlet that Rasmussen had sought for two weeks from one side of India to the other, might fly off before it could be positively identified and captured on film. Elation because the chunky, 9-inch-long owl that she was staring at was a species that had gone unseen by any scientist for 113 years. Seven stuffed skins in a handful of museums were all that seemed to remain of a species that several experts had crossed off as extinct.

Fortunately, the forest owlet was not only alive, but 'absurdly cooperative,' says Rasmussen, a museum specialist in the Division of Birds at the National Museum of Natural History. 'It just sat there,' she says, while she and a colleague videotaped it for half an hour before another bird finally chased it off. The next day, a second owlet, likely the first one's mate, revealed itself in the same patch of forest. Coming nose to beak with the long-absent species required days of difficult hunting along forest paths and stream beds. But before leaving for India, Rasmussen had already picked her way down another trail that led through a jungle of scientific deception. She had been in the final stages of preparing a field guide to birds of the Indian subcontinent (a project initiated by Smithsonian Secretary Emeritus S. Dillon Ripley), when she read an article that raised questions about the accuracy of bird records made by Col. Richard Meinertzhagen. A British soldier, spy and noted amateur ornithologist in the early part of this century, Meinertzhagen (who died in 1967) was long credited with creating one of the world's best private collections of Old World bird specimens. The 1993 article suggested, however, that labels on some of his birds were fraudulent. This was unsettling news for Rasmussen. There were more than a dozen kinds of birds for which Meinertzhagen was the only collector claiming to have found that species in India. 'I had to know whether to include all these taxa' in the field guide or rule them out as Indian birds', she says.

To find out, she visited Britain's Natural History Museum in London, where most of Meinertzhagen's collection of tens of thousands of birds now resides. Working with ornithologists there, she examined the colonel's unique India specimens. 'Each was either clearly fraudulent or highly suspicious,' Rasmussen says.

She discovered that Meinertzhagen had done some of his most successful bird hunting not in the wild but in the museum's specimen cabinets. He was known to boast of his collection's 'unique perfection,' she says. 'One of the reasons that it was uniquely perfect was because he was stealing the best specimens from other collectors' museum contributions.

With Robert Prys-Jones, head of the Bird Section at Britain's Natural History Museum, Rasmussen established that hundreds of Meinertzhagen specimens were birds he filched; some he restuffed and then relabeled with false information. Of all the ornithological treasures the colonel stole, the rarest was India's forest owlet. Cracking the case required sophisticated detective work.

Rasmussen had found that, of seven known specimens of the owlet in museums, only one was said to have been collected in this century—in 1914 by Meinertzhagen. Most of the others had been collected in the 1880s by James Davidson, a British official and bird enthusiast stationed in western India.

Working with ornithologist Nigel Collar of Bird Life International in England, Rasmussen examined the Meinertzhagen owlet at the British Museum. Both experts could see that original stitching and stuffing had been removed from the skin and that new stuffing had been inserted and the bird re-sewn. Closer study of the specimen and X-ray photographs of it revealed characteristic preparation touches unique to Davidson, a self-taught worker with one-of-a-kind methods for handling bird skins.

Fairly certain that Meinertzhagen's owlet actually had been collected by Davidson, the ornithologists still wanted more evidence. Even though the bird had been restuffed, Rasmussen remembers hoping 'maybe, just may be, there will be a fiber or something somewhere that will tie it to Davidson.' Luckily, there was.

Inside a wing, stuffed around a joint, there remained some raw cotton that had turned yellow from fat. Checking the wing of an owl of another species Davidson collected in India, the sleuths found what looked like similar cotton. They sent both samples to the Federal Bureau of Investigation in Washington, D.C., where forensic tests indicated that the two bits of cotton were virtually identical.

'That, along with other clues, just basically put the nail in the coffin,' Rasmussen says, noting the improbability that Meinertzhagen would have had access to the same kind of rough cotton Davidson used 30 years earlier. The Owlet was a previously unknown, fifth Davidson specimen, presumably stolen from Britain's Natural History Museum by Meinertzhagen and decades later returned to it as part of the colonel's rich collection.

Meanwhile, Rasmussen studied scientific literature on the owlet, including accounts of several searches for the living bird. She concluded that none of those owlet hunts had occurred in the four places where the bird had been collected. One well-intentioned but pointless search had focused on the area where Meinertzhagen claimed—falsely—that he had collected the bird.

If nobody had looked in the right places, maybe the owlet still existed, Rasmussen reasoned. In November 1997, she headed for India with Asian owl expert Ben King of the American Museum of Natural History in New York and with Virginia birder David Abbott. The owlet hunters concentrated on forests near sites where the bird had been collected by Davidson and others more than a century before.

Near the end of their stay, they were searching in foothills of the Satpura Range, northeast of Bombay. By 8:30 a.m. on Nov. 25, they had been in the forest for hours. It was hot, and Rasmussen was uncapping a water bottle when King quietly said. 'Look at that owlet.'

'And terror struck,' she recalls. She dropped the bottle. For a split second, she struggled to decide whether to aim her binoculars or video camera at the bird. 'It was like this huge decision—what am I going to do first?' But there was time to do both, as the forest owlet, missing no more after 113 years, sat tamely in the sun flicking its tail for 30 minutes.

Last summer, Rasmussen returned to India and revisited the discovery site. She relocated what she believes are both birds seen the previous November. She also obtained the first recording of the owlet's distinctive call—the species had been one of the last Indian birds whose vocalizations were unknown—and even watched one bird eat a lizard.

With support from the National Museum of Natural History's Office of Biodiversity Programs, Rasmussen also launched a project with India's Bombay Natural History Society to study the behaviour and ecology of *Athene blewitti*. In June 1998, she returned to the re-discovery site in India to do an emergency follow-up survey of the forest owlet. The survey, conducted primarily by the Bombay Natural History Society, resulted in the location of eight different individual owlets within a 30-mile area.

But nothing she learns about the species seems to top the thrill of finding the bird itself. 'It is certainly the most exciting bird related experience I've ever had,' she says.

'It was incomparable. And afterwards, we were all just grinning,' Rasmussen says, still smiling at the memory of the tail-wagging owlet that flew back from oblivion.

MICHAEL LIPSKE, *The Smithsonian*, 1999

Sources

Birds and the Human Mindscape

BIRDS IN THE WRITER'S IMAGINATION

Ryder, Arthur W. (trans.), *The Panchatantra*, Chicago: The University of Chicago Press, 1956.
Kipling, Joseph Rudyard, *The Jungle Book*, Macmillan, 1894.
Dewar, Douglas, *Bombay Ducks: An Account of Some of the Every-day Birds and Beasts Found in a Naturalist's Eldorado*, London: John Lane, 1906.
Fletcher, Thomas Bainbrigge and Charles McFarlane Inglis, *Birds of an Indian Garden*, Calcutta: Thacker, Spink & Co., 1936.

DELIGHTFUL DISTRACTIONS

Thackston, Wheeler M. (trans.), *The Jahangirnama: Memoirs of Jahangir, Emperor of India*, New York: Oxford University Press, 1999.
Khushwant Singh, *I Shall Not Hear the Nightingale*, London: John Calder, 1959.
Twain, Mark, *Following the Equator*, Hartford: The American Publishing Company, 1897.
Ahmed, Abul Kalam Muhiyuddin, *Ghubar-e-Khatir* (translated as *Sallies of Mind* by D.R. Goyal), Kolkata: Maulana Abul Kalam Azad Institute of Asian Studies, 2003.
Corbett, Jim, *Man-eaters of Kumaon*, Bombay: Oxford University Press, 1944.
Krishnan, M., 'Rescuing a Fledgeling' in Ramachandra Guha (ed.) *Nature's Spokesman: M. Krishnan and Indian Wildlife*, New Delhi: Oxford University Press, 2000.
Lokaranjan, R., 'Delightful Distractions', *Newsletter for Birdwatchers*, II (11), November 1971.
Stairmand, D.A., 'Castaway with Birds', *Newsletter for Birdwatchers*, February 1972 (reprinted in Volume V, No. 40, Sept.–Oct. 2000).

Barnes, Simon, *How to be a (Bad) Birdwatcher*, New York: Pantheon Books, 2005.
Gee, Edward Pritchard, *The Wildlife of India*, New York: E.P. Dullon & Co., 1964.

Sport, Entertainment, and Falconry

Hunting and Sport

Corbett, Jim, *The Temple Tiger and More Man-eaters of Kumaon*, Delhi: Oxford University Press, 1954 (rpt. 1988).
Ali, Sálim, *The Fall of a Sparrow*, Delhi: Oxford University Press, 1985.
Weaver, Mary Anne, 'Hunting with the Sheikhs', *The New Yorker*, 14 December 1992.

Entertainment

Thackston, Wheeler M. (trans.), *The Baburnama: Memoirs of Babur, Prince and Emperor*, Paperback Edition, New York: Modern Library, 2002.
Thackston, Wheeler M. (trans.), *The Jahangirnama: Memoirs of Jahangir, Emperor of India*, New York, Oxford University Press, 1999.
Sharar, Abdul Halim, *Lucknow: The Last Phase of an Oriental Culture* (edited and translated by E.S. Harcourt and F. Hussain), London: Paul Elek, 1975.
Aitken, Edward Hamilton, *A Naturalist on the Prowl*, London: W. Thacker & Co., 1917.
Darymple, William, *City of Djinns*, HarperCollins Publishers, 1994.

Falcons and Falconry

T.C. Jerdon, *'Falconinae'*, *The Birds of India*, Vol. 1, Calcutta: The Military Orphan Press, 1862.
Dharmakumarsinhji, R.S., *Reminiscences of Indian Wildlife*, Delhi: Oxford University Press, 1998.

Naturalists on the Prowl

Aitken, Edward Hamilton, *The Common Birds of Bombay*, Bombay: Thacker, Spink & Co., 1900.
Aitken, E.H., *A Naturalist on the Prowl*, London: W. Thacker & Co., 1917.

Dewar, Douglas, *Birds of the Indian Hills*, London: The Bodley Head, 1915.
Dewar, Douglas, *Bombay Ducks*, London: John Bale and The Bodley Head, 1906.
Bates, R.S.P., *Bird Life in India*, Madras: Madras Diocesan Press, 1931.
Osmaston, B.B, 'From the Diaries of B.B. Osmaston (1904–1907)', *Newsletter for Birdwatchers*, 41(4), 2001.
Fletcher, Thomas Bainbrigge and Charles McFarlane Inglis, *Birds of an Indian Garden*, Calcutta: Thacker, Spink & Co., 1936.
Donald, C.H., 'The Flight of Eagles', *Journal of the Bombay Natural History Society*, Vol. 50, 1952.

Natural History and Science

Thackston, Wheeler M. (trans.), *The Baburnama: Memoirs of Babur, Prince and Emperor*, Paperback Edition, New York: Modern Library, 2002.
Thackston, Wheeler M. (trans.), *The Jahangirnama: Memoirs of Jahangir, Emperor of India*, New York: Oxford University Press, 1999.
Hume, Allan Octavian and Charles Henry Tilson Marshall, *The Game Birds of India, Burmah, and Ceylon* Vols 1–3, London: John Bale, 1879–81.
Mason, C.W. and Harold Maxwell-Lefroy, 'The Food of Birds in India', *Memoirs of the Department of Agriculture in India, Entomological Series*, Calcutta: Thacker, Spink & Co., 1912.
'Hugh Whistler's Suggestions on How to Run a Bird Survey' in Sálim Ali's *The Fall of a Sparrow*, Delhi: Oxford University Press, 1985.
Haldane, John Burdon Sanderson, 'The Non-violent Scientific Study of Birds', *Journal of the Bombay Natural History Society*, Vol. 56, 1959.

Birdwatching and Beyond

Ali, Sálim, 'Stopping by the Woods on a Sunday Morning', *Express*, 11 November 1984, rpt. in *Newsletter for Birdwatchers*.
Ali, Sálim, *The Fall of a Sparrow*, Delhi: Oxford University Press, 1985.
Shivrajkumar (of Jasdan), Ramesh M. Naik, and K.S. Lavkumar, 'A Visit to the Flamingos in the Great Rann of Kutch', *Journal of the Bombay Natural History Society*, Vol. 57, 1960.
MacDonald, Malcolm, *Birds in my Indian Garden*, London: Jonathan Cape, 1960.
Gee, Edward Pritchard, *The Wildlife of India*, New York: E.P. Dullon & Co., 1964.

Melluish, R.A. Stewart, 'Notes from Madras', *Newsletter for Birdwatchers*, 5(3), 1965.
Ezekiel, Nissim, *The Exact Name*, New Delhi: Oxford University Press, 1965.
Gay , Thomas, 'An Evening at Pashan Lake , Poona', *Newsletter for Birdwatchers*, 43(2), 2003.
Jackson, Peter, 'A Day's Worth of Delhi Birds', *Newsletter for Birdwatchers*, 1971.
Kahl, Philip, 'The Courtship of Storks', *Natural History*, 80(8), 1971.
Gadgil, Madhav, 'Ornithology in Bandipur', *Newsletter for Birdwatchers*, 18(5), 1978.
Saiduzzafar, Hamida, 'Some Observations on the Apparent Decrease in Numbers of the Northern Roller or Blue Jay (*Coracius benghalensis*)', *Newsletter for Birdwatchers*, Vol. 24, 1984.
Gole, Prakash, 'The Pair beside the Lake', *The ICF Bugle*, 15(4), 1989.
Rahmani, Asad Rafi, 'The Greater Adjutant Stork', *Newsletter for Birdwatchers*, Vol. 29, 1989
Pfister, Otto, 'Cranes of Hanley', *Sanctuary Magazine*, 15(6), 1995.
Cocker, Mark, 'Rare Bird of the Mountains', *Guardian Weekly*, 16 June 1996.
Baskaran, S. Theodore, 'The Haflong Phenomenon' in *The Dance of the Sarus*, Delhi: Oxford University Press, 1999.
Whittaker, Zai, 'Misty Binoculars and Other Strategies for Survival among Birdwatchers', *Newsletter for Birdwatchers*, 39(6), 1999.

Personalities and Controversies

REMEMBERING SÁLIM ALI

Futehally, Zafar, 'Remembering Sálim Ali', *Sanctuary Magazine*, 15(5), 1995.
Bhushan, Bharat, 'The EmPee Saar in Andhra Pradesh', *Hornbill*, 1995: 26–9.

SÁLIM ALI ON OTHERS

Ali, Sálim, *The Fall of a Sparrow*, Delhi: Oxford University Press, 1985.
Meinertzhagen, Richard, '... To See Ourselves as Others See Us' in Sálim Ali's *The Fall of a Sparrow*, Delhi: Oxford University Press, 1985.

CONTROVERSIES

Katti, Madhusudan, 'Are Warblers Less Important than Tigers?' in *In Danger* (edited by Paola Manfriedi), Delhi: Ranthambore Foundation, 1997.
Ali, Sálim, *The Fall of a Sparrow*, Delhi: Oxford University Press, 1985.
Manakadan, R., J.C. Daniel, Asad Rafi Rahmani, M. Inamdar, and Gayatri Ugra, 'Standardized English Common Names of the Birds of the Indian Subcontinent—A Proposal', *Buceros*, 3(2), 1998.
Lipske, Michael, 'Forest Owlet Thought to be Extinct is Spotted Anew After 113 Years', *The Smithsonian*, Vol. 96, 1999.

Acknowledgements

The editor and publisher would like to place on record their gratitude to the following individuals/publishers for permissions to reproduce their pieces/extracts in this book. Copyright holders who could not be contacted earlier because of lack of information are requested to correspond with Oxford University Press, New Delhi.

The University of Chicago Press (Chicago) for 'The Heron and the Crab' and 'The Shrewd Old Gander' from *The Panchatantra* (translated from the Sanskrit by Arthur W. Ryder).

Khushwant Singh and John Calder Publishers, London for extracts from the novel *I Shall Not Hear the Nightingale*.

Oxford University Press, New York for extracts from *The Baburnama* (translated by Wheeler M. Thackston) and *The Jahangirnama* (translated by Wheeler M. Thackston).

Oxford University Press, Mumbai and Delhi for extracts from *Man-eaters of Kumaon* by Jim Corbett, *Reminiscences of Indian Wildlife* by R.S. Dharmakumarsinhji, *The Fall of a Sparrow* by Sálim Ali, *The Dance of the Sarus* by S. Theodore Baskaran, and 'Poet, Lover, Birdwatcher' by Nissim Ezekiel from *The Extact Name*.

Zafar Futehally, Editor, *Newsletter for Birdwatchers*, for 'Castaway with Birds' by D.A. Stairmand, 'From the Diaries of B.B. Osmaston (1904–1907)' by B.B. Osmaston, 'An Evening at Pashan Lake, Poona' by Thomas Gay, 'A Day's Worth of Delhi Birds' by Peter Jackson, 'Notes from Madras' by R.A. Stewart Melluish, 'Ornithology in Bandipur' by Madhav Gadgil, 'Stopping by the Woods on a Sunday Morning' by Sálim Ali, 'Some Observations on the Apparent Decrease in Numbers of the Northern Roller or Blue Jay' (*Coracius benghalensis*) by Hamida Saiduzzafar, 'The Greater Adjutant Stork' by Asad Rafi Rahmani, and 'Misty Binoculars and Other Strategies for Survival Among Birdwatchers' by Zai Whittaker.

William Dalrymple and HarperCollins for extracts from *City of Djinns*.

Paul Elek, London for extracts from Abdul Halim Sharar's *Lucknow: The Last Phase of an Oriental Culture*, edited and translated by E.S. Harcourt and F. Hussain.

The New Yorker, New York, USA for 'Hunting with the Sheikhs' by Mary Anne Weaver.

Bombay Natural History Society (BHNS), Mumbai for 'The Flight of Eagles' by C.H. Donald, 'The Non-Violent Scientific Study of Birds' by John Burdon Sanderson Haldane, 'A Visit to the Flamingos in the Great Rann of Kutch' by Shivrajkumar (of Jasdan), Ramesh M. Naik, and K.S. Lavkumar, 'The EmPee Saar in Andhra Pradesh' by Bharat Bhushan, and 'Common English Names of the Birds' of the Indian Subcontinent' by R. Manakadan, J.C. Daniel, Asad Rafi Rahmani, M. Inamdar, and Gayatri Ugra.

The Guardian, London for 'Rare Bird of the Mountains' by Mark Cocker.

E.P. Dullon & Co., New York for extracts from *The Wildlife of India* by E.P. Gee.

Jonathan Cape, London for extracts from *Birds in my Indian Garden* by Malcolm MacDonald.

Internal Crane Foundation, Wisconsin, USA for 'The Pair beside the Lake' by Prakash Gole.

Philip Kahl and *Natural History Magazine*, New York, USA for 'The Courtships of Storks' by Philip Kahl.

Sanctuary Asia, Mumbai for 'Cranes of Hanley' by Otto Pfister and 'Remembering Sálim Ali' by Zafar Futehally.

Ranthambore Foundation, New Delhi for 'Are Warblers Less Important than Tigers?' by Madhusudan Katti from *In Danger* (edited by Paola Manfriedi).

The Smithsonian, Washington D.C. for 'Forest Owlet Thought to be Extinct is Spotted Anew after 113 years' by Michael Lipske.

Further Reading

Aitken, Edward Hamilton (EHA), *A Naturalist on the Prowl: Or in the Jungle*, London: W. Thacker & Co., 1923.
____, *The Common Birds of India*, Bombay: Thacker & Co. Ltd., 1947.
____, *The Common Birds of Bombay*, Bombay: Thacker, Spink & Co., 1900.
Ali, Sálim, *The Book of Indian Birds*, Bombay: BNHS, 1941.
____, *Indian Hill Birds*, Bombay: Oxford University Press, 1949.
____, *Bird Study in India: Its History and its Importance*, New Delhi: Indian Council for Cultural Relations, 1979.
____, *Fall of a Sparrow*, Bombay: Oxford University Press, 1985.
____, *The Book of Indian Birds*, 12th and enlarged centenary edn., Bombay: BNHS and Oxford University Press, 1996.
Ali, Sálim and S.D. Ripley, *Handbook of the Birds of India and Pakistan*, Vols 1–10, Bombay: Oxford University Press, 1964–74.
____, *A Pictorial Guide to the Birds of the Indian Subcontinent*, Bombay: Oxford University Press, 1983.
____, *Handbook of the Birds of India and Pakistan*, Compact edn., Bombay: Oxford University Press, 1987.
Ara, Jamal, *Watching Birds*, New Delhi: National Book Trust, 1970.
Baker, Edward Charles Stuart, *The Indian Ducks and Their Allies*, Bombay: BNHS, 1908.
____, *Indian Pigeons and Doves*, London: Witherby & Co., 1913.
____, *The Game-birds of India, Burma, and Ceylon*, Vols 1–3, London: John Bale, 1921.
____, *Birds*, Vols 1–8, 'Fauna of British India' series, London: Taylor and Francis, 1921–30.
____, *The Nidification of Birds of the Indian Empire*, Vols 1–4, London: Taylor and Francis, 1932–5.
____, *Cuckoo Problems*, London: H.F. & G. Witherby, 1942.
Bates, R.S.P., *Bird Life in India*, Madras: Madras Diocesan Press, 1931.
Bates, R.S.P. and E.H.N. Lowther, *Breeding Birds of Kashmir*, Bombay: Oxford University Press, 1952.

Burg, C., Bruce Beehler, and S. Dillon Ripley, *Ornithology of the Indian Subcontinent 1872–1992: An Annotated Bibliography*, Washington DC: Smithsonian Institution, 1994.
Dave, K.N., *Birds in Sanskrit Literature*, Delhi: Motilal Banarsidass, 1985.
Dewar, Douglas, *Bombay Ducks: An Account of Some of the Everyday Birds and Beasts Found in a Naturalist's Eldorado*, London: John Lane and The Bodley Head, 1906.
____, *Birds of the Plains*, London: John Lane, 1909.
____, *The Indian Crow: His Book*, Madras: Higginbotham & Co. and London: Luzac & Co., 1911.
____, *Glimpses of Indian Birds*, London: John Lane, 1913.
____, *Birds of the Indian Hills*, London: The Bodley Head, 1915.
____, *A Bird Calendar for Northern India*, London: Thacker & Co., 1916.
____, *Himalayan and Kashmiri Birds*, London: John Lane, 1923.
____, *Indian Birds*, London: John Lane and The Bodley Head, 1923.
____, *Birds at the Nest*, London: John Lane, 1923.
____, *The Common Birds of India*, Vols 1–2, Calcutta and Simla: Thacker & Spink, 1923–5.
____, *Birds of an Indian Village*, Bombay: Oxford University Press, 1924.
____, *Indian Bird Life*, London: John Lane and The Bodley Head Limited, 1925.
____, *Game Birds*, London: Chapman & Hall, 1928.
____, *Indian Birds' Nests*, Bombay: Thacker & Spink, 1929.
Dharmakumarsinhji, R.S., *Birds of Saurashtra*, Bombay: Times of India Press, 1954.
Dharmakumarsinhji, R.S. and K.S. Lavkumar, *Sixty Indian Birds*, New Delhi: Ministry of Information and Broadcasting, 1972.
Finn, F., *How to Know the Indian Ducks*, Calcutta: Thacker, Spink & Co., 1901.
____, *Game Birds of India and Asia*, Calcutta: Thacker, Spink & Co., 1916.
____, *The Birds of Calcutta*, Calcutta: Thacker, Spink & Co., 1917.
____, *How to Know the Indian Waders*, Calcutta: Thacker, Spink & Co., 1920.
____, *Indian Sporting Birds*, London: F. Edwards, 1920.
____, *The Water Fowl of India and Asia*, Calcutta and Simla: Thacker, Spink & Co., 1921.
____, *Garden and Aviary Birds of India*, Calcutta: Thacker, Spink & Co., 1950.

Fletcher, T.B. and C.M. Inglis, *Birds of an Indian Garden*, Calcutta: Thacker, Spink & Co., 1936.
Futehally, Zafar (ed.), *India through its Birds*, Bangalore: Dronequill Publishers, 2007.
Ganguli, Usha, *A Guide to the Birds of the Delhi Area*, New Delhi: Indian Council for Agricultural Research, 1975.
Gaur, R.K., *Indian Birds*, New Delhi: Brijbasi Printers, 1994.
Grewal, Bikram, *Birds of the Indian Subcontinent*, Hong Kong: The Guidebook Company Limited, 1995.
Grewal Bikram, Bill Harvey, and Otto Pfister, *A Photographic Guide to the Birds of the Indian Subcontinent*, Singapore: Periplus, 2002.
Grimmet, R, T. Inskipp, and C. Inskipp, *Birds of the Indian Subcontinent*, UK: A&C Black, 1998.
Hume, A.O., *My Scrap Book or Rough Notes on Indian Oology and Ornithology*, Calcutta: C.B. Lewis, Baptist Mission Press, 1869.
———, *Contributions to Indian Ornithology*, London: L. Reeve & Co., 1873.
——— (ed.), *Stray Feathers*, Vols 1–12, A.O. Hume, 1873–83.
Hume A.O. and C.H.T. Marshall, *The Game Birds of India, Burmah, and Ceylon*, Vols 1–3, London: John Bale, 1879–81.
Hume A.O. and E.W. Oates., *The Nests and Eggs of Indian Birds*, London: R.H. Porter, 1889.
Jerdon, T.C., *The Birds of India: A Natural History*, Vols 1–3, Calcutta: The Military Orphan Press, 1862–4.
———, *The Game Birds and Wildfowl of India*, London: Military Orphan Press, 1864.
Kalpavriksh, *What's That Bird? A Guide to Birdwatching, with Special Reference to Delhi*, New Delhi: Kalpavriksh, 1991.
Kazmierczak, Krys, *A Field Guide to the Birds of the Indian Subcontinent*, New Haven: Yale University Press, 2000.
Kazmierczak, Krys and R. Singh, *A Bird Watchers Guide to India*, Delhi: Oxford University Press, 2001.
Kothari, A. and B.F. Chhapgar, (eds), *Salim Ali's India*, New Delhi: Oxford University Press and BNHS, 1996.
Lowther, E.H.N., *A Bird Photographer in India*, London: Oxford University Press, 1949.
MacDonald, M., *Birds in My Indian Garden*, London: Jonathan Cape, 1960.
———, *Birds in the Sun: Some Beautiful Birds of India*, London: D.B. Taraporevala Sons & Co. Pvt. Ltd., 1962.
Mason, C.W. and H. Maxwell–Lefroy, *The Food of Birds in India*, Calcutta: Thacker, Spink & Co., 1912.

Mookherjee, K., *Birds and Trees of Tolly*, Calcutta: Tollygunge Club, 1995.
Nugent, R., *The Search for the Pink-headed Duck*, Boston: Houghton Mifflin, 1991.
Oates, E.W. and Blandford, *Birds*, Vols 1–4, 'Fauna of British India' series, London: Taylor & Francis, 1889–98.
Osman, S.M., *Hunters of the Air: A Falconer's Notes*, New Delhi: WWF, 1991.
Pittie, Aasheesh, *A Bibliographic Index to the Ornithology of the Indian Region – Part I*, Hyderabad: S.M. Osman, 1995.
Rasmussen, Pamela C. and John C. Anderton, *The Birds of South Asia: The Ripley Guide*, Vols 1 and 2, Washington D.C. and Barcelona: Smithsonian Institution and Lynx Edicions, 2005.
Rangaswami, S. and S. Sridhar, *Birds of Rishi Valley and Renewal of their Habitats*, Andhra Pradesh: Rishi Valley Education Centre, 1993.
Ripley, S.D., *Search for the Spiny Babbler*, Boston: Houghton Mifflin, 1952.
____, *A Bundle of Feathers*, London: Oxford University Press, 1978.
____, *A Synopsis of the Birds of India and Pakistan*, Bombay: BNHS, 1982.
Tikader, B.K., *Birds of Andaman and Nicobar Islands*, Calcutta: Zoological Survey of India, 1984.
Urfi, A.J., *Birds, Beyond Watching*, Hyderabad: Universities Press, 2004.
Wan Tho, Loke, *A Company of Birds*, London: Michael Joseph, 1958.
Wedderburn, Sir William, *Allan Octavian Hume*, London: T. Fisher Unwin, 1912.
Whistler, Hugh, *Popular Handbook of Indian Birds*, London: Gurney & Jackson, 1928.
Woodcock, Martin, *Collins Handguide to the Birds of the Indian Subcontinent*, London: Collins, 1980.